T0145410

Advances in Intelligent Systems and Computing

Volume 839

Series editor

Janusz Kacprzyk, Polish Academy of Sciences, Warsaw, Poland
e-mail: kacprzyk@ibspan.waw.pl

The series "Advances in Intelligent Systems and Computing" contains publications on theory, applications, and design methods of Intelligent Systems and Intelligent Computing. Virtually all disciplines such as engineering, natural sciences, computer and information science, ICT, economics, business, e-commerce, environment, healthcare, life science are covered. The list of topics spans all the areas of modern intelligent systems and computing such as: computational intelligence, soft computing including neural networks, fuzzy systems, evolutionary computing and the fusion of these paradigms, social intelligence, ambient intelligence, computational neuroscience, artificial life, virtual worlds and society, cognitive science and systems, Perception and Vision, DNA and immune based systems, self-organizing and adaptive systems, e-Learning and teaching, human-centered and human-centric computing, recommender systems, intelligent control, robotics and mechatronics including human-machine teaming, knowledge-based paradigms, learning paradigms, machine ethics, intelligent data analysis, knowledge management, intelligent agents, intelligent decision making and support, intelligent network security, trust management, interactive entertainment, Web intelligence and multimedia.

The publications within "Advances in Intelligent Systems and Computing" are primarily proceedings of important conferences, symposia and congresses. They cover significant recent developments in the field, both of a foundational and applicable character. An important characteristic feature of the series is the short publication time and world-wide distribution. This permits a rapid and broad dissemination of research results.

More information about this series at http://www.springer.com/series/11156

Valentina Emilia Balas · Neha Sharma
Amlan Chakrabarti
Editors

Data Management, Analytics and Innovation

Proceedings of ICDMAI 2018, Volume 2

 Springer

Editors
Valentina Emilia Balas
Department of Automatics and Applied
  Software
Aurel Vlaicu University of Arad
Arad, Romania

Amlan Chakrabarti
Faculty of Engineering and Technology
A.K. Choudhury School of Information
  Technology
Kolkata, India

Neha Sharma
Adyogik Tantra Shikshan Sanstha's IICMR
Pune, Maharashtra, India

ISSN 2194-5357   ISSN 2194-5365 (electronic)
Advances in Intelligent Systems and Computing
ISBN 978-981-13-1273-1   ISBN 978-981-13-1274-8 (eBook)
https://doi.org/10.1007/978-981-13-1274-8

Library of Congress Control Number: 2018947293

This Springer imprint is published by the registered company Springer Nature Singapore Pte Ltd.
The registered company address is: 152 Beach Road, #21-01/04 Gateway East, Singapore 189721, Singapore

Preface

These two volumes constitute the Proceedings of the International Conference on Data Management, Analytics and Innovation (ICDMAI 2018) held from January 19 to 21, 2018, which was jointly organized by Computer Society of India, Div II and Pune Section, and Institute of Industrial and Computer Management and Research (IICMR), Pune. The conference was supported by the industry leaders like TCS, IBM, Ellicium Solutions Pvt. Ltd, Omneslaw Pvt. Ltd, and premier academic universities like Savitribai Phule Pune University, Pune; Lincoln University, Malaysia; Defence Institute of Advanced Technology, Pune. The conference witnessed participants from 14 industries and 10 international universities from 14 countries. Utmost care was taken in each and every facet of the conference, especially regarding the quality of the paper submissions. Out of 488 papers submitted to ICDMAI 2018 from 133 institutions, only 76 papers (15.5%) were selected for oral presentation. Besides quality paper presentation, the conference also showcased 04 workshops, 04 tutorials, 08 keynote sessions, and 05 plenary talk by the experts of the respective fields.

The volumes cover a broad spectrum of fields such as computer science, information technology, computational engineering, electronics and telecommunication, electrical, computer application, and all the relevant disciplines. The conference papers included in this proceedings, published post-conference, are grouped into four areas of research such as data management and smart informatics, big data management, artificial intelligence and data analytics, advances in network technologies. All four tracks of the conference were much relevant to the current technological advancements and had the Best Paper Award in each of their respective tracks. Very stringent selection process was adopted for the paper selection, from plagiarism check to technical chairs' review to double-blind review; every step was religiously followed. We are thankful to all the authors who the have submitted papers for keeping the quality of the ICDMAI 2018 conference at high levels. The editors would like to acknowledge all the authors for their contributions and also the efforts taken by the reviewers and session chairs of the conference, without whom it would have been difficult to select these papers. We have received important help from the members of the International Program Committee.

We appreciate the role of special sessions organizers. It was really interesting to hear the participants of the conference highlighting the new areas and the resulting challenges as well as opportunities. This conference has served as a vehicle for a spirited debate and discussion on many challenges that the world faces today.

We especially thank our Chief Mentor, Dr. Vijay Bhatkar, Chancellor, Nalanda University; General Chair, Dr. P. K. Sinha, Vice Chancellor and Director, Dr. S.P. Mukherjee International Institute of Information Technology, Naya Raipur (IIIT-NR), Chhattisgarh; other eminent personalities like Dr. Rajat Moona, Director, IIT Bhilai, and Ex-Director General, CDAC; Dr. Juergen Seitz, Head of Business Information Systems Department, Baden-Wuerttemberg Cooperative State University, Heidenheim, Germany; Dr. Valentina Balas, Professor, Aurel Vlaicu University of Arad, Romania; Mr. Aninda Bose, Senior Publishing Editor, Springer India Pvt. Ltd; Dr. Vincenzo Piuri, IEEE Fellow, University of Milano, Italy; Mr. Birjodh Tiwana, Staff Software Engineer, Linkedin; Dr. Jan Martinovic, Head of Advanced Data Analysis and Simulations Lab, IT4 Innovations National Supercomputing Centre of the Czech Republic, VŠB Technical University of Ostrava; Mr. Makarand Gadre, CTO, Hexanika, Ex-Chief Architect, Microsoft; Col. Inderjit Singh Barara, Technology Evangelist, Solution Architect and Mentor; Dr. Rajesh Arora, President and CEO, TQMS; Dr. Deepak Shikarpur, Technology Evangelist and Member, IT Board, AICTE; Dr. Satish Chand, Chair, IT Board, AICTE, and Professor, CSE, JNU; and many more who were associated with ICDMAI 2018. Besides, there was CSI-Startup and Entrepreneurship Award to felicitate budding job creators.

Our special thanks go to Janus Kacprzyk (Editor-in-Chief, Springer, Advances in Intelligent Systems and Computing Series) for the opportunity to organize this guest-edited volume. We are grateful to Springer, especially to Mr. Aninda Bose (Senior Publishing Editor, Springer India Pvt. Ltd) for the excellent collaboration, patience, and help during the evolvement of this volume.

We are confident that the volumes will provide state-of-the-art information to professors, researchers, practitioners, and graduate students in the areas of data management, analytics, and innovation, and all will find this collection of papers inspiring and useful.

Arad, Romania   Valentina Emilia Balas
Pune, India   Neha Sharma
Kolkata, India   Amlan Chakrabarti

Organizing Committee Details

Dr. Deepali Sawai is a Computer Engineer and is the alumnus of Janana Probodhini Prashala, a school formed for gifted students. She has obtained Master of Computer Management and doctorate (Ph.D.) in the field of RFID technology from Savitribai Phule Pune University. She is a certified Microsoft Technology Associate in database and software development. She has worked in various IT industries/organizations for more than 12 years in various capacities.

At present, her profession spheres over directorship with several institutions under the parent trust Audyogik Tantra Shikshan Sanstha (ATSS), Pune. At present, she is Professor and Founder Director, ATSS's Institute of Industrial and Computer Management and Research (IICMR), Nigdi, a Postgraduate, NAAC accredited institute affiliated to Savitribai Phule Pune University, recognized by DTE Maharashtra conducting MCA programme approved by AICTE, New Delhi; Founder Director, City Pride School, a NABET accredited school, affiliated to CBSE, New Delhi; Founder Director, ATSS College of Business Studies and Computer Applications, Chinchwad, Graduate Degree College affiliated to Savitribai Phule Pune University. Along with this, she is also carrying the responsibility as Technical Director of the parent trust ATSS and the CMF College of Physiotherapy conducting BPth affiliated to MUHS, Nasik, and MPth affiliated to Savitribai Phule Pune University.

She has so far authored six books for Computer Education for school children from grade 1 to 9 and three books for undergraduate and postgraduate IT students. She has conducted management development programs for the organizations like nationalized banks, hospitals, and industries. She has chaired national and international conferences as an expert. Her areas of interest include databases, analysis and design, big data, artificial intelligence, robotics, and IoT.

She has been awarded and appreciated by various organizations for her tangible and significant work in the educational field.

Prof. Dr. Neha Sharma is serving as Secretary of Society for Data Science, India. Prior to this, she worked as Director, Zeal Institute of Business Administration, Computer Application and Research, Pune, Maharashtra, India; as Dy. Director, Padmashree Dr. D. Y. Patil Institute of Master of Computer Applications, Akurdi, Pune; and as Professor at IICMR, Pune. She is an alumnus of a premier college of engineering affiliated to Orissa University of Agriculture and Technology, Bhubaneshwar. She has completed her Ph.D. from the prestigious Indian Institute of Technology (ISM), Dhanbad. She is Website and Newsletter Chair of IEEE Pune Section and has served as Student Activity Committee Chair for IEEE Pune Section as well. She is an astute academician and has organized several national and international conferences and seminars. She has published several papers in reputed indexed journals, both at national and international levels. She is a well-known figure among the IT circles of Pune and well sought over for her sound knowledge and professional skills. She has been instrumental in integrating teaching with the current needs of the industry and steering the college to the present stature. Not only loved by her students, who currently are employed in reputed firms; for her passion to mingle freely with every one, Neha Sharma enjoys the support of her colleagues as well. She is the recipient of **"Best PhD Thesis Award"** and **"Best Paper Presenter at International Conference Award"** at National Level by Computer Society of India. Her area of interest includes data mining, database design, analysis and design, artificial intelligence, big data, and cloud computing.

Acknowledgements

We, the editors of the book, Dr. Valentina Balas, Dr. Neha Sharma, and Dr. Amlan Chakrabarti, take this opportunity to express our heartfelt gratitude toward all those who have contributed toward this book and supported us in one way or the other. This book incorporates the work of many people all over the globe. We are indebted to all those people who helped us in the making of this high-quality book which deals with state-of-the-art topics in the areas of data management, analysis and innovation.

At the outset, we would like to extend our deepest gratitude and appreciation to our affiliations, Dr. Valentina Balas from the Department of Automatics and Applied Software, Faculty of Engineering, University of Arad, Romania; Dr. Neha Sharma from IICMR, Nigdi of Savitribai Phule Pune University, India; and Dr. Amlan Chakraborty from A.K. Choudhury School of IT, University of Calcutta, India, for providing all the necessary support throughout the process of book publishing. We are grateful to all the officers and staff members of our affiliated institutions who have always been very supportive and have always been companions as well as contributed graciously to the making of this book.

Our sincere heartfelt thanks goes to our entire family for their undying prayers, love, encouragement, and moral support and for being with us throughout this period, constantly encouraging us to work hard. "Thank You" for being our backbone during this journey of compilation and editing of this book.

About the Book

This book is divided into two volumes. This volume constitutes the Proceedings of the 2nd International Conference on Data Management, Analytics and Innovation 2018 or ICDMAI 2018, which was held from January 19 to 21, 2018, in Pune, India.

The aim of this conference was to bring together the researchers, practitioners, and students to discuss the numerous fields of computer science, information technology, computational engineering, electronics and telecommunication, electrical, computer application, and all the relevant disciplines.

The International Program Committee selected top 76 papers out of 488 submitted papers to be published in these two book volumes. These publications capture promising research ideas and outcomes in the areas of data management and smart informatics, big data management, artificial intelligence and data analytics, advances in network technologies. We are sure that these contributions made by the authors will create a great impact on the field of computer and information science.

Contents

Part II   Big Data Management

Part III   Artificial Intelligence and Data Analysis

About the Editors

Prof. Dr. Valentina Emilia Balas is currently Professor at the Aurel Vlaicu University of Arad, Romania. She is author of more than 270 research papers in refereed journals and international conferences. Her research interests are in intelligent systems, Fuzzy Control, soft computing, smart sensors, information fusion, modeling and simulation. She is Editor-in-Chief of International Journal of Advanced Intelligence Paradigms (IJAIP) and International Journal of Computational Systems Engineering (IJCSysE), Member in editorial boards for national and international journals, and serves as Reviewer for many International Journals. She is General Co-Chair to seven editions of International Workshop on Soft Computing Applications (SOFA) starting from 2005. She was Editor for more than 25 books in Springer and Elsevier. She is Series Editor for the work entitled Elsevier Biomedical Engineering from October 2017. She participated in many international conferences as General Chair, Organizer, session Chair, and Member in International Program Committee. She was Vice President (Awards) of IFSA International Fuzzy Systems Association Council (2013–2015), responsible with recruiting to European Society for Fuzzy logic and Technology (EUSFLAT) (2011–2013), Senior Member of IEEE, Member in Technical Committees to Fuzzy Sets and Systems and Emergent Technologies to IEEE CIS, and Member in Technical Committee to Soft Computing to IEEE SMC.

Prof. Dr. Neha Sharma is serving as Secretary of Society for Data Science, India. Prior to this, she worked as Director, Zeal Institute of Business Administration, Computer Application and Research, Pune, Maharashtra, India; as Dy. Director, Padmashree Dr. D. Y. Patil Institute of Master of Computer Applications, Akurdi, Pune; and as Professor, IICMR, Pune. She is Alumnus of a premier College of Engineering affiliated to Orissa University of Agriculture and Technology, Bhubaneshwar. She has completed her Ph.D. from prestigious Indian Institute of Technology (ISM), Dhanbad. She is Website and Newsletter Chair of IEEE Pune Section and served as Student Activity Committee Chair for IEEE Pune Section as well. She is an astute academician and has organized several national and international conferences and seminars. She has published several papers in reputed indexed journals, both at national and international levels. She is a well-known figure among the IT circles of Pune and well sought over for her sound knowledge and professional skills. She has been instrumental in integrating teaching with the current needs of the Industry and steering the college to the present stature. Not only loved by her students, who currently are employed in reputed firms; for her passion to mingle freely with every one, Neha Sharma enjoys the support of her colleagues as well. She is the recipient of "**Best Ph.D. Thesis Award**" and "**Best Paper Presenter at International Conference Award**" at national level by Computer Society of India. Her areas of interest include data mining, database design, analysis and design, artificial intelligence, big data, and cloud computing.

Prof. Dr. Amlan Chakrabarti is *ACM Distinguished Speaker*, who is presently *Dean Faculty of Engineering and Technology* and Director of the A.K.Choudhury School of Information Technology, *University of Calcutta*. He obtained M.Tech from University of Calcutta and did his *doctoral research at the Indian Statistical Institute, Kolkata*. He was *Postdoctoral Fellow at the School of Engineering, Princeton University*, USA, during 2011–2012. He is the recipient of *DST BOYSCAST Fellowship Award* in the area of engineering science in 2011, *Indian National Science Academy Visiting Scientist Fellowship* in 2014, *JSPS Invitation Research Award* from Japan in 2016, *Erasmus Mundus Leaders Award* from European Union in 2017, and *Hamied Visiting Fellowship of the University of Cambridge* in 2018. He is Team Leader of European Center for Research in Nuclear Science (CERN, Geneva) ALICE-India project for University of Calcutta and also a key member of the CBM-FAIR project at Darmstadt Germany. He is also Principal Investigator of the Center of Excellence in Systems Biology and Biomedical Engineering, University of Calcutta, funded by MHRD (TEQIP-II). He has published around 120 research papers in refereed journals and conferences. He has been involved in research projects *funded by DRDO, DST, DAE, DeITy, UGC, Ministry of Social Empowerment, TCS, and TEQIP-II*. He is Senior Member of IEEE, Secretary of IEEE—CEDA India Chapter, and Senior Member of ACM. His research interests are quantum computing, VLSI design, embedded system design, computer vision and analytics.

Part I
Data Management and Smart Informatics

An Empirical Study of Website Personalization Effect on Users Intention to Revisit E-commerce Website Through Cognitive and Hedonic Experience

Darshana Desai

Abstract Personalization is used as an emerging strategy to reduce information overload and attract users and leveraging business through online web portals in recent years. However, less attention is given to study what are different design aspects of web personalization and how it impacts on users' decision-making. To address this gap, this study draws on both stimulus–organism–response theory and information overload theory to propose a model for users' information processing and decision-making. Different personalization aspects induce cognitive and hedonic user's experience during interaction with websites which in turn generates satisfaction and effect on users' decision-making to revisit the personalized website. This research identifies personalization aspects used in e-commerce websites as information, navigation, presentation personalization, and proposed research model and validated it empirically. Using Exploratory Factor Analysis (EFA) supports the factors identified with model as information, navigation, presentation personalization, cognitive, hedonic experience, satisfaction, and intention to revisit the personalized website. Confirmatory Factor Analysis (CFA) result supports proposed model representing interrelation of constructs information, presentation, navigation, cognitive, hedonic experience, satisfaction, and intention to revisit. The model is tested with the data collected from personalized e-commerce website users. 547 out of 600 data from e-commerce website users were used for analysis and for testing the model. EFA of responses extracted seven factors information, presentation, navigation personalization, cognitive experience, hedonic experience, satisfaction, and intention to revisit. CFA confirms model with RMSEA, CFI, and NFI values near to 0.9 which indicates good model fit for e-commerce websites. Structural equation modeling results indicate correlation between personalization aspects, i.e., information, presentation, navigation personalization, and users' satisfaction and intention to revisit through cognitive and hedonic experience. Structural equation modeling technique result validates proposed model and reveals that different design aspects of personalized website design information, presentation, and navigation personalization play a vital role in forming user's positive cognitive

D. Desai (✉)
Department of MCA, Indira College of Engineering & Management, Pune, India
e-mail: desai.darshana@indiraicem.ac.in

© Springer Nature Singapore Pte Ltd. 2019
V. E. Balas et al. (eds.), *Data Management, Analytics and Innovation*,
Advances in Intelligent Systems and Computing 839,
https://doi.org/10.1007/978-981-13-1274-8_1

experience by inducing perceived usefulness, perceived ease of use, enjoyment and hedonic experience of control leading higher satisfaction level, and revisit of e-commerce website.

Keywords Web personalization · Information personalization · Navigation personalization · Presentation personalization · Cognitive experience Hedonic experience · Satisfaction · Perceived ease of use · Perceived usefulness Enjoyment · Control

1 Introduction

With the advent of Internet, website has invaluable source for information exchange for users and e-tailers. Today, every part of business and social media worldwide are using the website as an integral part of business to interact with the customer, brand promotions, marketing, after sales services, and support. Diversity of its users and complexity of web application lead to information overload and one-size-fits-all issue. Cognitive limitation of user information processing leads to lost users in the world of information and results in inefficiency in decision-making. Website personalization has emerged as an effectual solution to overcome this difficulty of information overload in recent years. Many firms are developing personalized websites by investing in the development of personalization tools to attract the users and retain the customers. E-commerce websites like Amazon.in, Flipkart.com, eBay. in, etc. provide personalization features, personalized offerings with categories of products, and services to attract and retain users. Previous research shows significant effect of perceived usefulness of personalized e-services [22] and users interest in personalized services [15], and indicated that various personalized services affect differently on customer satisfaction [8, 9]. Web personalization has become a pervasive phenomenon in a wide range of web applications, e.g., Internet banking, e-commerce, etc. Accordingly, a boom in research on real-world implementation of personalization features has been witnessed recently, and typically focusing on the impact of isolated, one-dimensional personalization features on users. It has been recognized that necessary and well-designed personalization features facilitate the effectiveness, perceived usefulness, perceived ease of use and efficiency as well as the feeling of enjoyment, control, and satisfaction while using a website. Such features have become increasingly diverse and multifaceted in Information System (IS) and Human–Computer Interaction (HCI) research. In light of this, and in view of a continuing gap in the contemporary literature, we would like to investigate different personalization aspects, the role played by these aspects of personalization used in e-commerce website design and how they impact the user intention to revisit or reuse the website. We would also like to study personalization design aspects of e-commerce websites and its impact on user information processing and aspects related to it. This paper is organized as follows: Section 2 discusses previous studies on various personalization dimensions. Section 3 represents research framework derived from previous studies and corresponding hypotheses. Section 4 describes research methodology, research design, and data collection with analysis. Section 5 summarizes the results of the data analysis with EFA, CFA, and SEM. Results are discussed with major findings, theoretical and practical contributions, limitations, and possible directions for future work in Sect. 6.

2 Literature Review

Personalization is the process of catering tailored content, website structure, and look and feel of website with presentation by identifying users' implicit and explicit needs. Personalization has been researched by large community of researchers from diverse fields; personalization research according to literature review is classified into three areas of research [27]. The first stream researches personalization technologies used like mining data, adaptable and adaptive personalization, and push technologies. The second stream researches user-centered personalization, e.g., users' working, privacy issues, and the application context. The third stream of research investigates presentation features which users personalize, and how the effectiveness of the website should be measured [7]. In different areas, personalization has been defined as a toolbox, a feature, or a process.

2.1 Personalization Dimension

In previous literature, there exist at least three perspectives in interpreting the effect of personalization: information and effort reduction, personal persuasion, and relationship building [24]. The relationship building perspective adopts the concept of relationship marketing and treats personalized services as a tool for building a close relationship between the sender and the receiver. Personalized messages intend to develop positive affection between the sender and the receiver. This feeling may include care, trust, and other related emotions. For instance, Komiak and Benbasat [1] proposed a trust-centered perspective in studying the adoption of personalized recommendation agents. Both cognitive trust and emotional trust have been found to influence the intention to adopt personalization agents. This finding indicates that personalized services induce individual's emotional process and give the user a sense of togetherness with the personalized service and its provider. In addition, Liang et al. [21] found that perceived care (an emotional factor) was more influential than transaction costs reduction (a rational factor) on the users' perceived usefulness with personalized services offered by online bookstores. These findings suggest that personalization may have significant affective influence on consumer.

Among all the issues pertaining to personalization, "what" to personalize is the most fundamental problem researched for the effective personalized website design. Different design aspects of personalization may have different impacts on users' information processing and decision-making. Moreover, the different roles played by different personalization features in website design have not been comprehensively investigated. Effective personalized website design is an important issue to be researched to meet the expectation and dynamic need of the users. Different design aspects of personalization impact differently on user's perception and fulfill different kinds of user requirements. However, in previous literature, studies often have focused on only one or more aspects of personalization, e.g., information personalization [1, 2, 4, 8, 9, 17, 27, 31] or visualization [7, 23] but little is researched on effectiveness of the design aspects of personalization. Few studies investigate the roles played by multiple dimensions of personalization [9, 31]. In fact, the existing literature has serious deficit in actionable guidance on personalization design issues and effective personalized web design. To address this research gap, this study comprehensively reviews literature on personalization and develops methodologically construct framework for personalized website design and test the impact of different aspects of personalization. Based on environmental psychology theory and TAM, this paper investigates the different roles played by dimensions of personalization, i.e., information personalization, presentation personalization, and navigation personalization.

3 Research Framework

This research aims to study, first, various personalization design aspects, i.e., information personalization, presentation personalization, and navigation personalization used in websites which are web stimuli, second, impact of personalization aspects (Web Stimuli) on hedonic, utilitarian state of user, third, its effect on user's behavioral response and satisfaction. Moreover, interactions among cognitive/hedonic experience, utilitarian/affective state, satisfaction, and intention to revisit are also taken into consideration, which is missing in prior literature.

The proposed research model is derived from the environmental psychology theory, S-O-R (Stimulus–Organism–Response) theory, information overload theory, Technology Acceptance Model (TAM), and information system success model. Definitions of different personalization design aspects are presented based on environmental psychology. Impact of different aspects of personalization effects on decision-making process is described with cognitive/hedonic and utilitarian experience of user like perceived ease of use, perceived usefulness, enjoyment, and control. User with positive hedonic and utilitarian experience has more satisfaction and is likely to revisit/reuse the personalized websites. More specifically, this study focuses on how user perceives personalization aspects and their influence in decision-making to reuse the website. Hypotheses are proposed to address the research questions (Fig. 1).

Eroglu et al. [10, 11] defined website stimuli, e.g., environmental cues in two different categories like low task-relevant and high task-relevant cues presented online. Low task-relevant cues are responsible to create a mood or an image for the

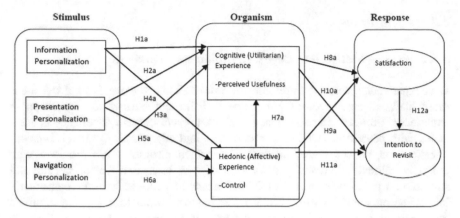

Fig. 1 Research framework for website personalization

online website with environment or esthetic view of website. High task-relevant cues comprise textual contents represented verbal or pictorial contents, whereas low task-relevant cues represent peripheral contents like color, background themes, typestyles, fonts, and images. Research shows that low task-relevant cues can lead to a more pleasant online shopping experience, and these cues do not directly influence the completion of the shopping task [10].

Personalization is the process of tailoring website by satisfying user's implicit and explicit need [7, 8]. The objective of web personalization is to deliver right content to users based on its individual implicit or explicit preferences at the right time to induce a favorable response to the personalized offerings and to increase user satisfaction to build loyalty for future interaction. Personalization is the process perceived to provide information/interface/navigation personalized to cater unique needs of each user. Information personalization is the extent to which information can be catered according to user's implicit or explicit requirement [7]. Users can specify their requirements of the information through customization choices to search or get recommendations from the website. Presentation personalization is the extent to which interface can be modified according to user implicit or explicit requirement (e.g., color, layout, background, themes, etc.). Navigation personalization is the extent to which navigation can be modified in according to user requirement (e.g., new tabs and reorganized the elements to new tabs). User can reorganize the website structure by creating new categories and move information into them or generating quick links.

Research in environmental psychology conceptualized the affective states along three dimensions [11], i.e., Pleasure, Arousal, and Dominance (PAD). Cognitive state refers to user internal mental processes and states including attitudes, beliefs, attention, comprehension, memory, and knowledge. User's cognitive or utilitarian and affective/hedonic states are induced by environmental stimuli and also influence response. Users experience utilitarian benefit with the relevant personalized information that reduces information search.

3.1   Hypotheses

3.1.1   Personalization and Cognitive/Utilitarian Experience

User experiencing perceived usefulness of information and ease of use of website is more likely to enjoy using e-commerce website and creates positive shopping experience. So researcher says that users' cognitive/utilitarian experience is associated with perceived usefulness, ease of use, and enjoyment. Content of personalization can be considered as degree to which customers are provided with uniquely tailored information in the form of text, look, and feel of website and structure on the basis of users' individual needs as gathered with the consumer's interaction on website visit [3, 20, 30]. Personalized content reduces the cognitive effort needed by the user in order to process information, proposed hypotheses as follows:

H1a: Users' Cognitive Experience is positively associated with Information personalization.

The perceived ease of use while interacting with website and personalized layout influences consumers' internal states and behavior [24]. Wang [9, 31, 32] posits that navigation personalization is positively associated with user's' cognitive state perceived usefulness and ease of use. Navigation personalization facilitates users with system initiated personalized structure that reduces users efforts of searching for information. Also, it provides quick links to minimize navigations, resulting in less cognitive load, user feel enjoyment, and increased cognitive experience with perceived ease of use and usefulness. User-initiated personalization can be produced by explicitly giving users choice of quick links and producing personalized website structure, and hypothesis can be proposed as follows:

H2a: Users' cognitive experience is positively associated with presentation personalization.

Modification of the interface to users' need helps, reduces information processing complexity, and facilitates the effectiveness and efficiency with which user can personalize a website [9]. When there are more choices in modifying the presentation feature, e.g., layout and background, the higher level of personalization will give more flexibility in alleviating the complexity. Therefore, more presentation personalization facilitates the user task effectively. Personalized interface induces positive cognitive feeling in user with improved esthetics, finds ease of use, and enjoys operating with the personalized system [23]. So researcher posits the following hypothesis:

H3a: Users' cognitive experience is positively associated with navigation personalization.

3.1.2 Personalization and Hedonic Experience

Personalization provided with the choices to users generates high level of perceived control and users' experience flow with personalization process [9, 16] is more likely to have comfort level and enjoy [9] the interaction with the website. So author postulates hypotheses.

H4a: Users' hedonic experience is positively associated with information personalization.

Website contents like the structure of information presentation and navigation positively persuade the consumer's perception of being in control during the online shopping episode in interaction [12]. So, the researcher proposes the following hypotheses:

H5a: Users' hedonic experience is positively associated with presentation personalization.

H6a: Users' hedonic experience is positively associated with navigation personalization.

Greater customer control of the shopping experience increased the pleasure of shopping [8, 11]. Users with a high level of perceived control during usage of personalized website are expected to feel increase in high comfort level with the activity. Thus, they would be more inclined to feelings of joy using the website more frequently [8, 9]. Studies in HCI also found that more control correlates with enjoyment [23]. Therefore, the following hypothesis is proposed:

H7a: Users' cognitive experience is positively associated with hedonic experience.

DeLone and McLean reported that user satisfaction has been widely adopted in practice as a substitute measure of information systems effectiveness [5]. So, author posits the following hypothesis:

H8a: Users' satisfaction is positively associated with cognitive experience using personalized website.

Prior research suggested that emotions mediate the impact of environment on user intention [19]. We expect the effects of using a web portal to be similar. If the users enjoy their experience in interacting with the web portal, they are more likely to visit the web portal again. Echoing TAM3 research study showed that the degree of perceived ease of use positively influences users' perception of usefulness and their intention to continue to use the website [29].

H9a: Users' intention to revisit is positively associated with cognitive experience using personalized website.

Research shows that accurate personalization process reduces information overload and increases user involvement with increased efficiency, performance, and satisfaction [7, 17, 22, 29]. User with positive hedonic experience of control with personalization features like user interface, information, and navigation over website with involvement using website is more satisfied and likely to revisit the personalized website. So, author proposes the following hypotheses:

H10a: Users' satisfaction is positively associated with hedonic experience (Control) using personalized website.

H11a: Users' intention to revisit website is positively associated with hedonic experience (Control) using personalized website.

DeLone and McLean identified satisfaction and usage of system to measure the information system success which is found as an antecedent of information and system quality [6]. DeLone and McLean [5, 6, 9] in updated IS success model states that user's intention to reuse the system is highly associated with satisfaction. So author proposes the following hypothesis:

H12a: Users' intention to reuse/revisit the personalized website is positively related to user satisfaction.

4 Research Methodology

The survey method is used for data collection to test the proposed research model/ framework. Author collected data from web users using both e-commerce websites having personalization features on their web portal. Selection of website was done with most popular and frequently visited e-commerce personalized websites for study. Besides dimensions of personalization (i.e., information personalization, presentation personalization, and navigation personalization), all other research variables are measured using multiple-item scale adapted from prior studies. Constructs of information personalization, presentation personalization, and navigation personalization are developed from previous literature that relates to the definition of personalization in our context. This research is descriptive research with qualitative nature of study as we investigate the effect of personalization on user's behavioral intentions and satisfaction. Data collection was done through online questionnaire form filling as well as manual form filled by respondents.

4.1 Data Collection and Analysis

Pilot study conducted with 50 online users (Students, business owners, IT professionals, and housewives) who have used e-commerce web portals. Select e-commerce websites were done for study as it adapts all the aspects of personalization features implemented in websites like Amazon.in, eBay.in, Flipkart.com, etc. Pilot study was conducted to first verify the reliability of the questionnaire items, and to determine if survey items needed to be elucidated or changed.

Responses from 50 users were collected through questionnaires by asking them about their general online shopping experiences with personalized websites, their perceptions, and attitudes toward different personalization aspects when using e-commerce websites. Non-probability sampling method, i.e., convenience sampling, was adopted as data collection method for the main survey for online users of e-commerce websites in India. The population for the study is the online users of e-commerce web portals in India. After the completion of pilot study in two stages, final questionnaire was developed with several revisions. Data were collected with 600 responses through online form filling both by mailing users using social media sites and also visiting users personally from all over India. Researcher adopted multi-item scales to measure user's cognitive/utilitarian and hedonic experience like perceived usefulness, perceived ease of use, enjoyment, and control which is adopted from previous literature [9, 31]. Construct satisfaction and intention to revisit is adopted from [5, 7, 8, 19]. All construct items use five-point Likert scale.

After collection of 600 responses from e-commerce website users in India, incomplete and inconsistent data from responses were cleaned in data screening process. After initial screening of data, further responses with less standard deviation (i.e., below 0.30) were also removed to get valid responses. Before proceeding with the final analysis, data were cleaned by removal of incomplete and inconsistent data from both responses of e-commerce website out of which 547 valid responses were used from e-commerce. Cleaned data were analyzed with tool SPSS 20.0 for Exploratory Factor Analysis (EFA) and Confirmatory Factor Analysis (CFA), the final model was tested with SPSS Amos 21.0 with Structural Equation Modeling (SEM) technique.

The Cronbach's alpha coefficient for assessing reliability of survey items (variables) and analysis result indicates that all survey items were in the range of 0.70–0.93 which shows high level of internal consistency for questionnaire items used of scales in this survey. According to Nunnally [25], reliability coefficients representing internal consistency of construct items which is 0.70 or more are considered internal consistency of scale constructs of survey items. Thus, all survey items in constructs in Table 1 were reliable and appropriate to use in an actual research study.

5 Results and Findings

For analysis of data, factor analysis technique is used to summarize data, to interpret the relationships and understand the patterns of variables. This technique is used to regroup the variables in set of clusters based on their shared variance. EFA is used to identify the number of factors with group of variables and named that factors or constructs. CFA is used to find interrelationship among constructs. CFA confirms hypotheses and uses path analysis diagrams to represent variables and factors,

Table 1 Cronbach alpha coefficient of constructs

Web portal	Constructs	No. of items	Cronbach's alpha
E-commerce questionnaire	Information personalization	6	0.777
	Presentation personalization	6	0.816
	Navigation personalization	5	0.767
	Utilitarian/cognitive experience (perceived ease of use, perceived usefulness, enjoyment)	9	0.892
	Hedonic experience(Control)	2	0.772
	Satisfaction	2	0.945
	Intention to revisit	3	0.989

Table 2 Age-wise responses of users e-commerce website

		Frequency	Percent
Valid responses	18–25	424	77.5
	26–35	81	14.8
	36–50	40	7.3
	Above 60	2	0.4
	Total	547	100.0

whereas EFA is used to uncover complex patterns by exploring the dataset and testing predictions. In this study, EFA is needed to explore different aspects or dimensions of personalization and items of satisfaction. Maximum likelihood method of extraction is used as it gives correlation between factors in addition to factor loadings, and Promax oblique rotation technique is used because it is relatively efficient in achieving a simple oblique structure. The larger the sample size, the smaller the loadings are allowed for a factor to be considered significant [26]. Factor loading score of variable above 0.32 is statistically significant for sample size above 300 [28]. The factor loadings in Table 2 of e-commerce websites show fairly desirable factor loadings above 0.32.

5.1   EFA for E-commerce Website

Kaiser-Meyer-Olkin measure of sampling adequacy plays an essential role in accepting the sample adequacy, KMO ranges from 0 to 1 value, and the accepted index is over 0.6. Results in Table 3 show KMO value 0.926 which is above 0.6 that depicts good sampling adequacy for our research (Fig. 2).

Table 4 states factor loadings of through pattern matrix generated with maximum likelihood extraction method and Promax rotation method. Pattern matrix result gives all the factors and their loadings with items with similarity in EFA. Appropriate name of the factors was given based on nature of the questions and measuring variables falling under each factor. EFA identified seven factors as

Table 3 KMO and Bartlett's test (e-commerce website)

Kaiser-Meyer-Olkin measure of sampling adequacy		0.926
Bartlett's test of sphericity	Approx. Chi-Square	12,420.300
	Df	496
	Sig.	0.000

Total Variance Explained

Factor	Initial Eigenvalues			Extraction Sums of Squared Loadings			Rotation Sums of Squared Loadings[a]
	Total	% of Variance	Cumulative %	Total	% of Variance	Cumulative %	Total
1	11.770	36.783	36.783	8.515	26.608	26.608	9.732
2	2.803	8.759	45.541	1.876	5.861	32.469	7.670
3	2.283	7.133	52.674	3.979	12.436	44.905	5.056
4	1.544	4.826	57.500	2.112	6.601	51.507	7.249
5	1.259	3.934	61.434	.812	2.538	54.044	6.149
6	1.077	3.366	64.800	1.152	3.599	57.643	5.793
7	1.005	3.140	67.940	.939	2.935	60.578	5.112
8	.785	2.453	70.393				
9	.742	2.318	72.711				

Fig. 2 Total variance explained for e-commerce websites (SPSS EFA result snapshot)

information personalization, navigation personalization, presentation personalization, cognitive\utilitarian experience, hedonic experience (control), satisfaction, and intention to revisit. Table 4 mentions factor loadings of variables with underlying constructs of e-commerce web portals' personalization design aspects and its interrelationship with users cognitive experience, control, satisfaction, and intention to revisit.

The communality estimate values represent estimated proportion of variance of the variable which is free of error variance which is shared with other variables in the matrix and common with all others together [15]. There are 20 (4.0%) nonredundant residuals with absolute values greater than 0.05. A good fit model has less than 50% of the nonredundant residuals with absolute values that are greater than 0.05 which is true for our result. After comparing the reproduced correlation matrix with the original correlation coefficients matrix, our result in Fig. 3 shows 4% of residual which shows good model fit of factors.

Figure 4 is SPSS EFA result snapshot which shows correlation matrix of all seven identified matrixes with good correlations among factors. Information, navigation, and presentation personalization are highly correlated with cognitive, hedonic experience, satisfaction, and intention to revisit.

Table 4 Factor loadings with e-commerce website

Constructs	Variables	Factor						
		1	2	3	4	5	6	7
Cognitive experience	ECPEU4	0.820						
	ECPEU1	0.820						
	ECENJ1	0.801						
	ECENJ2	0.772						
	ECPU1	0.661						
	ECPEU3	0.594						
	ECPEU2	0.555						
	ECPU3	0.473						
	ECPU2	0.463						
Presentation personalization	ECPP5		0.837					
	ECPP3		0.813					
	ECPP2		0.674					
	ECPP1		0.657					
	ECPP4		0.629					
	ECPP6		0.628					
Information personalization	ECIP2			0.813				
	ECIP3			0.755				
	ECIP4			0.717				
	ECIP6			0.633				
	ECIP1			0.631				
	ECIP5			0.627				
Intention to revisit	ECINT1				0.970			
	ECINT3				0.947			
	ECINT2				0.916			
Navigation personalization	ECNP2					0.838		
	ECNP1					0.706		
	ECNP3					0.674		
	ECNP4					0.619		
Satisfaction	ECSAT1						0.983	
	ECSAT2						0.914	
Control	ECCON1							0.875
	ECCON2							0.502
	Extraction method: maximum likelihood							
	Rotation method: Promax with Kaiser normalization							
	Rotation converged in seven iterations							

ECPP5	.003	.005	-.005	.005	.000	.006	.001	.019	-.029
ECPP6	.010	-.035	.013	.005	-.004	-.012	-.001	.024	-.007
ECSAT1	.000	.001	.000	.000	.000	.002	-.002	.003	.000
ECSAT2	.002	-.004	.001	.001	-.001	-.007	.005	-.009	.005

Extraction Method: Maximum Likelihood.

a. Reproduced communalities

b. Residuals are computed between observed and reproduced correlations. There are 20 (4.0%) nonredundant residuals with absolute values greater than 0.05.

Fig. 3 Non redundant residuals with e-commerce website (SPSS EFA result snapshot)

Factor Correlation Matrix

Factor	1	2	3	4	5	6	7
1	1.000	.609	.447	.664	.561	.576	.529
2	.609	1.000	.392	.419	.568	.434	.441
3	.447	.392	1.000	.248	.257	.260	.282
4	.664	.419	.248	1.000	.427	.531	.512
5	.561	.568	.257	.427	1.000	.330	.351
6	.576	.434	.260	.531	.330	1.000	.455
7	.529	.441	.282	.512	.351	.455	1.000

Extraction Method: Maximum Likelihood.
Rotation Method: Promax with Kaiser Normalization.

Fig. 4 Factor correlation matrix (SPSS EFA result snapshot)

5.2 CFA and SEM for E-commerce Website

Our result of CFA for e-commerce website shows minimum discrepancy which is chi-square divided by degree of freedom, i.e., CMIN/DF 2.393 which should be less than 5, so my parsimonious model is fit. All NFI, RFI, and TLI are nearer to 0.9 which is good. RMSEA is 0.051 which is less than 0.06, so the model is having good fit. The Root-Mean-Square Error of Approximation (RMSEA) is associated with the residuals in the model, and good fit model ranges from zero to one [14]. The research results of the model estimation are shown in figure which is less than 0.06 showing better model fit of CFA. The CFA showed an acceptable overall model fit, and hence, the theorized model fits well with the observed data. It can be concluded that the hypothesized factor CFA model fits the sample data very well.

Structural Equation Modeling (SEM) technique tests the models stating causal relationships between latent variables which are hypothesized. SEM of e-commerce website data shows that all the hypotheses are supported. This indicates that personalized e-commerce website has a positive effect on users' satisfaction and intention to revisit website through positive cognitive and hedonic experience (Fig. 5).

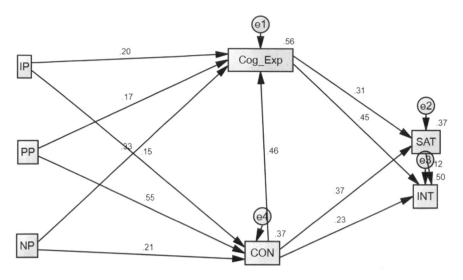

Fig. 5 SEM for personalized e-commerce website (SEM result snapshot from SPSS Amos)

6 Conclusion

This study addressed two related research questions: what and how personalization features have impacts on the cognitive and hedonic experience determining users' satisfaction and intention to continue to use a website. Our results suggest that different design aspects of personalization play a different role in this decision-making process. Users experience better enjoyment when the level of presentation personalization is perceived to be higher. Users also value information, presentation, and navigation personalization very much as it enhances the perceived usefulness, perceived ease of use of a website, enjoyment, and give users the experience of control. Among all the decision variables, cognitive experience with perceived ease of use, perceived usefulness, and enjoyment are found to be the most important antecedent factor determining the decision to continue using a website. Some of the findings in the are consistent with previous research, while others stand in contrast to other studies.

Presentation personalization adjusts the layout of user interface and provides content with good look and feel in the form of personalized themes, font, and background color generating ease of use and enjoy while browsing personalized e-commerce. Navigation personalization also makes the website easy to use by giving Internet users more flexibility and control. Result shows that information, presentation, and navigation personalization increase perceived usefulness, perceived ease of use, and enjoyment inducing positive cognitive experience with both e-commerce websites. Our result supports finding by Koufaris [17] that user experience positive impact on enjoyment and control as user may experience flow during the personalization process. Eroglu et al. [10, 11], Desai [9] identified that the presence of low task-relevant cues like look and feel of website positively affects the organism, e.g., pleasure, our results. Therefore, the presentation personalization can arouse the enjoyment [8, 11]. Personalization is a process that changes the website, information content, or distinctiveness functionality, interface of a system to increase its personal relevance to an individual user of website [1], and this finding is in line with our research findings.

Major findings of our research show that personalization reduces cognitive efforts of the user by personalized information provided which, in turn, decreases search time of user and increases efficiency. Also, relevant personalized information induces perceived usefulness with increased ease of use and enjoyment, user experience flow using personalized e-commerce, and social networking websites. Also, users feel satisfied with positive cognitive experience with personalized websites and likely to revisit the website, and this finding is consistent with similar findings in earlier research [11, 16, 31]. Tam and Ho [29] proposed conceptualization of web personalization and posit that the effectiveness of personalization is determined by the use of self-referent cues and the timely display of content relevant to the processing goal of the user. Such a conceptualization captures many of the functionalities of contemporary personalization agents such as adaptive content generation, customer profiling, web mining, and clickstream analysis. Earlier research considers control as important aspect of PAD emotional experience of pleasure, arousal, and dominance as intervening organismic state. User's positive cognitive and hedonic experience with personalization aspects lead to satisfaction [5, 9, 11, 32]. In accordance with previous research findings, this study finds that user with higher satisfaction is likely to revisit the personalized websites. Result in this research reveals that users, who experience satisfaction with personalization features through positive cognitive and hedonic experience, intend to return with personalized e-commerce websites.

Future Scope Research Future research can be conducted in several directions. First, different methodologies can be applied to cross-validate the findings in the current study. Longitudinal study can be done to study different roles of personalization features as user gains more experience over the period of time. Second, added dimensions of personalization from different viewpoints are also interesting and can be the subject of investigation, e.g., personalization strategies. Then, more mediating and moderating factors could also be taken into consideration.

References

1. Benbasat, I., & Komiak, S. (2006). The effects of personalization and familiarity on trust and a doption of recommendation agents. *MIS Quarterly, 30*(4), 941–960.
2. Blom, J. O., & Monk, A. F. (2003). Theory of personalization of appearance: Why users personalize their pcs and mobile phones. *Human-Computer Interaction, 18*(3), 193–228.
3. Chellappa, R. K., & Sin, R. G. (2005). Personalization versus privacy: An empirical examination of the online consumer's dilemma. *Information Technology and Management, 6*(2–3), 181–202.

4. Dabholkar, P. A., & Sheng, X. (2012). Consumer participation in using online recommendation agents: effects on satisfaction, trust, and purchase intentions. *The Service Industries Journal, 32*(9), 1433–1449. https://doi.org/10.1080/02642069.2011.624596.

5. DeLone, W. H., & McLean, E. R. (2013). The DeLone and McLean model of information systems success: A ten-year update. *Journal of Management Information Systems, 8*(4), 9–30. https://doi.org/10.1073/pnas.0914199107.

6. DeLone, W., & McLean, E. (1992). The quest for the dependent variable. *Information Systems Research, 3*(1), 60–95. https://doi.org/10.1287/isre.3.1.60.

7. Desai, D., & Kumar, S. (2015). Web personalization: A perspective of design and implementation strategies in websites. *Journal of Management Research & Practices.* ISSN No: 0976-8262.

8. Desai D. (2016). A study of personalization effect on users' satisfaction with ecommerce websites. *Sankalpa-Journal of Management & Research.* ISSN No. 2231-1904.

9. Desai D. (2017). *A study of design aspects of web personalization for online users in India.* Ph.D. thesis, Gujarat Technological University.

10. Eroglu, Sa, Machleit, Ka, & Davis, L. M. (2001). Atmospheric qualities of online retailing: A conceptual model and implications. *Journal of Business Research, 54*(2), 177–184. https://doi.org/10.1016/S0148-2963(99)00087-9.

11. Eroglu, S., Machleit, K., & Davis, L. (2001). Atmospheric qualities of online retailing. *Journal of Business Research, 54*(2), 177–184.

12. Eroglu, S. A., Machleit, K. A., & Davis, L. M. (2003). Empirical testing of a model of online store atmospherics and shopper responses. *Psychology and Marketing, 20*(2), 139–150. https://doi.org/10.1002/mar.10064.

13. Gie, Y. A., & Pearce, S. (2013). A beginner's guide to factor analysis: Focusing on exploratory factor analysis. *Tutorials in Quantitative Methods for Psychology, 9*(2), 79–94.

14. Hu, L. T., & Bentler, P. M. (1999). Cutoff criteria for fit indexes in covariance structure analysis: Conventional criteria versus new alternatives. *Structural Equation Modeling, 6*(1), 1–55.

15. Kamis, A., Marios, K., & Stern, T. (2008, March). Using an attribute-based decision support system for user-customized products online: an experimental investigation. *Mis Quarterly, 32*, 159–177.

16. Kobsa, A. (2007). Privacy-enhanced web personalization. *Communications of the ACM, 50*(8), 628–670. Retrieved from http://portal.acm.org/citation.cfm?id=1768197.1768222.

17. Koufaris, M. (2002). Applying the technology acceptance model and flow theory to online consumer behavior. *Information Systems Research, 13*(2), 205–223.

18. Kwon, K., & Kim, C. (2012). How to design personalization in a context of customer retention: Who personalizes what and to what extent? *Electronic Commerce Research and Applications, 11*(2), 101–116. https://doi.org/10.1016/j.elerap.2011.05.002.

19. Lai, J.-Y., Wang, C.-T., & Chou, C.-Y. (2008). How knowledge map and personalization affect effectiveness of KMS in high-tech firms. In *Proceedings of the Hawaii International Conference on System Sciences, Mauihi* (p. 355).

20. Lee, H.-H., Fiore, A. M., & Kim, J. (2006). The role of the technology acceptance model in explaining effects of image interactivity technology on consumer responses. *International Journal of Retail & Distribution Management, 34*(8), 621–644. https://doi.org/10.1108/09590550610675949.

21. Liang, T. P., Yang, Y. F., Chen, D. N., & Ku, Y. C. (2008). A semantic-expansion approach to personalized knowledge recommendation. *Decision Support Systems, 45*, 401–412. http://doi.org/10.1016/j.dss.2007.05.004.

22. Liang, T.-P., Li, Y.-W., & Turban, E. (2009). Personalized services as empathic responses: The role of intimacy. In *PACIS 2009 Proceedings*. Retrieved from http://aisel.aisnet.org/pacis2009/73.

23. Liang, T.-P., Chen, H.-Y., Du, T., Turban, E., & Li, Y. (2012). Effect of personalization on the perceived usefulness of online customer services: A dual-core theory. *Journal of Electronic Commerce Research, 13*(4), 275–288. Retrieved from http://www.ecrc.nsysu.edu. tw/liang/paper/2/79. Effect of Personalization on the Perceived (JECR, 2012).pdf.

24. Monk, A. F., & Blom, J. O. (2007). A theory of personalization of appearance: Quantitative evaluation of qualitatively derived data. *Behavior & Information Technology, 26*(3), 237–246. http://doi.org/10.1080/01449290500348168.

25. Montgomery, A. L., & Smith, M. D. (2009). Prospects for personalization on the internet. *Journal of Interactive Marketing, 23*(2), 130–137.

26. Nunnally, J. C. (1978). *Psychometric theory* (2nd ed.). New York: McGraw-Hill.

27. Oulasvirta, A., & Blom, J. (2008). Motivations in personalization behavior. *Interacting with Computers, 20*(1), 1–16. https://doi.org/10.1016/j.intcom.2007.06.002.

28. Tabachnick, B. G., & Fidell, L. S. (2007). *Using multivariate statistics* (5th ed.). Boston, MA: Allyn & Bacon/Pearson Education.

29. Tam, K. Y., & Ho, S. Y. (2006). Understanding the impact of web personalization on user information processing and decision. *MIS Quarterly, 30*(4), 865–890.

30. Thongpapanl, N., Catharines, & Ashraf, A. R. (2011). Enhance online performance through website content and personalization. *Journal of Computer Information Systems, 52*(1), 3–13.

31. Tsekouras, D., Dellaert, B. G. C., & Li, T. (2011). Content learning on websites: The effects of information personalization. SSRN eLibrary. Retrieved from http://papers.ssrn.com/sol3/papers.cfm?abstract_id=1976178.

32. Wang, Y., & Yen, B. (2010). The effects of website personalization on user intention to return through cognitive beliefs and affective reactions. In *PACIS 2010 Proceedings* (pp. 1610–1617).

33. Ying, W. (2009). The effects of website personalization on user intention to return through cognitive beliefs and affective reactions. The University of Hong Kong.

Hierarchical Summarization of Text Documents Using Topic Modeling and Formal Concept Analysis

Nadeem Akhtar, Hira Javed and Tameem Ahmad

Abstract Availability of large collection of text documents triggers the need for large-scale text summarization. Identification of topics and organization of documents are important for analysis and exploration of textual data. This study is a part of the growing body of research on large-scale text summarization. It is an experimental approach and it differs from earlier works in the sense that it generates topic hierarchy which simultaneously provides a hierarchical structure to the document corpus. The documents are labeled with topics using latent Dirichlet allocation (LDA) and are automatically organized in a lattice structure using formal concept analysis (FCA). This groups the semantically related documents together. The lattice is further converted to a tree for better visualization and easy exploration of data. To signify the effectiveness of the approach, we have carried out the experiment and evaluation on 20 Newsgroup Dataset. Results depict that the presented method is considerably successful in forming topic hierarchy.

Keywords Formal concept analysis · Latent Dirichlet allocation
Topic modeling

1 Introduction

With the explosion of web, there has been a huge increase in textual data. This textual data covers all domains. It could be in the form of emails, reports, blogposts, scientific articles, news, etc. As such, there is a large need for summarization of this textual data for further search, exploration and analysis. This has led to the

N. Akhtar (✉) · H. Javed · T. Ahmad
Department of Computer Engineering, ZHCET, AMU, Aligarh, India
e-mail: nadeemalkhtar@gmail.com

H. Javed
e-mail: hirajaved05@gmail.com

T. Ahmad
e-mail: tameemahmad@gmail.com

© Springer Nature Singapore Pte Ltd. 2019   21
V. E. Balas et al. (eds.), *Data Management, Analytics and Innovation*,
Advances in Intelligent Systems and Computing 839,
https://doi.org/10.1007/978-981-13-1274-8_2

development of several topic modeling programs. The topic modeling programs give a summary of text documents in the form of topics. Hierarchical summarization is a recent advancement in the area of text summarization. The hierarchically summarized information is more organized, coherent, collated, personalized, and interactive [1].

This work aims at automatic summarization of text documents by creating a hierarchy of topics. At the top of the hierarchy, the topics are generic while they become specific towards the bottom. The general topics have more number of documents associated with them at the upper level while the count decreases as one goes down the hierarchy. This work utilizes the supervised methodology of latent Dirichlet allocation (LDA) to generate topics and the unsupervised methodology of formal concept analysis (FCA) is used for hierarchy creation. The documents are taken as objects and topics generated by LDA are taken as their attributes. The novelty lies in the fact that the themes of the concept lattice will represent a topic associated with their corresponding documents. The inherent property of generalization and specialization of FCA is used. The result is a concept lattice which is further converted to a tree structure by eliminating the edges that cross each other. Thus, a tree-based summary of the entire document corpus is obtained. The experiment focuses at:

- Forming hierarchy of topics
- Discovering relationships between topics generated by LDA
- Labeling documents with topics
- Organizing documents in hierarchical structure.

1.1 Topic Modeling

Topic models are algorithms which infer topics from text. The main themes of the text are obtained through topic models. A topic is a list of words that occur in statistically meaningful ways [2]. Topic model programs assume that any document is composed by selecting words from possible topics. A topic modeling program aims at mathematically decomposing a text into topics which give rise to it. It iteratively follows this process until it settles on the most likely distribution of words into baskets which we call topics. In LDA, the topic structure from the documents is determined by computing the posterior distribution and the conditional distribution of the hidden variables when the documents are given [3]. The LDA technique is the most popular technique used for topic modeling. Our work uses MALLET [4] for topic modeling. Topic modeling is a useful concept when we have large document corpus and we cannot read documents to manually detect topics in corpus.

1.2 Formal Concept Analysis

FCA is a data analysis theory which was founded by Rudolf Wille [5]. FCA facilitates the discovery of concepts in datasets. The idea of FCA is to transform the explicit knowledge which is given in a data table called as formal context into the implicit knowledge. FCA is useful in the deduction of concepts having similar objects and common features. In FCA, the is-a or has-a relationship of objects and their attributes is converted to a generalization–specialization relationship between the formal concepts [6]. The resulting structure is a concept lattice which is depicted by a Hasse diagram yielding visualization for analyzing large datasets in a hierarchical form.

A (formal) context can be described as $K = (G, M, I)$, where G is a set of objects, M is a set of attributes, and the binary relation $I \subseteq G \times M$ describes the relationship of objects with their attributes. The derivation operator $(.)^I$ is defined for $A \subseteq G$ and $B \subseteq M$ as follows:

$$A^I = \{m \in M | \forall g \in A : gIm\};$$

$$B^I = \{g \in G | \forall m \in B : gIm\};$$

In a formal concept (A, B), A (known as extent of formal concept) is a set of objects and B (known as intent of formal concept) is a set of attributes. Here, $A^I = B$ is the set of attributes which are common to all objects of A and $B^I = A$ is the set of objects which share all attributes of B [7]. A concept (C, D) will be a sub-concept of (A, B) if $C \subseteq A$. Therefore, due to the fact that the formal concepts are ordered sets, they can be represented by Hasse diagrams in the form of a lattice. Thus with $C \subseteq A$ and no concept (E, F) with $C \subseteq E \subseteq A$, (C, D) will be placed directly below (A, B) in the concept lattice. This gives rise to generalization–specialization property in FCA because if B is the property related to a group of objects A, then a property related to the subset of A will be a specialized property as it satisfies the subset of objects.

2 Related Work

Numerous methods have been used and a lot of algorithms have been proposed for topic identification. The different methods include classification techniques, clustering techniques, matrix factorization, graph-based techniques, probabilistic techniques, and recently FCA. Bengel et al. [8] introduced the concept of automatic topic identification in chat rooms using classification techniques. Cataldi et al. [9] classified short narrative from scientific documents by using the concept of LDA. Titov and McDonald [10] used extended methods like LDA and PLSA to incite multigrain topics in modeling online reviews. Several hierarchical clustering algorithms have been implemented for the organization of textual data. The algorithm in [11] is based on a constrained agglomerative clustering framework that organizes documents in hierarchical trees. In [12], the basis vectors and matrix projections by a nonnegative matrix factorization approach were applied to identify topics in individual electronic mail messages. Hierarchical Dirichlet process has been introduced in [13]. In this process, topics are arranged in a hierarchy through the nested CRP. Vo and Ock [14] use graphical approach for the detection of topics. Topics are obtained from the structure of graph using community detection approach. The drawback in graph-based approaches is that further processing of graphs is carried out for identifying communities or topics. Topics are intuitively hierarchical so representing them in a hierarchical structure is preferred [6, 15]. Cigarrán et al. [6] utilized the FCA theory to detect topics in tweets by analyzing the terms of tweets and the tweets which generated them. In our work, the formal concepts denote the conceptual representations of topics and the documents that have originated them.

3   Proposed Work

In our work, we have proposed a novel idea of generating a topic hierarchy based on FCA. This idea could be understood by an example. Take for example the formal context of Table 1. Here, the documents are taken as objects and the topics they fit in are taken as attributes.

The lattice of Fig. 1 depicts a topic hierarchy. There are eight concepts in the above example. The concepts numbered "0" and "7" are irrelevant for inferring topics. Each concept consists of a pair depicting a topic and the corresponding documents of that topic. The property of a concept lattice is that the attributes have a generalization–specialization property. Hence, the topics at upper level are general topics while at lower levels they are specialized. So a topic will have more documents associated with it while the subtopic will contain the subset of the documents. For example, if in concept number 2 the topic is "Computer" and documents 1 and 3 belong to it, then Topics 2 and 3 could be "software" and "hardware" respectively with document 3 belonging to software while document 2 belonging to hardware. The topic hierarchy framework is shown in Fig. 2.

Table 1 An example of formal context

	Topic 1	Topic 2	Topic 3	Topic 4	Topic 5	Topic 6
Doc 1	x		x			
Doc 2				x	x	
Doc 3	x	x				
Doc 4				x		x

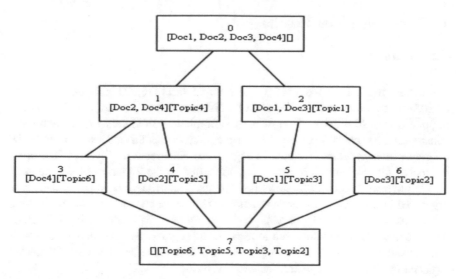

Fig. 1 Concept lattice corresponding to the formal context of Table 1

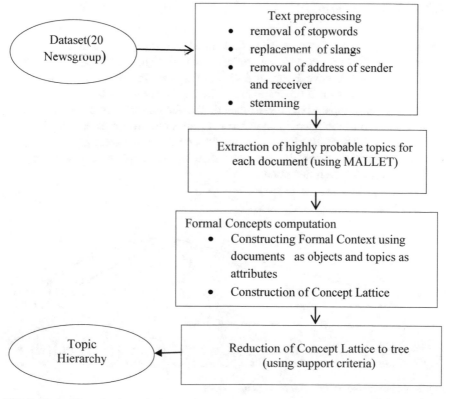

Fig. 2 Topic hierarchy framework overview

4   Experiment and Results

4.1   Dataset

The dataset used in this work is 20 newsgroup dataset [16]. This dataset comprises of 20,000 news documents taken from 20 different usenet newsgroups. The data is organized into 20 different groups, each symbolizing a different topic. Each topic consists of 1000 news documents. The topic hierarchy of this dataset is taken as the reference hierarchy and is shown in Fig. 3 and elaborated in Table 2. The topics numbered from 1.1 to 5.3 in the table correspond to the 20 topics of the dataset.

We have worked on 2000 news documents. The original newsgroup dataset is organized in 20 folders. We have retained 100 news documents in each folder. We have considered each folder as a single document to extract topics using LDA. The news documents are preprocessed to remove stopwords, mentions, and addresses of senders and receivers and stemming is performed on the reduced dataset. Slangs are replaced by their corresponding meaningful terms.

4.2   Topic Generation

For generating topics, topic modeling tool called Mallet [4] is used. Mallet uses the technique of LDA for generating topics. Two files, tutorial_keys.txt and tutorial_composition.txt, have been utilized. tutorial_keys.txt gives a text document outputting the topics for 2000 documents. The number of topics is set to 25 and top 20 words for each topic have been extracted. The file tutorial_composition.txt gives the composition of each document in terms of all topics in percentage. Each of the documents consists of 100 original news documents. After getting the topic proportion for each document, a threshold of 0.11 was applied to get the highly probable topics for each document.

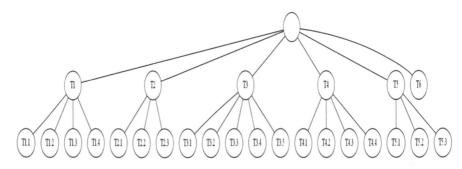

Fig. 3 Reference hierarchy of 20 newsgroup dataset

Table 2 Topic information corresponding to the reference hierarchy of Fig. 3

Topic No.	Topic
1	Recreation
2	Religion
3	Computer
4	Science
5	Talk
6	Miscellaneous (misc.forsale)
1.1	rec.autos
1.2	rec.motorcycles
1.3	rec.sport.baseball
1.4	rec.sport.hockey
2.1	talk.religion.misc
2.2	alt.atheism
2.3	soc.religion.christian
3.1	comp.graphics
3.2	comp.os.ms-windows.misc
3.3	comp.sys.ibm.pc.hardware
3.4	comp.sys.mac.hardware
3.5	comp.windows.x
4.1	sci.crypt
4.2	sci.electronics
4.3	sci.med
4.4	sci.space
5.1	talk.politics.misc
5.2	talk.politics.guns
5.3	talk.politics.mideast

4.3   Topic Hierarchy Computation

After finding the characteristic topics for each document, a formal context is prepared by considering documents as objects and topics as attributes. The formal context is shown in Fig. 4.

It is difficult to understand the hierarchical organization of topics in a lattice, so the lattice has to be reduced to a tree. The primary difference between a tree and a lattice is that in a tree each child node has exactly one parent while in lattice a child node can have more than one parent. So, the lattice is converted to a tree by selecting a single parent. This is done by removing multiple edges between a child and its many parents and retaining only one parent. In this work, the tree is obtained using the support criteria.

For the formal context given by $K = (G, M, I)$, suppose $B \subset M$, the support count of the attribute set B in K is

```
                              8 17 11 19 18 15 5 9 16 4 13 12 22 6 10 20 23 21 1 3 14 2 24 7
alt.atheism                   1  1  0  0  0  0 0 0  0 0  0  0  0 0  0  0  0  0 0 0  0 0  0 0
comp.graphics                 0  0  1  1  1  0 0 0  0 0  0  0  0 0  0  0  0  0 0 0  0 0  0 0
comp.os.ms-windows.misc       0  0  0  1  0  1 0 0  0 0  0  0  0 0  0  0  0  0 0 0  0 0  0 0
comp.sys.ibm.pc.hardware      0  0  0  0  0  0 1 0  0 0  0  0  0 0  0  0  0  0 0 0  0 0  0 0
comp.sys.mac.hardware         0  0  0  1  0  0 1 1  0 0  0  0  0 0  0  0  0  0 0 0  0 0  0 0
comp.windows.x                0  0  0  0  1  0 0 0  1 0  0  0  0 0  0  0  0  0 0 0  0 0  0 0
misc.forsale                  0  0  0  0  0  0 0 1  0 0  0  0  0 0  0  0  0  0 0 0  0 0  0 0
rec.autos                     0  0  0  1  0  0 0 0  0 1  1  0  0 0  0  0  0  0 0 0  0 0  0 0
rec.motorcycles               0  0  0  0  0  0 0 0  0 1  1  0  0 0  0  0  0  0 0 0  0 0  0 0
rec.sport.baseball            0  0  0  0  0  0 0 0  0 0  1  1  0 0  0  0  0  0 0 0  0 0  0 0
rec.sport.hockey              0  0  0  0  0  0 0 0  0 0  1  1  0 0  0  0  0  0 0 0  0 0  0 0
sci.crypt                     0  0  0  0  0  0 0 0  0 0  0  1  1 0  0  0  0  0 0 0  0 0  0 0
sci.electronics               0  0  0  1  0  0 0 0  0 0  0  0  0 1  0  0  0  0 0 0  0 0  0 0
sci.med                       0  0  0  1  0  0 0 0  0 0  0  0  0 0  1  0  0  0 0 0  0 0  0 0
sci.space                     0  0  0  0  0  0 0 0  0 0  0  0  0 1  0  0  0  0 0 0  0 0  0 0
soc.religion.christian        0  1  0  0  0  0 0 0  0 0  0  0  0 0  0  1  0  0 0 0  0 0  0 0
talk.politics.guns            0  0  0  0  0  0 0 0  0 0  0  0  0 0  0  0  1  1 0 0  0 0  0 0
talk.politics.mideast         0  0  0  0  0  0 0 0  0 0  0  0  0 0  0  0  0  1 1 0  0 0  0 0
talk.politics.misc            0  0  0  0  0  0 0 0  0 0  0  0  0 0  0  0  0  0 1 0  0 1  0 0
talk.religion.misc            0  1  0  0  0  0 0 0  0 0  0  0  0 0  0  0  1  0 0 0  0 0  0 1
```

Fig. 4 Formal context of documents and topics

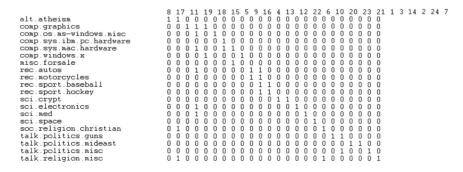

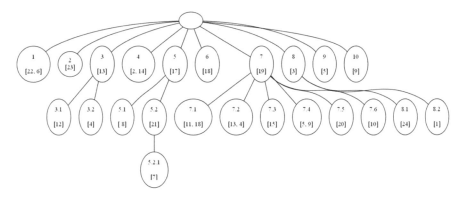

Fig. 5 Resulting topic hierarchy

$$\Phi = \text{Card } (B')/\text{Card (G)}$$

where Card stands for cardinality [17]. For a concept with multiple parent nodes, the node with the highest support is selected as the parent node. In other words, the subtopic will be attached to that parent topic which has large number of documents. The formal context of Fig. 4 is utilized to obtain the topic hierarchy shown in Fig. 5 and its elaboration in Table 3.

A total of 23 nodes are obtained in the hierarchy. Each node corresponds to a topic. The numbers in the square bracket in Fig. 5 are topic numbers obtained through LDA. The top 5 keywords of these topics are taken as topic name in the hierarchy. The concept of FCA gives a combination of attributes and objects. So each of the topics is associated with its documents.

Table 3 Topic information corresponding to the reference hierarchy of Fig. 5

Topic No.	Topical words	Documents
1	Key encryption privacy anonymous security Mail data computer technology users	sci.crypt
2	Space NASA orbit shuttle earth	sci.space
3	Writes good back day bad	rec.autos, rec.motorcycles, rec.sport.baseball, rec.sport.hockey
4	Human dead women policy peace Armenian Turkish Armenians Israeli people	talk.politics.mideast
5	People time point fact things	talk.religion.misc, alt.atheism soc.religion.christian
6	Information number set based list	comp.windows.x, comp.graphics
7	Work writes time make university	comp.sys.mac.hardware, rec.autos, sci.electronics, comp.graphics, sci.med, comp.os.mswindows.misc
8	Law state rights make public	talk.politics.misc, talk.politics.guns
9	Drive disk hard card drives	comp.sys.mac.hardware, comp.sys.ibm.pc.hardware
10	Apple price mac computer mail	comp.sys.mac.hardware, misc.forsale
3.1	Team play game year players	rec.sport.baseball, rec.sport.hockey
3.2	Car defeat day cars bike	rec.autos, rec.motorcycles
5.1	God atheists system atheism religious	alt.atheism
5.2	God Jesus Bible Christ Church	soc.religion.christian, talk.religion.mis
5.2.1	Man moral words good word	talk.religion.misc
7.1	Graphics color software image polygon Information number set based list	comp.graphics
7.2	Writes good back day bad Car defeat day cars bike	rec.autos
7.3	Windows file size memory system	comp.os.mswindows.misc
7.4	Drive disk hard card drives Apple price mac computer mail	comp.sys.mac.hardware

(continued)

Table 3 (continued)

Topic No.	Topical words	Documents
7.5	Banks Gordon pain disease intellect	sci.med
7.6	Power radar current low circuit	sci.electronics
8.1	War states south secret president	talk.politics.misc
8.2	Gun guns control crime weapons	talk.politics.guns

5   Evaluation

The reference hierarchy has six general topics. These are recreation, talk, religion, computer, science, and miscellaneous. There are five individual hierarchies corresponding to these topics, i.e., five topics have subtopics. The topic miscellaneous (misc.forsale) has no subtopic. In our analysis, the number of topics at level 1 which have subtopics is 4. This quite resembles the base hierarchy. The topics computer and science are merged in topic 7. All 20 documents appear in some topic or the other. Therefore, the document coverage is 100%.

The hierarchy is evaluated with respect to the documents of resulting hierarchy. Precision, recall, and F-measure are used to evaluate the hierarchy. Precision is the percentage of documents found in the topic that are relevant. Recall gives the percentage of the relevant documents that have been found in the hierarchy. The overall accuracy of the method is specified by F-measure which is the harmonic mean of precision and recall. Each of the six general topics of reference hierarchy is used to evaluate the hierarchy. This hierarchical evaluation is taking place by analyzing the hierarchy at level 1.

Recreation   Topic number 3 in the resulting hierarchy consists of the documents related to recreation. This topic is labeled as "writes good back day bad" as mentioned in Table 3. This topic is specialized further to two other topics which are at level 2. These are "team play game year players" and "car defeat day cars bike". The topic "team play game year players" consists of rec.sport.baseball and rec.sport.hockey as documents while the topic "car defeat day cars bike" contains rec. autos and rec.motorcycles as documents. This hierarchy at topic 3 has all the subtopics corresponding to recreation in the reference hierarchy. The precision of this is 1 because all four documents belong to the same subject matter. The recall is also 1 as all the four documents of this subject matter are obtained when compared with the reference hierarchy. F-measure is 1.

Religion The hierarchy at topic 5 contains three documents talk.religion.misc, alt. atheism, and soc.religion.christian, all related to religion. Hence the precision is 1. This signifies 100% convergence to the subtopics related to this subject matter in the reference hierarchy. So, the recall is also 1. F-measure is 1.

Science Topic number 7 has two topics related to science out of its 6 subtopics. The precision comes out to be 0.33333 while recall for the documents of is 2/4, i.e., 0.5 in this sub-hierarchy. The F-measure is 0.399997.

Computer For hierarchical evaluation, topic number 7 is considered. It has three documents related to computer which further branch and extend to level 2. The precision is therefore 3/6, i.e., 0.5 while the recall is 3/5, i.c., 0.6 for the subject matter computer. The F-measure is 0.545454.

Talk Two topics out of the three related to this subject matter are grouped under topic number 8. The precision at this part of hierarchy is 1 since both the documents belong to same subject matter. The recall is 0.6666666. F-measure is 0.8.

Miscellaneous There is one topic misc.forsale in the reference hierarchy. In the proposed hierarchy, the documents related to this subject are appearing under the topic "apple price mac computer mail" along with the documents of the topic comp. sys.mac.hardware. The precision is therefore 0.5 while recall is 1. F-measure is 0.66666.

The overall hierarchical precision is 0.722222. Recall is 0.7944444 and the F-measure comes out to be 0.7566136.

5.1 Comparison with hlda (Hierarchical Latent Dirichlet Allocation)

Hierarchical LDA (hlda) gives a topic hierarchy. It is based on the technique of LDA. The hierarchy obtained by running hlda on our dataset is shown in Fig. 6 and elaborated in Table 4.

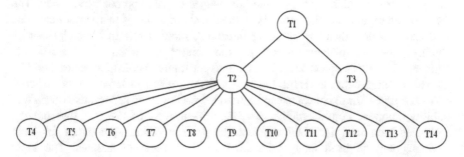

Fig. 6 Topic hierarchy of hlda

Table 4 Topic information corresponding to hlda hierarchy in Fig. 6

Topic No.	Topic
T1	Writes people time good make
T2	Car day defeat cars writes
T3	Space NASA orbit shuttle earth
T4	Drive disk card hard computer
T5	God Jesus Bible Christ Christian
T6	Windows file graphics software version
T7	God atheists atheism system religious
T8	File window entry program widget
T9	Year game players baseball runs
T10	Play team hockey game players
T11	Gun guns crime control weapons
T12	Armenian people Armenians Turkish
T13	War south states secret nuclear
T14	Key encryption privacy anonymous security

hlda gives 14 topics when the hierarchy is set to 3 levels. It gives the summary of the entire document corpus similar to the way obtained by our approach. Our approach has two advantages over hlda. First, hlda is not grouping the corpus according to the subject matter. The co-related topics do not show parent–child relationship here. In our approach, the distinct topics are separated at the first level. In the subsequent levels, the subtopics are someway related to the parent topic. Second, our methodology gives a way of arranging the documents. The subtopics have a subset of the documents of original topic.

6 Conclusion

A method is proposed for obtaining topic hierarchy in this work. The values of precision, recall, and F-measure show that the proposed approach is quite successful in forming topic hierarchy and organizing documents within these topics. Comparison with hlda highlights the advantages of this approach over hlda. The levels of the topic hierarchy depends on the threshold value, if more topics are used to describe a document by reducing threshold, more levels in hierarchy can be obtained, this completely depends on the dataset for which the hierarchy is retrieved. This work opens new research lines for future. Topic hierarchies could be formed on large datasets. These hierarchies would result in large number of concepts in lattice. These concepts can be reduced on the basis of various criteria like stability, support, confidentiality, etc. Experiments can be done to find out the optimized parameter for obtaining topic hierarchies. Other approaches for topic modeling can be used along with FCA to form topic hierarchies. The topic hierarchies could be obtained along multilingual dimensions. Moreover, this work is based on static topic modeling scheme, and the topic hierarchies could be obtained for real-time scenario.

References

1. Christensen, J. M. (2015). *Towards large scale summarization* (Doctoral dissertation).
2. http://programminghistorian.org/lessons/topic-modeling-and-mallet.
3. Blei, D. M. (2012). Probabilistic topic models. *Communications of the ACM, 55*(4), 77–84.
4. *MALLET Homepage.* (2017). *Mallet.cs.umass.edu.* Retrieved December 7, 2017, from http://mallet.cs.umass.edu/.
5. Uta priss HomePage. (2017). Retrieved December 7, 2017, from http://www.upriss.org.
6. Cigarrán, J., Castellanos, Á., & García-Serrano, A. (2016). A step forward for topic detection in Twitter: An FCA-based approach. *Expert Systems with Applications, 57,* 21–36.
7. Kuznetsov, S., Obiedkov, S., & Roth, C. (2007, July). Reducing the representation complexity of lattice-based taxonomies. In *International Conference on Conceptual Structures* (pp. 241–254). Berlin: Springer.
8. Bengel, J., Gauch, S., Mittur, E., & Vijayaraghavan, R. (2004, June). Chattrack: Chat room topic detection using classification. In *International Conference on Intelligence and Security Informatics* (pp. 266–277). Berlin: Springer.
9. Cataldi, M., Di Caro, L., & Schifanella, C. (2010, July). Emerging topic detection on Twitter based on temporal and social terms evaluation. In *Proceedings of the Tenth International Workshop on Multimedia Data Mining* (p. 4). ACM.
10. Titov, I., & McDonald, R. (2008, April). Modeling online reviews with multi-grain topic models. In *Proceedings of the 17th International Conference on World Wide Web* (pp. 111–120). ACM.
11. Zhao, Y. (2010, November). Topic-constrained hierarchical clustering for document datasets. In *International Conference on Advanced Data Mining and Applications* (pp. 181–192). Berlin: Springer.
12. Berry, M. W., & Browne, M. (2005). Email surveillance using non-negative matrix factorization. *Computational & Mathematical Organization Theory, 11*(3), 249–264.
13. Griffiths, D. M. B. T. L., & Tenenbaum, M. I. J. J. B. (2004). Hierarchical topic models and the nested Chinese restaurant process. *Advances in Neural Information Processing Systems, 16,* 17.
14. Vo, D. T., & Ock, C. Y. (2015). Learning to classify short text from scientific documents using topic models with various types of knowledge. *Expert Systems with Applications, 42*(3), 1684–1698.
15. Carpineto, C., & Romano, G. (2004). *Concept data analysis: Theory and applications.* Hoboken: John Wiley & Sons.
16. Newsgroup Dataset HomePage. (2017). [online] Available at: http://qwone.com/~jason/20Newsgroups/. Accessed December 7, 2017.
17. Melo, C., Le-Grand, B., Aufaure, M. A., & Bezerianos, A. (2011, July). Extracting and visualising tree-like structures from concept lattices. In *2011 15th International Conference on Information Visualisation (IV)* (pp. 261–266). IEEE.

Green Information and Communication Technology Techniques in Higher Technical Education Institutions for Future Sustainability

Kavita Suryawanshi

Abstract Information and communication technology (ICT) plays a crucial role in education sector. But ICT has negative impact if it is not used effectively. ICT needs to be used efficiently by keeping environment sustainability in mind to protect our mother Earth. The remarkable climate change in current years is a pointer that the Earth is sick. The entire world is responsible to save our environment. India has acknowledged the principle of sustainable development as part of their development policy. The Green or sustainable ICT practices are desirable to be followed by everyone for future sustainability. Every individual in the field of ICT is required to be a trendsetter for ICT sustainability, which is becoming an emerging research field. The objective of the paper is to recommend Green ICT techniques to be followed by higher education institutions. This paper provides introductory definitions of Green or sustainable ICT. It explains the importance of Green ICT. The study followed a qualitative methodological approach and primary data were collected through questionnaire and interview. These research findings show that Green ICT practices and awareness is less among the students of the selected education institutions in Pune in India. Based on the result analysis, the author has suggested techniques to improve Green ICT awareness in higher education institutions in order to solve some of the most challenging problems related to future sustainability.

Keywords Sustainable or Green ICT (GICT) · Green ICT techniques
Higher technical education institutions (HEI)

K. Suryawanshi (✉)
Dr. D. Y. Patil Institute of Master of Computer Applications,
Savitribai Phule Pune University, Akurdi, Pune, Maharashtra 411044, India
e-mail: kavita1104@yahoo.com

© Springer Nature Singapore Pte Ltd. 2019
V. E. Balas et al. (eds.), *Data Management, Analytics and Innovation*,
Advances in Intelligent Systems and Computing 839,
https://doi.org/10.1007/978-981-13-1274-8_3

1   Introduction and Literature Review

The entire world has been changing rapidly due to highly developed technological changes and globalization. In a very short period of time, information and communication technology (ICT) has become one of the essential requirements of day-to-day life of modern society. Many countries have been considering the basic proficiency and concepts of ICT as part of the core of education. UNESCO aims to ensure that all countries are having access to the most excellent learning facilities essential to prepare youth to contribute to the country [1].

Recently, Green IT was defined by professionals with few objectives such as to include the smallest quantity of harmful resources, to be energy efficient during their lifecycle, and to be used with the lowest result on the environment and human health [2].

Sustainable or Green ICT is a novel approach of using ICT connected to the ecological problems. The increasing changes in the ICT sector demand elucidation to conserve energy utilization in the field of ICT. These outcomes are called Green ICT [3].

GICT is defined as a method to promote ICT professionals to think about ecological harms and discover solutions to them [4]. The international energy agency (IEA) has to perform essential job in the implementation of Gleneagles Plan of Action on atmosphere transform, hygienic energy, and sustainable growth [5]. It is a necessity and huge requirement to think about future sustainability [6]. Recent studies on Green ICT define it with diverse characteristics and perceptions. Definitions made and proposed by Mingay [7], Molla [8], Vanessa [9], Mohamad [10], and Murugesan [2] core on manufacturing and commerce aspects in terms of resource utilization, e-waste creation, and carbon footprint.

The author studied and described Green ICT as below based on literature review. Green ICT is a revolutionary technique of applying ICT that comprises of policies and practices for sustenance of ICT [11]. To summarize, Green or sustainable ICT is simply an effective use of ICT in view of its application.

The rest of the paper proceeds as follow: first, the importance of sustainable ICT is presented and the research methodology is described in brief. Later, the result analysis and methods to be followed to improve Green ICT awareness among all stakeholders of higher education institutions are discussed, followed by conclusion.

2   Importance of Green ICT

ICT is progressively playing an important role as an enabling means for the delivery of efficient and effective government services. The universities are implementing the use of ICT for additional proficient and cutthroat processes both in delivery of services and in managerial procedures. The commencement of information and communication technologies (ICTs) is varying the way universities work. However, it has been reported by Gartner Research Group that ICT emissions are anticipated to rise by 60% by the year 2020 [12].

The wastage from electronic goods consist of venomous materials like cadmium and lead in the circuit boards and in monitor cathode ray tubes (CRTs), mercury in switches, etc. effects on physical condition and environment. The resources have been identified to be hard to reprocess in an ecologically sustainable way even in developed countries [13].

The ozone coating is getting thinner due to massive quantity of carbon discharge, which is dangerous for human life [14]. Currently, the world is suffering from environmental problems mostly due to global warming. It is need of the hour to deal with the climate changes of nature as the result of increasing atmospheric amalgamation of carbon dioxide (CO_2) and that an individual action was responsible for this rise in CO_2. This leads to undesirable social, economical, and ecological effects. At the same time, there is a huge need to pursue energy-efficient methods to diminish energy consumption and ultimately to optimize the use of natural resources.

Today, some of the places in India are suffering from power load sharing whereas, during use of ICT, many of us are not worried about energy consumption, generation of hazardous e-waste, and so on. Why people do not understand that the energy is precious, as it has been made from scared natural resources like coal, water, etc. which need to be preserved for our next generation. All human beings need to act as a motivator. ICT has been recognized by India as a key enabler for sustainable socioeconomic development and pillars of ICT needing green solutions.

The Green ICT mainly consists of sustenance of ICT for protecting energy for current ICT only as well as protecting Earth from harmful carbon emissions, which is the main reason for global warming. In view of this, the universities and colleges need to implement additional sustainable techniques toward the implementation of ICT. It is observed that the rate of expansion of higher education institutions and enrollment of the students coursewise are extremely rising. Similarly, the application of ICT in the field of education is tremendously increasing. A report of the United Nations predicted a very high hike in e-waste generation [15].

Global warming and climate change issues have brought the entire world closer together as part of social responsibility [16]. Now the sustenance of the planet and each creature on it has become need of the hour. To achieve this, there is a social demand for an innovative approach in using ICT. Green ICT recommends making use of energy-efficient materials and moderate energy expenditure [17]. In view of this, it is important to focus on Green ICT implementation at higher education institutions keeping future sustainability in mindset.

3   Research Methodology

This section presents the design of the research conducted to find out the GICT awareness among all the stakeholders of the HEI. This study is primarily focused on awareness and usage of Green ICT practices followed by higher education institutions in Pune district. In view of this, primary data are collected from the stakeholders of higher education institutions affiliated to different universities in Pune district in India. A survey method is employed and a structured questionnaire is designed to measure Green ICT awareness and practices followed by the respondents categorized into Directors, Faculty, and Students. Both open-ended and close-ended questions have been employed for eliciting desired information from respondents. A pilot study was conducted at the end of the exploratory phase. The feedback received at the end of the pilot study was used to improve the questionnaire before the survey is carried out. The Cronbach alpha reliability test has been used to measure the pilot survey data.

The survey questionnaire is developed based on the literature reviews. The researcher has collected primary data in the fieldwork conducted at selected 47 higher technical education institutions (HEI) in Pune district in India. The questionnaire consists of various data attributes like Green ICT meaning, awareness of GICT practices and barriers in the implementation of Green ICT [18], etc. The collected data are properly analyzed with the use of SPSS (statistical package for the social sciences) tool of version 16.0. The research study has used statistical techniques such as averages, percentages, median, proportion, and tabulation [19].

The data are represented in the form of frequency tables and charts in order to promote better understanding and interpretations drawn out of the data analysis [20].

4   Analysis and Discussions

The participants were asked about their opinion on the understanding of Green ICT as an innovative way of using ICT related to the environment protection and sustainability of ICT in future. The responses from participants following in categories namely Strongly Agree, Agree, Partially Agree, and Disagree are collected from all respondents, which are shown cumulatively in Fig. 1.

From Fig. 1, it is observed that almost all directors, faculties, and students of professional institutes are in agreement with Green ICT as an innovative way of using ICT.

The understanding of what is meant by Green ICT is not only important but also to know the awareness about Green ICT practices and initiatives among directors, faculties, and students of selected professional institutes is also essential. Table 1 demonstrated the responses on the same.

It is seen in Table 1 that there is a moderate difference observed among directors (19.57%) and faculties (30.50%) in terms of unawareness about Green ICT practices.

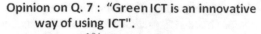

Opinion on Q. 7 : "Green ICT is an innovative
way of using ICT".

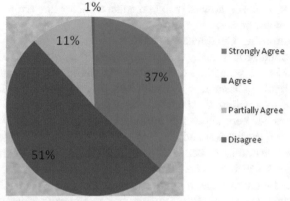

Fig. 1 Opinion of all respondents on Green ICT

Table 1 Green ICT practices awareness

Respondent category	Green ICT practices awareness (%)	Green ICT practices unawareness (%)
Director	80.43	19.57
Faculty	69.50	30.50
Student	40.46	59.54

However, 59.54% of students are not aware about Green ICT practices or initiatives, so there is a noticeable difference among their awareness. It can be concluded that the awareness of GICT is more among the directors and faculties as compared with the students of selected professional institutes.

5 Techniques to Improve Green ICT Awareness

This section provides the methods to be followed to improve Green or sustainable ICT awareness among all the stakeholders of higher education institutions such as faculty, staff, student, and top management based on rigorous literature review of secondary data, primary data, and web-based survey of few HEI in India. The techniques to improve the awareness and practices of Green ICT are listed as follows:

A. **Obligatory Green ICT Program Course**

It is the necessity of time to increase not only the awareness of green practices but also to motivate to be green in the approach of designing, implementing, and disposing ICT resources. The youth of the nation are now pursuing their higher education from various universities of nation. Based on rigorous literature review, it is suggested that all universities in India have to offer green computing program or Green ICT practices course as a mandatory curriculum to educate students on how to sustain the use of ICT while reducing its negative impact on environment.

B. **Green Computing Certification (GCC)**

The most popular IT literacy course in Maharashtra state of India is MS-CIT that started in the year 2002 and made more than 9 million learners in information technology literate. In line with this, it is suggested that its high time to start Green Computing Certification (GCC) course to educate as many academicians or learners about the importance of green practices to be implemented during application and discarding of ICT in everyday life to save Earth from global warming issue. GCC certification should demonstrate that an individual has specific green computing knowledge, including awareness of sustainable computing, energy-efficient methods, and Green ICT practices as well as laws and regulations concerning with environmental issues and protection.

C. **Green Procurement Procedure**

The implementation of Green ICT at education institutions ensures the sustainability of the ICT resources. The study has discovered green parameters so that the educational institutes can reduce, reuse, and recycle infrastructure [21]. It is recommended that there is a strong need to give more attention for increasing green procurement practices by higher education institutes. Further, I have suggested few guidelines for green procurement procedure and described them as follows:

   i. Timely financial assistance should be supported by top management authority of higher education institutes.
   ii. ICT's environmental impacts should be considered significantly serious.
   iii. Reuse and recycling policy should be strictly followed by extending the lifecycle of all ICT equipment.
   iv. Only energy certified and five energy star rated ICT equipment should be purchased.
   v. In implementation of Green ICT strategies, initial capital investment is usually high. Therefore, appropriate budget should be approved to ensure green procurement.

D. **Reward or Testimonial for Best Green Practices**

Due to collaborative and knowledge management sharing among educational institutes, it can be seen that sharing of best green practices would change and upscale the education system of world and therefore, it is suggested to appreciate the same for motivation among the higher education institutions. Green ICT is an organization-wide continuous process and it is not a product that can be implemented overnight. It needs behavioral change and zest to sustain ecology. To facilitate continuous green practices improvement in the institute, it is further recommended to present motivational award or incentives to all categories of stakeholders of the institutes for those who have efficiently adopted green practices. Also, recognitions or testimonials like Green University, Green Institute, Green Campus, and Green Teacher should be declared after successfully followed green practices by concerned authority based on laid down green policies and green strategies or innovative practices.

E. **Webinar and Online Education**

The webinar as green practice would reduce carbon releases from students/staff/faculty traveling activities. Each professional institute should choose webinar/video conferencing to lessen journey and ultimately carbon footprint will diminish. Online education mechanisms, sharing of online classrooms, and tutorials can provide significant advantage in terms of reduced infrastructure, and therefore diminished carbon footprint [22].

The higher education institutions and universities in better off places generally have the competent faculties to tutor and guide students. There is an urgency to change the mindset of students toward ecologically responsible youth. This research recommended organizing as much virtual conferences and webinar as regular green practice by using cloud services.

F. **Renewable Energy Sources**

Renewable energy sources like solar should be effectively used for ICT in institutes since every daytime world receives more energy from Sun which should be efficiently utilized. As government has initiated renewable energy policy for the nation according to Green India Mission, all the institutes should be implanted with solar panels or any other renewable energy source as part of this green practice. Necessary research and development to be made on renewable energy in order to cut down their application cost. Also, appropriate subsidy should be provided by concerned government authority so that there will be an increase in the number of subscribers for renewable energy.

G. **Green Promotion and Awareness Program**

The Green ICT awareness of students or faculties is highly dependent on training provided by higher education institutions. Green ICT is a transition, which takes time and the end objective is to lower operational cost, increase optimum utilization of ICT resources, and reduce energy consumption [23]. The author here suggested the change in the mindset of academicians on green technology through effective promotion and awareness program. There is a need to enhance green technology research and development by developing skilled and competent human resources. It is also suggested that green practice like carpooling or use of bicycle to be preferred on large scale to reduce carbon footprint.

6 Conclusion

This paper has an academic contribution considering the emerging trends in the field of computer and information technology. These research findings show that Green ICT practices awareness is less among the students of the selected education institutions in India. In this paper, I have suggested the strategies to increase the awareness of GICT practices among higher education institutions. This paper is an attempt to motivate all students to adopt and think in terms of Green ICT not only for conserving energy but also for obtaining a cost-effective solution. Also, we can certainly build the sustainable future for our next generation.

References

1. Blurton, C. (1999). *UNESCO's world communication and information report*. On-line available at www.unesco.org/education/educprog/wer/wer.html.
2. Murugesan, S. (2007, August). Going Green with IT: You're responsibility toward environmental sustainability. *Cutter Consortium Business-IT Strategies Executive Report, 10*(8).
3. Paruchuri, V. (2009). Greener ICT: Feasibility of successful technologies from energy sector. In *ISABEL* (pp. 1–3).
4. Chai-Arayalert, S., & Nakata, K. (2011). The evolution of Green ICT practice: UK higher education institutions case study. In *IEEE International Conference on Green Computing and Communications, United Kingdom* (pp. 220–225).
5. International Energy Agency. (2005, July 13). *Open Energy Technology Bulletin*, no. 27. [online available] http://www.iea.org/impagr/cip/archieved_bulletines/issue_no27.html. Accessed December 2011.
6. James, P., & Hopkinson, L. (2009). *Sustainable ICT in further and higher education*. A Report for the Joint Information Services Committee (JISC), SusteIT, London, UK.
7. Mingay, (2007). *Green IT: The new industry shock wave* (p. 2007). USA: Gartner.
8. Molla, A. (2008). GITAM—A model for the adoption of Green IT. In *Proceeding of 19th Australian Conference on Information Systems, Australia* (pp. 658–668).
9. Molla, A., Vanessa, C., Brain, C., Hepu, D., & Say, Y. T. (2008). E-readiness to G-readiness: Developing a green information technology readiness framework. In *Proceeding of 19th Australian Conference on Information Systems*, 3–5 December 2008, Christchurch, Australia (pp. 669–678).
10. Mohamad, T. I., Alemayehu, M., Asmare E., & Say, Y. T. (2010). Seeking the Green in Green IS: A spirit, practice and impact perspective. In *Proceeding of 14th Pacific Asia Conference on Information Systems PACIS, Taiwan* (pp. 433–443).
11. Suryawanshi, K., Narkhede, S. (2013). Green ICT implementation at educational institution: A step towards sustainable future. In *Innovation and Technology in Education (MITE), 2013 IEEE International Conference in MOOC* (pp. 251–255), 20–22 December 2013. https://doi.org/10.1109/mite.2013.6756344.
12. Gartner Research Group. *GISFI report*. Online available at www.gisfi.org/.../GISFI_GICT_201109.pdf.
13. Kandhari, R., Sood, J., & Bera, S. (2010, May 16–31). IT's underbelly. *Down to Earth (E-Magazine), 19*(1).
14. Kendhe, O. (2011, January). Basics of Green IT and India perspective. *CSI Communications: Green Computing, 34*(10), 11–15.

15. Tom, Y. (2010, February 22). E-waste a growing problem for China and India. www. computing.co.uk.
16. Sheehan, M. C., & Smith, S. D. (2010, April). Powering down: Green IT in higher education. Research Study, EDUCAUSE Center for Applied Research, Washington.
17. Meurant, R. C. (2011). Facing reality: Using ICT to go green in education. In *Communications in Computer and Information Science, 1 (Vol. 150)*: *Ubiquitous Computing and Multimedia Applications* (pp. 211–222).
18. Suryawanshi, K., & Narkhede, S. (2015). Green ICT for sustainable development: A higher education perspective. *Procedia Computer Science, 70*, 701–707. https://doi.org/10.1016/j. procs.2015.10.107. ISSN: 1877-0509.
19. Kothari, C. R. (2004). *Research methodology, methods and techniques* (2nd ed).
20. Gupta, S. C., & Kapoor, V. K. (2007). *Fundamentals of mathematical statistics* (11th ed, Reprint). Sultan Chand and Sons.
21. Suryawanshi, K., & Narkhede, S. (2014). Green ICT at higher education institution: Solution for sustenance of ICT in future. *International Journal of Computer Applications (IJCA), 107*(14), 35–38. https://doi.org/10.5120/18823-0237. ISSN: 0975-8887.
22. Alla, K. R., & Chen, S. D. (2017, May). Strategies for achieving and maintaining Green ICT campus for Malaysian higher education institutes. *Advanced Science Letters, 23*(5), 3967–3971.
23. Nana, Y. A., Amevi, A., & Nii, Q. (2016, March). Encouraging Green ICT implementation strategies in polytechnic education in Ghana. *International Journal of Applied Information Systems (IJAIS)*. ISSN: 2249-0868.

A Comparative Study of Different Similarity Metrics in Highly Sparse Rating Dataset

Pradeep Kumar Singh, Pijush Kanti Dutta Pramanik
and Prasenjit Choudhury

Abstract Recommender system has been popularly used for recommending products and services to the online buyers and users. Collaborative Filtering (CF) is one of the most popular filtering approaches used to find the preferences of users for the recommendation. CF works on the ratings given by the users for a particular item. It predicts the rating that is not explicitly given for any item and build the recommendation list for a particular user. Different similarity metrics and prediction approaches are used for this purpose. But these metrics and approaches have some issues in dealing with highly sparse datasets. In this paper, we sought to find the most accurate combinations of similarity metrics and prediction approaches for both user and item similarity based CF. In this comparative study, we deliberately instill sparsity of different magnitudes (10, 20, 30 and 40%) by deleting given ratings in an existing dataset. We then predict the deleted ratings using different combinations of similarity metrics and prediction approach. We assessed the accuracy of the prediction with the help of two evaluation metrics (MAE and RMSE).

Keywords Recommender systems · Collaborative filtering · Similarity metrics
Prediction approaches · Rating

P. K. Singh (✉) · P. K. D. Pramanik · P. Choudhury
Department of Computer Science and Engineering, National Institute
of Technology Durgapur, Durgapur, West Bengal, India
e-mail: pksingh300689se@gmail.com

P. K. D. Pramanik
e-mail: pijushjld@yahoo.co.in

P. Choudhury
e-mail: prasenjit0007@yahoo.co.in

© Springer Nature Singapore Pte Ltd. 2019
V. E. Balas et al. (eds.), *Data Management, Analytics and Innovation*,
Advances in Intelligent Systems and Computing 839,
https://doi.org/10.1007/978-981-13-1274-8_4

1   Introduction

Recommender System (RS) is an information filtering tool that helps the users to find the desired services from the ever-increasing list of online stuffs. RSs adopt various filtering approaches such as Content-Based (CB), Collaborative Filtering (CF) and Hybrid Filtering (HF) to extract the meaningful information. Out of these, CF is the most used approach that exercises user–item rating matrix information for the recommendation process. Rating is a measurement scheme through which users convey their preferences. That is why we generally check out the user rating for considering some item to purchase, a movie to watch or to book a hotel. Some popular movie recommendation engines like Netflix, MovieLens [1], IMDb, etc. use user's rating for the recommendation [2]. Considering the user's ratings, these platforms generate the list of top-rated movies in terms of genre, current trending, most popular movies and most popular movies by genre. A lot of researches [3–6] have been conducted using rating dataset (collected from Netflix, MovieLens and IMDb) to improve the performance of CF for the movie recommendation. In CF, ratings have been collected using two ways: explicitly and implicitly. In an explicit way, users directly give their feedback as a form of ratings [7]. Pandora and YouTube use thumbs up/thumbs down rating for the recommendation, and Amazon follows star system to collect rating. These systems observe the behaviour of users, for example, keeping track of what a user clicks on a particular page for converting number of clicks into ratings in implicit rating collection.

But what if a product has got no or very few ratings? This scenario can misguide a user from being recommended with the best or as per the choice. For precise recommendation, these missing ratings are to be predicted on the basis of which the recommendation is done.

There are two types of promising algorithms used in prediction of missing ratings. First is user-based CF and another is item-based CF. The major challenge of these types of the algorithm is sparsity. The aim of this paper is to provide a comparative analysis of different Similarity Metrics (SMs) on the basis of prediction accuracy applied on a dataset with different degrees of sparseness.

The highlights of this paper are summarized below which has also been pictorially portrayed in Fig. 1.

- The dataset for the experiment is collected from MovieLens.
- The highly sparse data are removed from the dataset to get a low sparse dataset.
- This low sparse dataset is converted to high sparse data by intentionally deleting some given ratings.
- Then, the original values of those deleted ratings are predicted using different combinations of SMs and prediction approaches.
- And finally, the accuracy of the prediction has been assessed.

The rest of this paper has been organized as follows. Section 2 discusses the related works. Section 3 presents the mathematical equations of different SMs of user-based and item-based CF algorithm. Section 3 also elaborates the

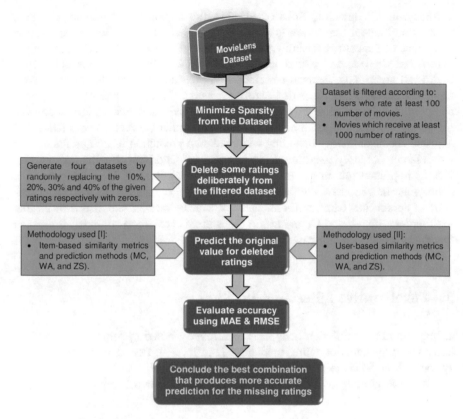

Fig. 1 Highlights of this paper

mathematical equations of different prediction approaches. Section 4 portrays the experimental results of our comparative analysis. And finally, in Sect. 5, we conclude the paper with an indication of future work.

2 Related Works

Collaborative filtering uses the known ratings to predict the recommendation list of users. Cold start problem and high sparse data are the major challenges of this filtering technique. SMs used in CF techniques do not predict accurately due to the presence of these challenges on the dataset. To minimize the limitation of the Pearson correlation for cold start items, a hybrid algorithm used, and Sun et al. [8] have proposed a novel approach for collaborative filtering. Dimensionality reduction technique has been used to alleviate the sparsity problem in CF [9, 10]. Hu and Lu [9] have provided a Hybrid Predictive Algorithm with Smoothing (HSPA), to build the predictive model in order to ensure robust to data sparsity and prediction accuracy. In this direction, Bokde et al. [11] illustrate an item-based collaborative filtering on Mahout framework to reduce the problem of scalability and sparsity in CF. Value of item attribute similarity has been used to predict the missing rating in item-based CF [12] and experimental results showed that it predicts better than traditional algorithms. Pearson correlation and cosine similarity have been investigated in movie rating prediction algorithm with CF [13]. Sarwar et al. [14] have analysed item-based similarity algorithm that provides better performance than user-based similarity algorithm. They use cosine similarity, Pearson correlation and adjusted cosine similarity including weighted sum prediction technique for finding the different similarity measures on item-based CF algorithm. Neighbour selection has the important role in generating the recommendation list. In order to find the appropriate neighbours, Bilge and Kaleli [15] have compared the traditional correlation approaches with multi-dimensional distance metrics and also showed that multi-criteria item-based CF algorithm gives more accurate recommendations than the single-criterion rating-based algorithm.

3 Collaborative Filtering Algorithm

CF algorithm uses different SMs in user similarity based systems and item similarity based systems for rating prediction [14, 16, 17]. For item similarity based systems, these SMs are as follows.

Cosine Similarity (CS): Cosine similarity between two items i and j is computed as

$$\text{sim}(i,j) = \cos(i,j) = \frac{i \cdot j}{||i||^2 * ||j||^2} \tag{1}$$

where '·' represents the dot product of the two items.

Adjusted Cosine Similarity (ACS): It is calculated using

$$\text{sim}(i,j) = \frac{\sum_{u \in U} \left(R_{u,i} - \bar{R}_u \right) \left(R_{u,j} - \bar{R}_u \right)}{\sqrt[2]{\sum_{u \in U} \left(R_{u,i} - \bar{R}_u \right)^2} \sqrt[2]{\sum_{u \in U} \left(R_{u,j} - \bar{R}_u \right)^2}} \tag{2}$$

where $R_{u,i}$, $R_{u,j}$ are the ratings of user u on items i and j, respectively. $\bar{R}_u$ shows the average rating of user u.

Pearson Correlation (PC): Similarity using PC, in item-based recommender systems, is

$$\text{sim}(i,j) = \frac{\sum_{u \in U} \left(R_{u,i} - \bar{R}_i \right) \left(R_{u,j} - \bar{R}_j \right)}{\sqrt[2]{\sum_{u \in U} \left(R_{u,i} - \bar{R}_i \right)^2} \sqrt[2]{\sum_{u \in U} \left(R_{u,j} - \bar{R}_j \right)^2}} \tag{3}$$

where $\bar{R}_i$ and $\bar{R}_j$ are the mean rating value of item i and item j, respectively.

Spearman Correlation (SC): It is another way of calculating similarity by using rank of those rating in place of actual rating value [16]. The equation becomes

$$\text{sim}(i,j) = \frac{\sum_{u \in U} \left(k_{u,i} - \bar{k}_i\right)\left(k_{u,j} - \bar{k}_j\right)}{\sqrt[2]{\sum_{u \in U} \left(k_{u,i} - \bar{k}_i\right)^2} \sqrt[2]{\sum_{u \in U} \left(k_{u,j} - \bar{k}_j\right)^2}} \tag{4}$$

where $k_{u,i}$ denotes the rank of the rating of item i of user u and $k_{u,j}$ shows the rank of the rating of item i of user u. $\bar{k}_i$ and $\bar{k}_j$ calculate the average rank of item i and item j.

Jaccard Similarity (JS): Jaccard similarity between two items only considers the total number of common ratings between them [18]. It does not consider the absolute ratings of the items. It is formulated by

$$\text{sim}(i,j) = \frac{|R_i| \cap |R_j|}{|R_i| \cup |R_i|} \tag{5}$$

where $|R_i|$ and $|R_j|$ denote the total number of ratings of item i and item j, respectively.

Euclidean Distance (ED): Euclidean distance between two items is found to be

$$\text{sim}(i,j) = \sqrt[2]{\frac{\sum_{u \in U_{i,j}} \left(r_{i,u} - r_{j,u}\right)^2}{|U_{ij}|}} \tag{6}$$

where r_{iu} and r_{ju} are the ratings of items i and j given by user u. $|U_{ij}|$ denotes the number of users who rate both the items i and j.

Manhattan Distance (MD): After minor modification in ED, MD is formulated as

$$\text{sim}(i,j) = \frac{\sum_{u \in U_{ij}} \left(r_{iu} - r_{ju}\right)}{|U_{ij}|} \tag{7}$$

Mean Squared Distance (MSD): Similarity between item i and item j using MSD is computed as

$$\text{sim}(i,j) = \frac{\sum_{u \in U_{ij}} \left(r_{iu} - r_{ju}\right)^2}{|U_{ij}|} \tag{8}$$

3.1 Rating Prediction

RSs predict the rating of an item for further recommendation using similarity value with different prediction approaches [14, 19–21]. These methods are as follows.

Mean Centering (MC): The equation becomes as the following in the prediction of rating using similarity value with mean centering approach:

$$\hat{r}_{ui} = \bar{r}_i + \frac{\sum_{j \in N_u(i)} \text{sim}(i,j)(r_{ju} - \bar{r}_j)}{\sum_{j \in N_u(i)} |\text{sim}(i,j)|} \tag{9}$$

where $\hat{r}_{ui}$ depicts the predicted value of item i of user u.

Weighted Average (WA): The equation to compute a rating estimate in an item-based system is given as follows:

$$\hat{r}_{ui} = \frac{\sum_{j \in N_u(i)} \text{sim}(i,j)r_{ju}}{\sum_{j \in N_u(i)} |\text{sim}(i,j)|} \tag{10}$$

Z-Score (ZS): For item-based systems, the equation of Z-score becomes

$$\hat{r}_{ui} = \bar{r}_i + \sigma_i \frac{\sum_{j \in N_u(i)} \text{sim}(i,j)(r_{ju} - \bar{r}_j)/\sigma_i}{\sum_{j \in N_u(i)} |\text{sim}(i,j)|} \tag{11}$$

where σ_i is the standard deviation of rating of item i.

For user similarity based systems, the above equations (SMs and prediction methods) can be obtained by mutually exchanging i with u and j with v.

4 Experimental Results

After obtaining the explicit rating dataset from MovieLens [1], conversion has been performed into user–movie matrix, which contains 7120 users and 14,027 movies. This user–movie matrix contains the ratings information of user to a particular movie on the scale of 0.5–5 with 0.5 increment. Each selected user gives rating to minimum of 20 movies. Sparsity of this dataset is 98.95%, and calculated as

$$\text{Sparsity} = \frac{\text{number of zeros in user–movie matrix}}{\text{number of given ratings}} * 100 \tag{12}$$

For minimizing the sparsity, we consider only those users who give ratings to at least 100 movies and those movies which receive at least 1000 number of ratings. Sparsity level of modified dataset becomes 57.03%. We generate missing ratings artificially by deleting some given ratings. After creating missing ratings, we predict these missing ratings using different SMs (user-based and item-based SMs) and various prediction methods. Mean Absolute Error (MAE) and Root-Mean-Squared Error (RMSE) have been used for the comparison of accuracy of rating prediction of different SMs and prediction methods. The equations for calculating MAE and RMSE values [14, 22] are given below:

$$\text{MAE} = \frac{\sum_{i=1}^{N} |p_i - \hat{q}_i|}{N} \tag{13}$$

$$\text{RMSE} = \sqrt[2]{\frac{\sum_{i=1}^{N} (p_i - \hat{q}_i)^2}{N}} \tag{14}$$

where $<p_i - \hat{q}_i>$ represents each rating–prediction pair and N is the total number of rating prediction pair.

ED, MD and MSD give approximately same results, so we consider only ED in this comparative study in the place of ED, MD and MSD. Two types of similarity algorithm are used in CF for recommendation. These are as follows:

A. **Item-based CF Algorithm**

Item-based CF algorithms use item SMs for the prediction of missing ratings in the recommendation process. How different prediction approaches are used to find the most accurate item-based SM is discussed in this section. Figures 2, 3, 4, 5, 6, 7, 8 and 9 illustrate the comparative analysis of item-based SMs and rating prediction approaches.

Fig. 2 Comparison of different similarity metrics on MAE values

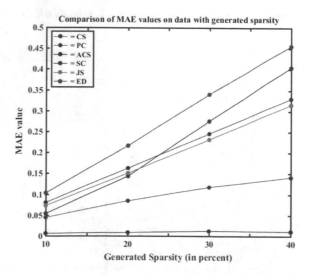

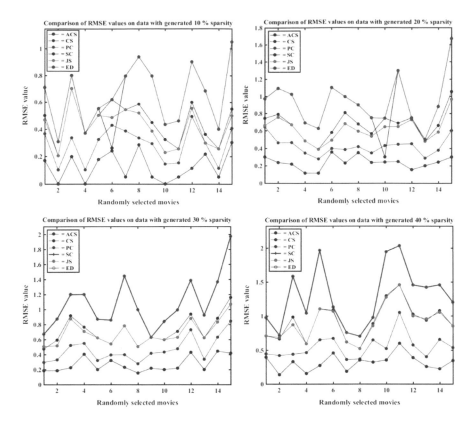

Fig. 3 Comparison of different similarity metrics on RMSE values

Fig. 4 Comparison of
different similarity metrics on
MAE values

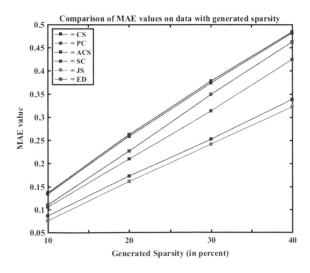

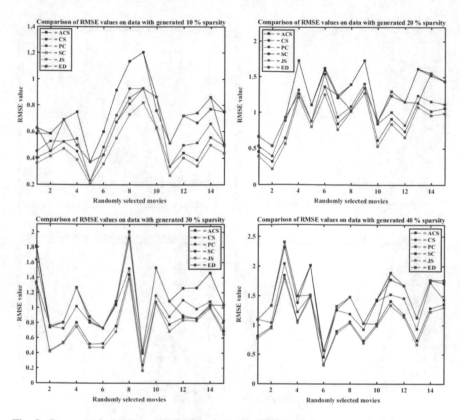

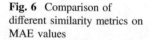

Fig. 5 Comparison of different similarity metrics on RMSE values

Fig. 6 Comparison of different similarity metrics on MAE values

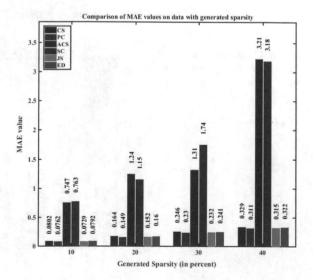

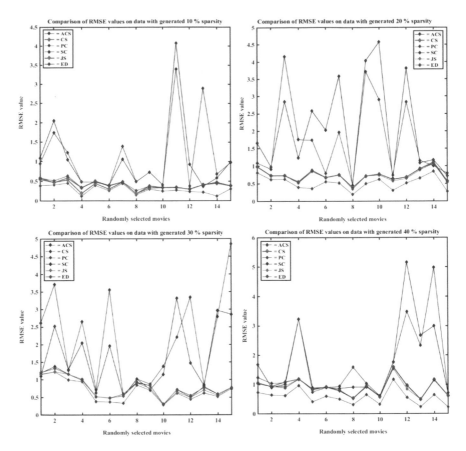

Fig. 7 Comparison of different similarity metrics on RMSE value

Fig. 8 Comparison of
different similarity metrics on
MAE values

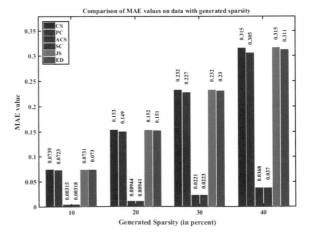

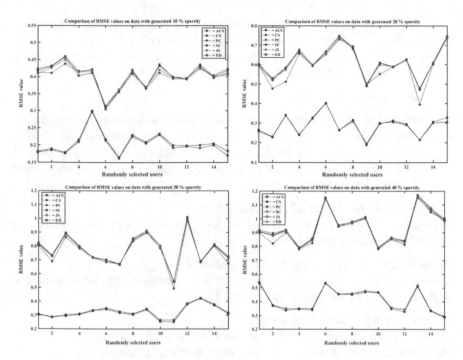

Fig. 9 Comparison of different similarity metrics on RMSE values

Using Mean Centering Prediction Approach

MAE values generated at 10, 20, 30 and 40% of sparsity: Fig. 2 shows the MAE values using different SMs with MC approach. ACS provides low MAE value on comparing to the other SMs and also provides the slow nature of the increase of MAE value at the different levels of sparsity.

RMSE values generated at 10, 20, 30 and 40% of sparsity: Fig. 3 shows the RMSE values of randomly selected movies at different sparse datasets.

ACS using MC predicts more accurate on comparing other SMs using MC as shown in Figs. 2 and 3.

Using Weighted Average Prediction Approach

MAE values generated at 10, 20, 30 and 40% of sparsity: Prediction of missing rating using JS with WA provides less error on different generated datasets.

RMSE values generated at 10, 20, 30 and 40% of sparsity: Fig. 5 clarifies that JS using WA gives the comparatively low value of RMSE.

Figures 4 and 5 illustrate the best similarity metric using WA in the prediction of missing rating for recommendation, JS.

Using Z-Score Prediction Approach

MAE values generated at 10, 20, 30 and 40% of sparsity: PC has got the low value of MAE on the generated four sparse datasets as shown in Fig. 6.

RMSE values generated at 10, 20, 30 and 40% of sparsity: Fig. 7 shows the RMSE value comparison on predicted ratings of randomly selected movies.

Figures 6 and 7 depict that PC is the most accurate similarity metric when the ZS prediction approach is used for prediction of rating.

B. User-based CF Algorithm

User-based CF algorithm uses user similarity in the prediction of missing ratings for the recommendation. Below, how different prediction approaches are used to find the most accurate user-based SMs is discussed. Figures 8, 9, 10, 11, 12 and 13 provide the comparative analysis of prediction of missing ratings using user-based SMs and prediction methods.

Using Mean Centering Prediction Approach
Figures 8 and 9 clarify that ACS and SC have low MAE and RMSE values on different generated sparse datasets.

Using Weighted Average Prediction Approach
ED and JS work comparatively better than other user-based SMs when prediction approach is WA as shown in Figs. 10 and 11.

Using Z-Score Prediction Approach
ED and JS are the most accurate user-based SMs in the prediction of missing rating for the recommendation when we use ZS as a prediction approach as shown in Figs. 12 and 13.

The summarized experimental results of the comparative analysis for SMs and prediction approaches based on the accuracy of missing rating predictions are shown in Table 1.

ACS with MC, JS with WA and PC with ZS are the best combinations of similarity metric and prediction approach for item similarity based collaborative filtering systems. And for the user similarity based collaborative filtering, the best pair of similarity metric and prediction approaches are ACS/SC with MC, ED/JS with WA and ED/JS with ZS.

Fig. 10 Comparison of different similarity metrics on MAE values

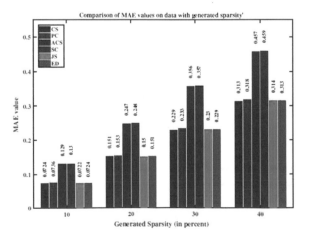

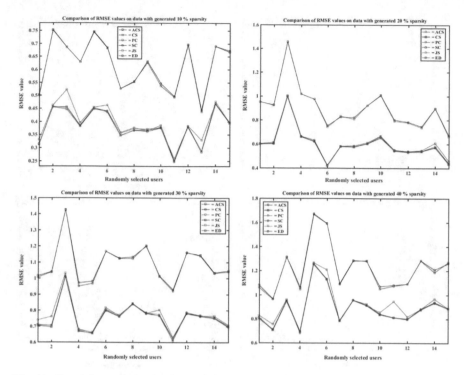

Fig. 11 Comparison of different similarity metrics on RMSE values

Fig. 12 Comparison of different similarity metrics on MAE values

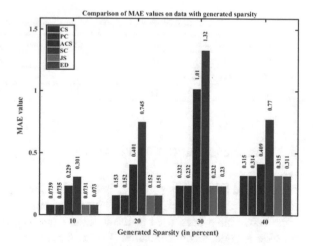

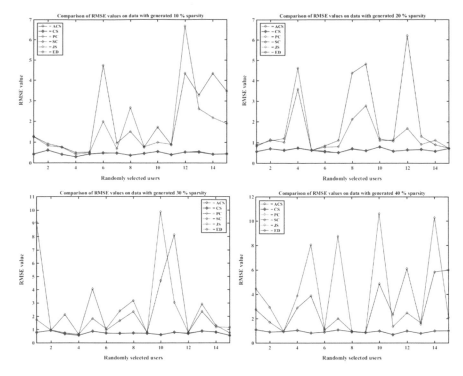

Fig. 13 Comparison of different similarity metrics on RMSE values

Table 1 Best combination of similarity metrics and prediction approaches

Prediction approaches	Item-based similarity metrics	User-based similarity metrics
Mean centering (MC)	ACS	ACS & SC
Weighted average (WA)	JS	ED & JS
Z-score (ZS)	PC	ED & JS

5   Conclusion and Future Work

A good deal of research works has been devoted to improve the performance of CF algorithms for recommendations. Various similarity metrics and prediction approaches have been used in the literature for prediction of unknown ratings to provide a good recommendation. Our comparative study shows the best combination of SMs and prediction approaches in terms of accuracy, which employed on different sizes of own generated sparse dataset.

Future works include the analysis of performance of various parametric rating datasets which will be collected from different domains and to propose a prediction algorithm that performs much better in respect of user and item similarity based collaborative filtering algorithms.

References

1. MovieLens|GroupLens, https://grouplens.org/datasets/movielens/. Last retrieved July 16, 2016.
2. Top 10 movie recommendation engines—CNET, https://www.cnet.com/news/top-10-movie-recommendation-engines/. Last Retrieved July 7, 2017.
3. García-Cumbreras, M. Á., Montejo-Ráez, A., & Díaz-Galiano, M. C. (2013). Pessimists and optimists: Improving collaborative filtering through sentiment analysis. *Expert Systems with Applications: An International Journal, 40*(17), 6758–6765.
4. Boratto, L., & Salvatore, C. (2014). Using collaborative filtering to overcome the curse of dimensionality when clustering users in a group recommender system. In *16th International Conference on Enterprise Information Systems*.
5. Said, A., Fields, B., Jain, B. J., & Albayrak, S. (2013). User-centric evaluation of a K-furthest neighbor collaborative filtering recommender algorithm. In *16th ACM Conference on Computer Supported Cooperative Work and Social Computing*.
6. Pirasteh, P., Jung, J. J., & Hwang, D. (2014). Item-based collaborative filtering with attribute correlation: A case study on movie recommendation. In *Intelligent Information and Database Systems: 6th Asian Conference*.
7. Jawaheer, G., Szomszor, M., & Kostkova, P. (2010). Comparison of implicit and explicit feedback from an online music recommendation service. 1st International Workshop on Information Heterogeneity and Fusion in Recommender Systems.
8. Sun, D., Luo, Z., & Zhang, F. (2011). A novel approach for collaborative filtering to alleviate the new item cold-start problem. In *11th International Symposium on Communications and Information Technologies*.
9. Hu, R., & Lu, Y. (2006). A hybrid user and item-based collaborative filtering with smoothing on sparse data. ICAT Workshops.
10. Sarwar, B. M., Karypis, G., Konstan, J. A., & Riedl, J. T. (2000). Application of dimensionality reduction in recommender system—A case study. ACM WEBKDD Workshop.
11. Bokde, D. K., Girase, S., & Mukhopadhyay, D. (2015). An item-based collaborative filtering using dimensionality reduction techniques on mahout framework. CoRR.
12. Puntheeranurak, S., & Chaiwitooanukool, T. (2011). An item-based collaborative filtering method using Item-based hybrid similarity. In *2nd International Conference on Software Engineering and Service Science*.
13. Fikir, O. B., Yaz, I. O., & Özyer, T. (2010). A movie rating prediction algorithm with collaborative filtering. In *International Conference on Advances in Social Networks Analysis and Mining*.
14. Sarwar, B., Karypis, G., Konstan, J., & Riedl, J. (2001). Item-based collaborative filtering recommendation algorithms. In *10th International Conference on World Wide Web*.
15. Bilge, A., & Kaleli, C. (2014). A multi-criteria item-based collaborative filtering framework. In *11th International Joint Conference on Computer Science and Software Engineering*.
16. Bobadilla, J., Hernando, A., Ortega, F., & Abraham, G. (2012). Collaborative filtering based on significances. *Information Sciences, 185*(1), 1–17.
17. Xu, J., & Man, H. (2011). Dictionary learning based on Laplacian score in sparse coding. In *Machine Learning and Data Mining in Pattern Recognition—7th International Conference*.

18. Liu, H., Hu, Z., Mian, A. U., Tian, H., & Zhu, X. (2014). A new user similarity model to improve the accuracy of collaborative filtering. *Knowledge-Based Systems, 56,* 156–166.
19. Wu, J., Chen, L., Feng, Z., Zhou, M., & Wu, Z. (2013). Predicting quality of service for selection by neighborhood based collaborative filtering. *IEEE Transactions Systems, Man, and Cybernetics: Systems, 43*(2), 428–439.
20. Herlocker, J. L., Konstan, J. A., Borchers, A., & Riedl, J. (1999). An algorithmic framework for performing collaborative filtering. In *22nd Annual International ACM SIGIR Conference on Research and Development in Information Retrieval.*
21. Herlocker, J. L., Konstan, J. A., & Riedl, J. (2002). An empirical analysis of design choices in neighborhood-based collaborative filtering algorithms. *Information Retrieval, 5,* 287–310.
22. Yang, X., Guo, Y., Liu, Y., & Steck, H. (2014). A survey of collaborative filtering based social recommender systems. *Computer Communications, 41,* 1–10.

Multi-document Summarization and Opinion Mining Using Stack Decoder Method and Neural Networks

Akshi Kumar and Sujal

Abstract Availability of data is not the foremost concern today; it is the extraction of relevant information from that data which requires the aid of technology. It is to help millions of users arrive at the desired information as quickly and effortlessly as possible. Document summarization and opinion-based document classification can effectively resolve the well-known problem of information overload on the Web. Summarization is about finding the perfect subset of data which holds the information of the entire set. In this paper, first we studied and evaluated three methods of generating summaries of multiple documents, namely, K-means clustering, novel-graph formulation method, and the stack decoder algorithm. The performance analysis emphasized on time, redundancy and coverage of the main content, was conducted along with the comparison between respective ROUGE scores. Next, hybrid architecture was proposed using a Stack decoder algorithm for creating automated summaries for multiple documents of similar kind, which were used as the dataset for analysis by a recursive neural tensor network to mine opinions of all the documents. The cross-validation of the generated summaries was done by comparing the polarity of summaries with their corresponding input documents. Finally, the results of opinion mining of each summary were compared with its corresponding documents and were found to be similar with few variations.

Keywords Document summarization · Document classification
Neural networks · Opinion mining · Stack decoder

A. Kumar · Sujal (✉)
Department of Computer Engineering, Delhi Technological University,
New Delhi, India
e-mail: Sujal.dce@gmail.com

A. Kumar
e-mail: akshikumar@dce.ac.in

© Springer Nature Singapore Pte Ltd. 2019
V. E. Balas et al. (eds.), *Data Management, Analytics and Innovation*,
Advances in Intelligent Systems and Computing 839,
https://doi.org/10.1007/978-981-13-1274-8_5

61

1 Introduction

A busy individual's time is often spent in searching for relevant information from the extensive volume of material as represented in Fig. 1, which needs to be properly investigated. These extended activities are quite overwhelming, just to have the relevant information in front of you. Automatic Summarization helps in creating summaries of extensive documents with just the key points. Account variables such as length, writing style and syntax are used to generate clear and concise summaries. Summarization of document automatically reduces the size of the passage and/or paragraph which is given as input to the document summarization system. This reduced text is sufficient to convey the main meaning or the important jest of the paragraph. This is a tedious task to be done if done manually. To search for the main information conveying a portion of the entire document manually not only takes a large amount of time but also requires skill to extract the important information. This calls for the need of an automatic software or system which can reduce the complexity of this job and save our time tremendously by automating the entire process. Gong and Liu in paper [1] presented two text-summarization methods. These methods create generic text summaries using ranking. These sentences are extracted from the original documents according to the rank of the sentences. In the first method, the relevance of the sentences is measured using "information retrieval methods". The second method uses a different approach. This method uses "Latent Semantic Analysis (LSA)" technique to find relevance of the sentences. The summary produced by using any of these methods work by ranking system. The ranking system generates the most important contents of the documents which is also non-redundant. This extraction of main information from the document automatically is the main idea behind the development of a tool called document summarizer. It is an application of machine learning and data mining. Search engines are an example; others include summarization of documents, image collections, and videos.

A clear and concise summary should only include the most imperative information which is coherent to the viewer. This is a challenging task to be achieved because of the variety in the types of information retrieved and also the uncertainty of user's needs. Hence, the most meaningful subsets are selected with the help of

Fig. 1 Summary for a single document

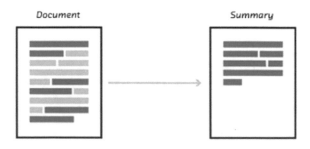

Fig. 2 Summarizing multiple documents

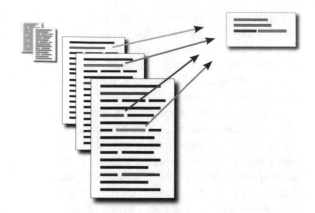

features by describing the weight/importance of those subsets in a given document. Summaries contain the highest weighted subsets or sentences.

The summary generated by the document summarizer helps the user to understand the document in very less amount of time. This is because the main idea of the whole passage or paragraph is expressed in very few lines. By reading only this limited information the whole idea of the document is conveyed to the user and thus helps in saving a considerable amount of his/her time. Today there is ample of information available in World Wide Web (WWW). So, taking out information from internet is no more an issue. But the major difficulty faced today is to process this large amount of information in a limited amount of time. A lot of work has to be done in a limited time constraint. This calls for a need to develop an automatic system. The document summarizer solves this issue gracefully.

When seeking information, users want to arrive at the desired information as quickly and effortlessly as possible. The overall aim of this paper is to help such users experience the solution to their trivial problem by suitable technologies. Here, our focus is on Multi-document summarization which involves exact same procedure as auto summarization with a slight difference of incorporating more than a single document for similar context. The same is shown in Fig. 2. It helps the professionals to familiarize with a huge cluster of data in a short span of time.

Additionally, searching out a relevant data that meets our requirements is a tedious and time consuming task. We can face two kinds of problems:

1. Searching a relevant document corresponding to our search.
2. Absorbing maximum amount of information from that bulk data source.

Here, we are working on such a technique that will resolve both of these problems, i.e., time and information.

Moreover, if the summary generated above is classified as positive or negative based on the sentiment, it would further ease the pain of retrieving the information. Hence, increase the overall efficiency. Applying Sentimental Analysis on the summarized documents helps in knowing the nature of that particular topic whether it is positive or negative. Since the documents summarized are taken from various sources, they hold a lot of different opinions. Also, the personal biases as seen while doing sentiment analysis on each blog is not there. This gives us the wider opinion about the topic. Document Summarization saves both a lot of time and money. The task to go through thousands of words is reduced. It gives us a crisp and brief note about what the document is all about.

Moreover, a new area of research which is gaining popularity recently is web page summarization. Web page summarization is carried out by using two methods: Content-Based Summarization and Context-Based Summarization. The content-based summarization as well as context-based summarization does not take into account the dynamic nature of the web documents. They both treat web documents as static documents whose content and attribute does not change with time. The methods of summarization of web pages as proposed in [2] are content-based techniques. The only difference is that it also takes into account the dynamic nature of the web pages. Web page summarization algorithms are proposed by Alguliev et al. in paper [3]. These algorithms extract the most relevant features from Web pages for improving Web classification approaches' efficiency. Further there is a need to improve the content of the knowledge that is extracted from web classification. In the paper [4] two text-summarization methods are proposed which provides a summary of the web pages. In the first method, "Luhn's significant-word selection" technique is used. The second method uses "LSA" as a technique to generate summary.

However, applying opinion mining on summarized content from multiple documents can be useful in many ways

- National Surveys.
- While making Government policies.
- Law firms to make a crisp summary of the whole case which includes the stories from both the parties.
- Improving the Education System. It will prove very useful to the students preparing for tough exams. Also the keen readers who want to keep themselves updated with all the happenings around the world.
- Good for Conferences. Sometimes while reading so much, people tend to either lose interest or distract themselves. The basic message of the conference can be lost. So to avoid this problem, document summarization is very important.

2 Literature Survey

We have studied three techniques to generate automated summaries from a collection of similar-documents.

- Stack Decoder Algorithm: the sentences with maximum importance are extracted to form the summary. Rather than taking a greedy approach in selecting sentences, solution is obtained by selecting the summary closest to the optimal solution according to our constraints.

- Graph decoder: it is a novel graph based formulation where cliques found in the sentence graph helps in generating summaries. These cliques are further undertaken to create summaries from these vertices. The absolute solution is selected on the basis of their importance scores.
- *K*-Means Clustering: the semantic score obtained using *K*-means clustering algorithm is used as a pivot to construct clusters of available documents. These clusters when processed further to obtain clusteroids, makes up the summary.

The challenges in the summarization of document are presented in the paper [5]. After studying these challenges, new algorithms are proposed. The efficiency of the proposed new algorithms depended on the "content-size" and the "target document's context". An analysis of news article summaries is presented by Goldstein et al. in [6]. The news article summaries are extracted by "Sentence Selection". The ranking of the sentences is determined by using a method of weighted combination. The attributes that are taken into consideration while deriving the rank are "statistical features" and "linguistic features". The statistical features are determined using the information retrieval methods. On the other hand, the linguistic features are determined by analyzing the news summaries. In this method, a modified version of "precision-recall curves" is used.

In the paper [7], an approach for extracting relevant sentences from the original document is presented which help in forming the summary. In this approach, local properties and the global properties of sentences are combined to generate the summary of the document. The local and global properties of the sentence are combined to generate the ranking of the sentences which help in extracting the top-ranked sentences from the document. The algorithm for the same is presented in the paper. The local property of the sentences consists of clusters of words which are significant in each sentence. The global property of the sentence is the relation of the sentences in a document. Hu et al. in paper [8] presented two text-summarization techniques: "modified corpus-based approach (MCBA)" and "LSA-based Text Relationship Maps (LSA + TRM)". MCBA is a trainable summarizer. It consists of various kinds of document features like position, keyword, similarity to title which help in generating the summaries of the document. The second method uses LSA. It uses the "semantic matrix" and "semantic sentence representation" to construct a semantic TRM which summarizes the document.

The paper [9] proposed a method for summarization of text documents of China. The distinct feature of the produced summary is that it can comprise of the main contents of various topics any number of times. *K*-Means Clustering algorithm and a novel clustering analysis algorithm are used to identify the different topics of the selected document.

In paper [10], semantic hypertext links are generated using the method of "information retrieval". Machine learning approaches are used to apply the concept of automatic generation of link to summarization of text. A text summarization tool is developed for implementation of the system.

An automatic general purpose text-summarization tool has been developed. The techniques used are the ones used to:

Table 1 Summary
generation time

Method	Time (s)
Stack decoder	0.78
Clustering based	1.62
Graph decoder	6.42

(a) Generate links between multiple documents
(b) Generate links between different passages of a single document.

On the basis of these patterns, the structure of text is characterized. Further, this understanding of the text structure can be applied to passage-wise extraction for text summarization.

Mitray et al. [11] also present an effective summarization method of text documents. Moreover, article [12] proposes another method for summarization in terms of minimum redundancy and maximum document coverage. The number of clusters is a parameter related to the complexity of the cluster structure. Paper [13], an assessment of a hierarchical method of summarization of text documents that aids to view summaries of documents of Web from mobile devices is shown. This method does not need the documents to be specifically in HTML because it automatically calculates it in a structure that is hierarchical. Presently it is used to summarize articles of news sent to a Web mail account in a simple text format. Summarization that is hierarchical in nature functions in two phases. First, the weight of each sentence is computed and then, the sentences are ranked accordingly. In the second phase, a tree is built from all sentences in such a way that the sentence with the maximum weight is the root and, provided any sentence of weight w present at depth d, sentences above that depth have a weight higher than w, while the weight of the rest of the sentences is below.

Multi-document summarization poses a number of new challenges over single document summarization. The problems identified in it are repetitions or contradictions across input documents and determining which information is important enough to include in the summary [14, 15]. Barzilay and Elhadad in paper [16] define five topic representations which are based on topic themes. Also, authors Diao and Shan [14] presented a multi-web page summarization algorithm which uses graph-based ranking algorithm.

The time taken for generating a summary for the Stack Decoder method is proved to be the least among other methods (Table 1).

2.1 Comparison of Approaches

- Stack decoder method gives results quicker and has a better ROUGE score compared to other two approaches.
- In K-Means Clustering method approach, the position of clusters and the clusteroids changes and sets points to the nearest clusteroid. Different coherent

summaries can be produced by reordering the sentences but it will not bring much difference to the ROUGE scores and processing time.

- In Graph-based method, finding cliques is an NP-hard problem. The time complexity for finding maximal clique in worst case is $O(3n/3)$. Dense graphs are therefore impossible to be worked upon.

The techniques of sentiment analysis help us in mining the opinions presented by each document in terms of positive, negative, or neutral. Twitter data is generally used to determine the feelings of individuals regarding a particular subject. The analysis of tweets using neural networks can answer the desired question. Even the corresponding reasons can be learnt, by extracting the exact words of their statements. For example, we could have easily found out how people were feeling about demonetization in India. If someone includes apposite reasons for their consent/discontent, we can even find why people are feeling that way. This demonstrates advancement in comparison to the mainstream concept of surveys and countless man-hours for market research. Lexalytic text-mining tools can easily analyze the content and save enormous amount of time and wealth.

Neural Networks are powerful and computational approaches which are considered as information processing models inspired by our biological brain. The entire structure of these networks is based on the human mind where a single neural unit is connected to many others. These neural units comprises of functions with even a threshold value in some cases, which combines all the input values together. They are self-trained instead of being explicitly programmed and excel in areas where traditional programs face much difficulty.

Neural networks are structured in a cube-like design which makes use of the process of back-propagation to obtain key results. The weights of "front" neural units are reset by forward stimulation. Analogous to machine learning, neural networks play key roles in problems which requires the concept of "learning from data". Natural language processing, speech recognition and vision are some examples which are hard to solve by traditional programming and therefore are aided by the powerful system of neural networks.

Recursive neural tensor networks (RNTNs) are neural nets which are instrumental in the applications of natural language processing. Each of their nodes has a tree structure with a neural net. They require some external components such as Word2vec. In order to analyze the content with neural nets, words are represented as continuous vectors of parameters. These word vectors hold important information even about the surrounding words such as its usage, context and other semantic information. RNTNs can be used to determine the polarity of words groups and also for boundary segmentation. Word vectors serve as the basis of sequential classifications. Further they are grouped into sub-phrases and these phrases are then pooled into one sentence, as shown in Fig. 3. This sentence can be then classified by sentiment or other metrics. Moreover, Deeplearning4j implements both Word2vec and recursive neural tensor networks.

Fig. 3 Constituency-based parse tree

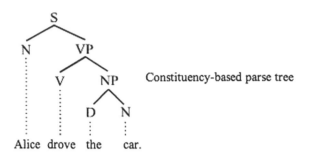

Constituency-based parse tree

Word vectorization serves as the initial step towards constructing a functioning RNTN and it can be achieved by an algorithm named Word2vec. It converts the corpus into vectors which are further mapped in a vector space in order to measure the cosine distance between them. In comparison to NLP, it has a separate pipeline through which it creates lookup tables which supplies word vectors at the time of processing different sentences. On the other hand, NLP's pipeline will ingest sentences, tokenize them and further tag them as POS.

RNTN use constituency parsing to group words into larger sub-phrases in the sentence, in order to organize sentences. Machine learning plays a crucial role in its implementation as it allows the observation of additional linguistic computations in the content. Trees are created by parsing the sentences. Further, the trees are binarized (making sure that each parent has two child leaves). With the root at the top, sentence trees follow a top-down approach as shown in Fig. 3.

3 Architecture

Preprocessing of data is an imperative step in data mining. During the process of collecting data, the collection methods cannot be totally accurate which generally results in out of range values as well as incorrect combinations and missing values to name a few inaccuracies. Working on such data corpus can lead to inaccurate and misleading results. Therefore, the quality of input dataset needs to be carefully checked before implementation.

If there is much redundant and irrelevant information present, then the information retrieved would be not useful. The unreliable data stuffed with noise should be preprocessed using suitable techniques like normalization, cleaning, transformation, feature extraction, and selection, etc. The most common words in a language are considered as "stop words". And stemming is the process of reducing words to their word stems-generally a written word form. Many algorithms are studied to perform stemming the one applied here is suffix-stripping algorithm. A stemming algorithm reduces words like "fishing", "fished", and "fisher" to their root word, "fish". On the contrary, "argue", "argued", "argues", "arguing", and "argus" reduces to the stem "argu". In this work, stemming is done in addition to the removal of stop-words from the dataset, which are filtered out before the processing step. Any group of words can be chosen as stop-words for a given undertaking. Most common words such as "the", "is", "a" which occur very frequently are removed to improve performance.

3.1 Document Representation

A document can be represented in numerous ways such as a bag of words, where these words are assumed to appear in any order independently. This model is commonly used in text-mining and information retrieval. The count of words in the bag differs from the mathematical definition of the set. Each and every word represents a dimension in the resulting data-space. Similarly, we have represented our document as a set of sentences and each sentence represents a vector in the vector space. The frequency of each term is used as its weight, which represents the number of times that term appeared in the document. It represents the importance of the respective term. In order to reduce the complexity of the document, it has to be transformed from the full text version to sentence vectors which describe the magnitude and direction of each sentence.

The entire procedure of vector space model shown in Fig. 4 can be divided to various steps. Initially, all the terms are extracted from the document. This is referred as document indexing. Then, these indexed terms are weighted in order to enhance the retrieval of document to the user effectively. Various schemes are used to differentiate the sentences in terms of their weights. Most of them assume that the importance of a term is proportional to the number of sentences that term appears in. Let $D = \{S_1, S_2, \ldots, S_n\}$ be a set of sentences occurring in document D. According to the model of vector space, sentences can be denoted as a vector of features with corresponding weights. Features are obtained from the words which

Fig. 4 Vector space model

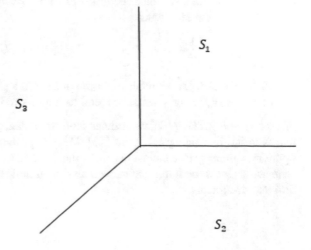

Fig. 5 Cosine similarity
between two documents

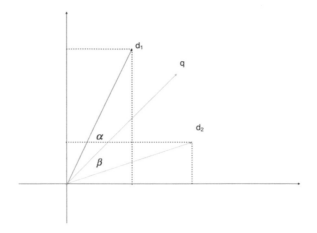

appear in that document, D. Each sentence S_i of document D can be represented as $(w_{1i}, w_{2i}, \ldots, w_{ni})$, where w_{ni} is the numeric weight for the sentence S_i, and n is the number of words in D.

Similarity Score Cosine similarity (as shown in Fig. 5) is the cosine value of angles between the vectors which are represented by the word-frequency vectors of any two sentences. It is given by:

$$\sin = (A \cdot B)/(||A||||B||) \tag{1}$$

Sentence Importance Score The importance scores assigned to each sentence depends on certain parameters. They are listed below

TF-IDF SUM Importance of words present in a given sentence represents the goodness of that sentence. TF-IDF is a straightforward yet powerful heuristic used to rank the words of a sentence according to their importance. It is the summation of TF_IDF scores for each word.

$$\text{tf}.\text{idf}(t,d) = (1 + \log(\text{tf}\_(t,d))) * \log(N/(df\_t))$$

SENTENCE LENGTH Number of words present in a given sentence represents its length. Generally, longer sentences tend to hold more information.

SENTENCE POSITION News articles tend to contain most of the important sentences in the first paragraph itself [15]. Generally, the ending paragraphs contain opinions, summary of conclusions about the document. Therefore, the position of a sentence seems to be important as well as a good indicator of its importance across different documents.

NUMERICAL LITERALS COUNT It is used to count the occurrence of numerical literals in a sentence. Sentences often contain statistical information in terms of numeric.

UPPER CASE LETTER COUNT The count of upper case letters can be useful in identifying entities which are generally represented in upper case.

3.2 Stack Decoder Algorithm

1. Since stack decoder formulation can test multiple summaries with different lengths, it was used to generate summaries close to the global optimal.
2. An overall score is calculated for the potential summaries according to the scores of their subsets. Finding the summary with the best overall importance score is the main objective. Here, score of summary is calculated as the sum of scores of individual sentences.

$$\text{imp (summary)} = \sum X \epsilon \text{ sentences imp(sentence)}$$

3. The set of all the sentences of input documents along with their importance scores serve as the input to summary generation technique.
4. The decoder has one stack for each length up to the maximum_length, a total of maximum_length + 1 stacks. Also an additional stack for summaries with greater length than maximum_length.
5. Priority queue is used to maintain the solutions of each computation. Every stack holds the best summary for a specific length. Since, the length of priority queue is limited; there will be at most stacksize different hypotheses for every stack, i.e., the obtained solution is best possible summary for that length.
6. The algorithm examines every set of sentences in a particular stack and then attempts to extend that solution with the sentences from original input document sentences.
7. Some important aspects:

   (a) A threshold value is decided as the minimum_length for sentences.
   (b) To tackle the problem of redundancy, a sentence which is considered to be added to an existing solution is added only if the similarity is below a certain value.
   (c) The maximum number of solutions maintained in the priority queue is restricted to stacksize for avoiding exponential blow-up.

3.3   Algorithm

The priority queue has a maximum of stacksize number of solution of summaries. All the sentences are placed at their respective length positions along with the (score,id) pair in this queue. Here, Fig. 6 is used to delineate the stack decoder algorithm with pseudo code and Fig. 7 lays out the entire flow of direction of the implementation.

Sentences are selected with their respective calculated sentiment scores. Picking a single stack at a time (after initialization phase), appropriate solution is selected

```
for i = 0 to maximum_length
{   for every solution in Stack[i]
    {   for every s in Sentences
        {
            newlen = maximum_length+1
            if (i+length(s) <= maximum_length)
                newlength = i+length(s)
                if (similarity < threshold)
                    new_solution = solution U {s}
                else
                    next
                score = importance(new_solution)
                Insert new_solution score into priority queue stack[newlength]
        }
    }
}
return best solution in stack[maximum_length]
```

Fig. 6 Stack decoder algorithm

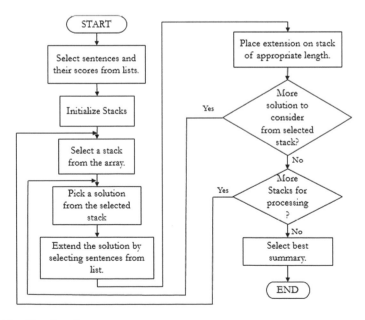

Fig. 7 Algorithm flowchart

from that stack and this solution is extended by further selecting other sentences from the list. After this step, this extended solution is placed in a stack of its respective length. If there are more sentences to be picked up from this stack, this process is repeated. Otherwise, we check if there are other such stacks available to be processed. If not, the best summary is picked and represented as a solution to the provided set of input; else the process is repeated for each of the stack.

After describing the algorithm and its flow, we should understand the system architecture of the system used specifically for document summarization, as shown in Fig. 8. Initiating by scanning the set of documents which needs to be summarized, these documents are firstly pre-processed for some checks in terms of stop-words, terminators, etc. Then, sentiment scores are calculated for each word separately to know the importance of each sentence of the document. These scores are calculated on the basis of some factors already defined in the previous sections of this paper. After successfully calculating the scores, ideal sentences which produce the most optimal summary are selected. Finally, the best solution or summary from each set is picked as the final solution.

Moreover, the system architecture for document classification is shown in Fig. 9. Our training set involved both positive and negative data from which various word features were also extracted. Training our recursive neural tensor network with this dataset and further classifying the extracted features from our input data gives us the polarity of our summary.

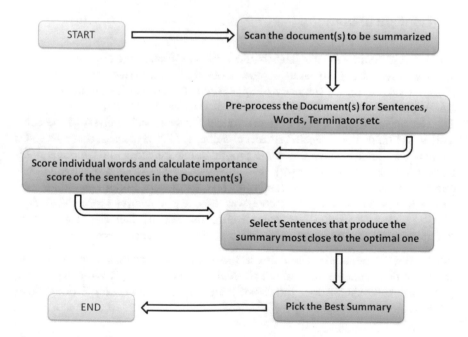

Fig. 8 System architecture (document summarization)

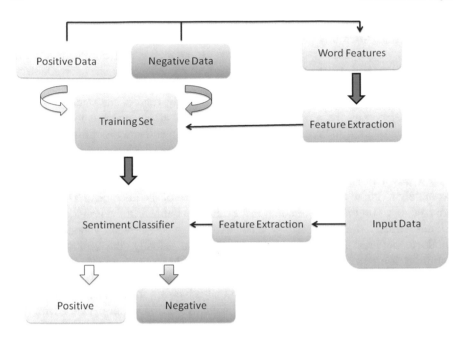

Fig. 9 System architecture (document classification)

4 Implementation

Dataset and Platform We have carried out our implementation of document summarization using datasets from Document Understanding Conference (DUC) 2004. A concise summary of approximately 100 words (number of words can be altered) was generated from the documents of DUC 2004.

The Recursive Neural Tensor Network model for sentiment analysis of the summaries was trained on a labeled dataset containing 7121 sentences. Stack Decoder method was implemented in JAVA programming language on Netbeans IDE. Recursive Neural Tensor Network was implemented in Python programming language on Linux OS. For preprocessing of text documents, Porter's algorithm was used whereas cosine similarity score was used for similarity score calculation. Finally sentence position, sentence length, number of literals, TF-IDF, Top K Important words and upper case were used for importance score calculation.

Functional Components The following serves as the functional components for automatic document summarization: (i) Sentence separation, (ii) Word separation, (iii) Stop-words elimination, (iv) Word-frequency calculation, (v) Scoring algorithm, (vi) Summarizing.

5   Result and Analysis

Experiments were done on the DUC 2004 data and various importance measures, similarity score were used to generate summaries.

Input Text Document:

Output Text Document:

Sentiment score was calculated for each of the input data and the corresponding summaries using Recursive Neural Tensor Networks. The scale for "sentiment score" used for sentiment analysis (Fig 10; Table 2).

The graph in Fig. 11 shows the comparison of sentiment scores of original documents and their summaries. It is clearly seen that their respective polarities are similar with some variations in File 4. The graph in Fig. 11 shows the comparison of sentiment scores of original documents and their summaries. It is clearly seen

Fig. 10 Sentiment score scale

Table 2 Comparison of sentiment scores

	Original	Summary
File 1	1.8	0.12
File 2	1.7	0.25
File 3	0.9	2.03
File 4	1.23	4.51
File 5	1.6	3.4

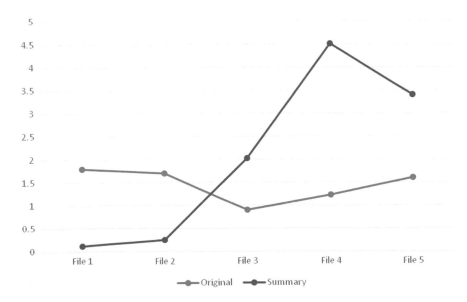

Fig. 11 Line graph (sentiment score)

that their respective polarities are similar with some variations in File 4. Hence, our verification methodology validates multi-document summarization using stack decoder method.

6 Conclusion

Multi-document summarization can efficiently draft the summary of input documents but it also has certain limitations such as poor coherence among the sentences, low coverage, redundancy and inaccurate extraction of essential sentences. Moreover, sentiment analysis of the summarized documents also has some limitations.

- Language-Specific Limitations: Words have different meanings; Language is a barrier, Sarcasm statements, Ex: Irony, Comparative and Complex statements, Short forms and many representations of the words, Co-relationship between the sentences.
- Technological Limitations: Extraction of Entities, Cannot identify the root cause of the review
- Data Specific Limitations: Noisy data, Videos and Images, Large data sets and lots of spam, Fails to classify with respect to others perspective, Knowledge of the data.

As the amount of textual content available electronically increases substantially, it becomes tougher for users to obtain data that is actually of their interest. Consequently methods of automatic document summarization are becoming progressively significant. Summarization of text document is a problem of shortening a source document into a shorter form and simultaneously preserving its data content. Stack decoder method is a good method to get started with the summarization as it is quicker and have fewer complexes to implement compared to other methods. It also provides better ROUGE scores compared to others. This paper [17] helped us to understand the different aspects of project with the help of detailed analysis. Using the combined system of document summarization and sentiment analysis, we can easily analyze and mine multitude of data in a time-efficient manner. Building upon cosine similarity seems to be most logical because of its speed and efficiency. Synset expansion which is based upon the POS tags of words can used before the application of cosine similarity to be more successful than just cosine similarity. We can reorder the sentences to generate more coherent summaries. Avoidable parts of the sentences can be removed by compression techniques in addition to the improvement in dataset, to attain better results.

Acknowledgements We would like thank our fellow classmates Vivek Thusu and Tanika Raghuvanshi for their insights and assistance during the course of this work.

References

1. Gong, Y., & Liu, X. (2001). Generic text summarization using relevance measure and latent semantic analysis. In *Proceedings of the 24th annual international ACM SIGIR Conference on Research and development in Information Retrieval* (pp. 19–25).
2. Alguliev, R. M., & Aliguliyev, R. M. (2005). Effective summarization method of text documents. In *The 2005 IEEE/WIC/ACM International Conference on Web Intelligence (WI'05)* (pp. 264–271). IEEE.
3. Alguliev, R. M., Alyguliev, R. M., & Bagirov, A. M. (2005). Global optimization in the summarization of text documents. *Automatic Control and Computer Sciences, 39*(6), 42–47.
4. Delort, J. Y., Bouchon-Meunier, B., & Rifqi, M. (2003). Enhanced web document summarization using hyperlinks. In *Proceedings of the Fourteenth ACM Conference on Hypertext and Hypermedia* (pp. 208–215). ACM.
5. Sun, J. T., Shen, D., Zeng, H. J., Yang, Q., Lu, Y., & Chen, Z. (2005). Web-page summarization using clickthrough data. In *Proceedings of the 28th annual international ACM SIGIR conference on Research and Development in Information Retrieval* (pp. 194–201).
6. Goldstein, J., Kantrowitz, M., Mittal, V., & Carbonell, J. (1999). Summarizing text documents: Sentence selection and evaluation metrics. In *Proceedings of the 22nd Annual International ACM SIGIR Conference on Research and Development in Information Retrieval* (pp. 121–128). ACM.
7. Harabagiu, S., & Lacatusu, F. (2005). Topic themes for multi-document summarization. In *Proceedings of the 28th Annual International ACM SIGIR Conference on Research and Development in Information Retrieval* (pp. 202–209). ACM.
8. Hu, P., He, T., Ji, D., & Wang, M. (2004). A study of Chinese text summarization using adaptive clustering of paragraphs. In *The Fourth International Conference on Computer and Information Technology, 2004. CIT'04* (pp. 1159–1164). IEEE.
9. Jatowt, A. (2004). Web page summarization using dynamic content. In *Proceedings of the 13th International World Wide Web conference on Alternate track papers & posters* (pp. 344–345). ACM.
10. Kruengkrai, C., & Jaruskulchai, C. (2003). Generic text summarization using local and global properties of sentences. In *Proceedings IEEE/WIC International Conference on Web Intelligence, 2003, WI 2003* (pp. 201–206). IEEE.
11. Mitray, M., Singhalz, A., & Buckleyyy, C. (1997). Automatic text summarization by paragraph extraction. *Compare, 22215*(22215), 26.
12. Shen, D., Chen, Z., Yang, Q., Zeng, H. J., Zhang, B., Lu, Y., & Ma, W. Y. (2004). Web-page classification through summarization. In *Proceedings of the 27th Annual International ACM SIGIR Conference on Research and Development in Information Retrieval* (pp. 242–249). ACM.
13. Yeh, J. Y., Ke, H. R., Yang, W. P., & Meng, I. H. (2005). Text summarization using a trainable summarizer and latent semantic analysis. *Information Processing and Management, 41*(1), 75–95.
14. Diao, Q., & Shan, J. (2006). A new web page summarization method. In *Proceedings of the 29th Annual International ACM SIGIR Conference on Research and Development in Information Retrieval* (pp. 639–640). ACM.
15. Lesk, M. (1986). Automatic sense disambiguation using machine readable dictionaries: How to tell a pine cone from an ice cream cone. In *Proceedings of the 5th Annual International Conference on Systems Documentation* (pp. 24–26). ACM.
16. Barzilay, R., & Elhadad, N. (2002). Inferring strategies for sentence ordering in multidocument news summarization. *Journal of Artificial Intelligence Research, 17,* 35–55.
17. Sripada, S., Kasturi, V. G., & Parai, G. K. (2005). Multi-document extraction based summarization. CS224, Final Project, Stanford University.

Data Mining Technology with Fuzzy Logic, Neural Networks and Machine Learning for Agriculture

Shivani S. Kale and Preeti S. Patil

Abstract Farmers countenance failure as the crop cultivation decisions by farmers always depend on current market price as the production sustainability processes are not taken into consideration. So there should be some platform which guides the farmer for taking correct decision depending on their need, environment, and changing seasons. The system proposes Marathi calendar using nakshatras which guide farmer for crop cultivation decision. It aims to create methodologies to strengthen the farmers' economic conditions by providing informed decisions. The methodology used for the system specially uses data mining to generate expert decision along with the fuzzy logic, machine learning to give decisions appropriately to farmer for cultivation of expected crops.

1 Introduction

Agriculture theaters a vital role in India's financial system. Exploration in farming is intended for the sake of increased crop production at cheap expenditures and with amplified yield. Not only final product (crop produced) should be acceptable but also processes to develop that product should also be sustainable. Today's need is to train the farmer with sufficient as well as useful techniques necessary for farming. The farming usually depends on weather conditions and monsoon predictions. Each specific day in each and every season, according to farming practices has its own tasks to be worked out and if it is followed correctly it will consequence in increase yield of production.

S. S. Kale (✉)
Department of Computer Science and Engineering, KIT's College of Engineering,
Kolhapur, India
e-mail: shivanikale33@gmail.com

P. S. Patil
Department of Information & Technology, D. Y. Patil College of Engineering,
Akurdi, Pune, India
e-mail: dr.preetipatil.kit@gmail.com

© Springer Nature Singapore Pte Ltd. 2019
V. E. Balas et al. (eds.), *Data Management, Analytics and Innovation*,
Advances in Intelligent Systems and Computing 839,
https://doi.org/10.1007/978-981-13-1274-8_6

The proposed work includes significant data from experienced farmers, the infallible and experienced procedure of farming, requirement of fertilizers, category of crops to be cultivated depending on farmers land condition, weather conditions, etc., to give suggestions to farmers for correct crop cultivation.

The proposed system analyzes methods such as fuzzy logic, neural networks to appropriately give decisions to farmer for cultivation of expected crops.

The system will provide the farmer with concrete model of farming which will be the perfect solution of sustainable agriculture.

2   Literature Survey: (See Acknowledgements)

From the table we can get to know that the production of crops is decreased tremendously over years. So the computer science and its algorithm can help the farmer to take informed decisions depending on different parameters such as climate, area, soil parameters, etc. Informed decision to farmers helps to take correct decisions to get increased yield for respective crop. Following are the authors who have studied on different aspects of the agriculture and provided some good solution (Table 1).

1. **Sawaitul et al.**, wrote about the effects of climatic conditions on agriculture. He has used prominent factors for weather prediction. He has checked the effect of variation of the different climatic parameters on the weather condition. The author has used algorithm such as back propagation and artificial neural networks [1].
2. **Somvanshi et al.**, did the analysis using Box Jenkins and ANN algorithm for rainfall prediction for agriculture purpose [2].
3. **Verheyen et al.**, discussed that different methodology of data mining are generally applied for calculation of farm land distinctiveness. The author has used K-Mean Algorithm for land classification by application of Global Positioning System Technology [3].

Table 1 % Growth rates in the Production of crop

Crops	1980–1990	1990–1996	1996–2005
Paddy	3.62	1.58	−0.07
Wheat	3.57	3.31	0.00
Jowar	0.27	−2.64	−4.09
Bajara	0.01	−1.51	1.68
Ragi	−0.10	0.19	−2.26
Pulses	1.49	−0.66	−0.83
Oil seeds	5.46	3.90	−0.88
Sugarcane	2.47	2.96	−1.95

Referred from Dr. Subhash Palekar's Book

4. **Jagielska et al.**, illustrate solicitations to farming associated extents. They focused and discussed the parameter crop yield as very crucial factor for farmer. Earlier, the crop production prediction was calculated by the facts known to farmer as well as his experience on precise land, yield, and weather circumstance. They added discussion regarding statistics such as likelihood in concepts of probability, relationship rating with respect to fuzzy set concepts [4].
5. **Tellaeche et al.**, identifying pest in accurate crop growing. Author has reviewed different software applications with respect to weeds revealing and finding accurate spray pattern for respective crop and related weed. They have proposed a system for giving decisions to farmers to identify the correct pattern. The algorithms used for the expert system are KNN, Fuzzy logic techniques [5].
6. **Veenadhari** The author has taken into consideration the Bhopal district of Madhya Pradesh for the study. They analyzed effect of climatic conditions with respect to rabi as well as kharif crop yield. The technique used is decision tree analysis. The study is undertaken for soyabean and wheat crop. The effects of rain, temperature, humidity, etc., are studied. So the study shows that paddy crops are mostly affected by rainfall and wheat crop is having dependency shown for temperature parameter mostly [6].
7. **Shalvi and De Claris** In this paper the author stated that Bayesian network is a popular application frequently used for agriculture databases. The system designed for farming purpose uses most popular methodology of Bayesian network. Finally they concluded that Bayesian Networks are realistic and proficient [7].
8. **Chinchulunn et al.**, The paper categorize the data mining algorithms KNN rules into two classes, i.e., (1) Coaching (2) Testing. They proposed that fewer points need to be considered in first class [8].
9. **Rajagopalan and Lal** The comparative analyses of data mining technologies and their applications are considered for agricultural databases. For example they took KNN for application on daily precipitations simulation, etc. [9].
10. **Veenadhari et al.**, The author has discussed SVM methodology for categorization of data into two classes [10].
11. **Tripathi et al.**, They have shown the application of Support Vector Machine based algorithms on future weather prediction. They verify the results for weather alteration on precipitation over India [11].

As nearly all the ES (expert system) concentrates on exacting features of yield similar to insect, weeds, compost, etc. So all the expert systems and technology available for farmers are giving solutions to problems but they are scattered among different applications. So farmers are not getting ONE STOP solution. Similarly the expert systems are designed are either for particular area or for according to some climatic parameters. So farmer needs expert system which may possibly answer nearly all the questions and suggests the better solution for their existing situation.

Also none of the system helps the new grower about the knowledge of farming concept

(1) What to grow when?
(2) What types of land required for which crops?
(3) Which type of fertilizers is required for specific crop?
(4) What can be taken as mixed crop? etc.

The literature survey provides that no recent work endow the farmer with the appropriate decision of crop cultivation throughout the life cycle of crop. Every expert system designed until is addressing only one problem as a system. So the proposed system will provide farmer with solutions to almost every problem to take informed decisions regarding crop cultivation, type of crop pattern, when to cultivate what, when to do spraying for crops and requirement of resources and remedies for the types of diseases.

The proposed expert system is using Google map GPS for getting the land area of farmer and depending on the attributes of land and other parameters outputs will be given to the farmer. The proposed system uses Fuzzy logic adaptive techniques such as adaptive neuro-fuzzy inference systems (ANFIS) and fuzzy subtractive clustering. The feed forward back propagation artificial neural network, this methodology has been used in our proposed system for modeling and forecasting of various crop yield, type of crop cultivation, etc., on the basis of various predictor variables, i.e., input by farmer, viz. type of soil, season, nakshtra, insecticide, fertilizer, rainfall.ANN with zero, one, and two hidden layers have been considered (Fig. 1).

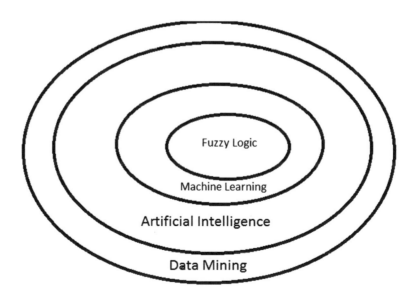

Fig. 1 Relationship between data mining algorithms

3   Proposed Work

In any sector management is important aspect. Agriculture field is also management related field. In agriculture Informed decisions are taken by the farmers based on the knowledge experience he had. Nowadays due to complex, complicated, computational world, it is required to adapt exact activity or agriculture process for cultivation of crop. Also it is required to make sure that no farmer survives on one crop pattern. So we suggest that farmer should use multiple crop patterns as per season in all types of monsoon regions of India reason for cultivation and profit. So we used data mining along with Artificial intelligence, Machine learning and fuzzy logic for well-being of crop and overall agriculture sustainability.

The three terms we used are as follows and can be described as Fig. 1

(I) Artificial Intelligence

In this we proposed all farming activities are designated in such pattern so that it is automation of crop pattern. For the given selected crop AI will take inputs such as

- Climate needed for that crop, according to climate—Atmospheric temperature, humidity, sunlight, etc. These are the direct inputs taken from the local weather forecasting department of India
- Soil required for that crop (TYPES OF SOIL)
- Plowing date required for that crop in various range
- Harvesting date required for that crop in various range
- Insecticides bites for that crop
- Fungicide bites for that crop
- Fertilizers for that crop
- Water requirement for that crop.

All the inputs are considered by AI for any type of crop. The simple algorithm of AI can take this information as input for producing proper crop decision. The inputs are also taken by machine learning algorithm at highest level by using ANN.AI uses more static programming instruction (Fig. 2).

(II) Machine Learning application in agriculture [12]

Full exploitation of data mining technology can be done by machine learning. Machine learning actually processes the data for example study of pattern recognition in agriculture field for particular crop or suitable cropping pattern. It also makes comparisons to give most suitable crop pattern that can be produced. Machine learning uses dynamic program instruction as well as static programming instructions too. Machine learning takes changing inputs such as change in temperature value which in turn makes changes into

- Harvesting time
- Spraying patterns of Insecticide and Pesticide

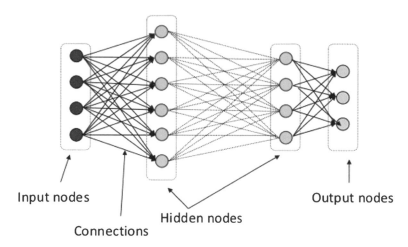

Input nodes

Connections

Hidden nodes

Output nodes

Fig. 2 Application of ANN, AI and machine learning and fuzzy in agriculture

- Water Requirement
- Fertilizer Usage-If temp is high farmer should not increase fertilizer usage for that period.

While using machine learning level in farming we can use more ANN to increase relationship between interconnected group nodes. So that back propagation can be applied more rigorously and by adjusting the network to reflect the exact decision. ANN is really helpful for making pesticide and fungicide spraying decision. We can also make use of Image processing and other tools which are not the subject of this paper.

(III) Fuzzy application in farming [12]

The fuzzy logic technique can be applied for agriculture to check feasibility of activity. We can also use fuzzy for evaluation of agriculture activity based on economic viability of the cropping pattern. In machine learning it may happen that depending on the data and activity input it will produce a single restricted output solution. It will not provide flexibility of solutions to be applied but fuzzy along with data can give us more than one solution flexibility in following aspects

- Inter cropping decision (Provide Options-refer Fig. 3)
- Insecticide and Fungicide determination decision(Less Expensive and Exact)
- Type of water input process determination
- Soil preparation decision
- Handling Control and decision making complex models.

Fig. 3 Mix cropping pattern

4 Result

There is no secret to success. It is the result of perfection, hard work, learning from failure, loyalty, and persistence. Productivity is never an accident. So for better crop yield and correct informed decision by farmer can be done using Machine learning algorithm. The following table shows the varying parameters input to agriculture and different algorithms we have used with the probable result (Fig. 4).

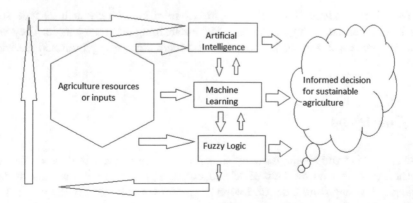

Fig. 4 Data flow of data mining for sustainable agriculture

Table 2 Result in different phases of data mining depending of inputs given

Farming input parameters	AI	Machine learning	Fuzzy	Result	Example
Temperature	Climate region	Suitable crop	Feasibility and economic viability	– Tropical crops – Hill Crops – Input usage	Coconut to Apple
Soil	Type of soil	Suitable crop	Feasibility and economic viability	Mix cropping pattern	Black cotton soil, Laterite. Sugarcane and cotton
Humidity	Type of climate region	Suitable crop	Feasibility and economic viability	– Crop Protection spray – Mix crop pattern	Kharip or rubby
Sunlight	Hot/cold region	Suitable crop	Feasibility and economic viability	– Amount of photosynthesis evaluation – Productivity	Increases 2.5 gm^2 foot food production
Above table inputs with machine learning in data mining technology gives following informed decision					
Fertilizer	Cropping pattern depending on temperature	Age of the crop and exact selection of fertilizer	Availability of water and climatic condition	– Guidance for fertilizer selection	NPK, micronutrients
Pesticide	Temperature, humidity and product stage	Organic composition mixture suggestion	Feasibility check	Increased productivity and crop protection	Insecticide and fungicide
Water requirement	Water resource as input	Per liter requirement	Type of irrigation	Irrigation system recognition	Drip, Sprinklers, Flood water
Human hours	Permanent and temperature combination	Temporary labor time schedule	Type of skilled and unskilled labor feasibility	Skilled and unskilled Human resource requirement	Human resources Per hour basis

The diagram given shows that the inputs from agriculture such as type soil, water, fertilizer, etc., are given to AI, Machine learning and Fuzzy logic. They will also get processed and exchanged amongst them for getting informed decision (Table 2).

5 Conclusion

Data mining is more about "farming" than it is about hunting. It is about cultivation decision making with the help of related output nodes. The real meaning of data mining is learning from the past, live in the present, and believe in the future. It is hard to prepare this type of model because interdisciplinary knowledge of Agriculture and climatology, geology, fluid engineering, agronomy and data mining is required for that but it can be handled all type of data by systematic approach of programmer or researcher.

So the paper has focused the possible inputs, outputs and usage of different algorithms to give correct decision to farmers for increased crop yield.

Now a day agriculture activity or decisions are very complex and not proper foundation for that by using data mining we can do that properly.

Acknowledgements Survey on Data Mining Techniques in Agriculture M. C. S. Geetha Assistant Professor, Dept. of Computer Applications, Kumaraguru College of Technology, Coimbatore, India. International Journal of Innovative Research in Computer and Communication Engineering (An ISO 3297: 2007 Certified Organization) Vol. 3, Issue 2, February 2015

References

1. Sawaitul, D. S., Prof Wagh, K. P., & Dr Chattur, P. N. (2012). Classification and prediction of future weather by using back propagation algorithm an approach. *International Journal of Emerging Technology and Advanced Engineering, 2*(1), 110–113.
2. Somvanshi, V. K., et al. (2006). Modeling and prediction of rainfall using artificial neural network and arima techniques. *The Journal of Indian Geophysical Union, 10*(2), 141–151.
3. Verheyen, K., Adriaens, D., Hermy, M., & Deckers, S. (2001). High resolution continuous soil classification using morphological soil profile descriptions. *Geoderma, 101*, 31–48.
4. Jagielska, I., Mattehews, C., & Whitfort, T. (1999). An investigation into the application of neural networks, fuzzy logic, genetic algorithms, and rough sets to automated knowledge acquisition for classification problems. *Neurocomputing, 24*, 37–54.
5. Tellaeche, A., BurgosArtizzu, X. P., Pajares, G., & Ribeiro, A. (2007). *A vision-based hybrid classifier for weeds detection in precision agriculture through the Bayesian and Fuzzy k-Means paradigms. In Innovations in Hybrid Intelligent Systems* (pp. 72–79). Berlin: Springer.
6. Veenadhari, S. (2007). *Crop productivity mapping based on decision tree and Bayesian classification* (Unpublished M. Tech thesis) submitted to Makhanlal Chaturvedi National University of Journalism and Communication, Bhopal.
7. Shalvi, D., & De Claris, N. (1998, May). Unsupervised neural network approach to medical data mining techniques. In *Proceedings of IEEE International Joint Conference on Neural Networks*, (Alaska), (pp. 171–176).
8. Chinchulunn, A., Xanthopoulos, P., Tomaino, V., & Pardalos, P. M. (1999). Data Mining Techniques in Agricultural and Environmental Sciences. *International Journal of Agricultural and Environmental Information Systems, 1*(1), 26–40.
9. Rajagopalan, B., & Lal, U. (1999). A K-nearest neighbor simulator for daily precipitation and other weather variable. *Water Resources, 35*, 3089–3101.
10. Veenadhari, S., Dr Misra, B., & Dr Singh, C. D. (2011, March). Data mining techniques for predicting crop productivity—A review article. *International Journal of Computer Science and Technology IJCST, 2*(1).
11. Tripathi, S., Srinivas, V. V., & Nanjundiah, R. S. (2006). Downscaling of precipitation for climate change scenarios: a support vector machine approach. *Journal of Hydrology, 330*(3), 621–640.
12. Murmua, S., & Biswasa, S. (2015). Application of fuzzy logic and neural network in crop classification: A review. In *ScienceDirect International Conference on Water Resources, Coastal and Ocean Engineering (Icwrcoe 2015) Aquatic Procedia 4* (pp. 1203–1210).

6-Tier Design Framework for Smart Solution as a Service (SSaaS) Using Big Data

Maitreya Sawai, Deepali Sawai and Neha Sharma

Abstract The data is floating everywhere as a result of availability of smart connected devices, and an attempt to decipher this ocean of data has made big data analytics a buzzword today. After industrial revolution, Internet revolution, we are witnessing third revolution, i.e., the Internet powered by IoT and Big Data. Extracting the value from this huge data and making the data useful through Big Data Analytics is the need of the hour. The world as a whole is facing many problems or challenges in multiple areas towards extracting valuable insight from the pile of digital data which is heterogeneous and distributed. The paper suggests a 6-tier framework using Text Mining, Natural Language Processing and Data Analytics to suggest Smart Solution(s) as a Service.

Keywords Big data · Big data analytics · Big data as a service
Cloud technologies · IoT

1 Introduction

With the invent of new technologies like Mobile Technologies, IoT and varied applications of social media, a huge amount of data of mottled data types is getting generated every day at a very fast rate. The traditional data processing and analysis of structured data using Relational Database Management Systems (RDBMS) and

M. Sawai
Department of Computer Science, Pimpri Chinchwad College of Engineering,
Akurdi, Pune, India
e-mail: maitreyasawai333@gmail.com

D. Sawai · N. Sharma (✉)
Institute of Industrial and Computer Management and Research, Savitribai Phule
Pune University, Pune, Maharashtra, India
e-mail: nvsharma@rediffmail.com

D. Sawai
e-mail: deepalisawai@gmail.com

© Springer Nature Singapore Pte Ltd. 2019
V. E. Balas et al. (eds.), *Data Management, Analytics and Innovation*,
Advances in Intelligent Systems and Computing 839,
https://doi.org/10.1007/978-981-13-1274-8_7

data warehousing no longer satisfy the challenges of Big Data. This big data, does not offer any value in its unprocessed state. Application of right set of tools can pull powerful insights from it. This huge, exponentially growing data, can drive decision making to a great extent only if it is analyzed in a systematic way, realizing the correlations and hidden patterns. For data analysis, the most important thing is to find the keywords, which are nothing but the condensed versions of documents and short forms of their summaries. For comparing, analyzing and mapping the information, extracting the keywords is the important task. There are many keyword detection and extraction algorithms available, like ranking algorithm called HITS [1] to directed graphs representing source documents. Mihalcea and Tarau [2] have applied the PageRank algorithm suggested by Brin and Page [3] for keyword extraction using a simpler graph representation (undirected unweighted graphs), and showed that their results compare favorably with results on established benchmarks of manually assigned keywords. Mihalcea and Tarau, are also using the HITS algorithm for automatic sentence extraction from documents represented by graphs built from sentences connected by similarity relationships. According to Gonenc Ercan and Ilyas Cicekli, encouraging results can be obtained by a keyword extraction using lexical chains [4]. Sufficient work has been done by various researchers to perform search operation on the web, which is the basis for data analysis. However, these techniques may be used for effective mapping of problems and the "discovered" knowledge along with prescriptive analysis can resolve many difficulties faced across the world in diverse fields.

The objective of this paper is suggest a design of a framework for solution as a service. The framework attempts to bring scholars, researchers, scientists, various domain experts, administrators, and government officials, ministers on a common platform to organize and channelize the information flow to solve many challenges faced by the world. This framework will bring together three entities viz. the one who is facing a problem in some vertical or domain, second is the one who has got the solution for such problem or have experience in solving similar problems or has an idea and knowhow. The third entity is data which may be useful for solving the challenge, which would be extracted from the freely available big data. The rest of the paper is organized as follows: Sect. 2 reviews related literature, Sect. 3 presents a brief design and working of the framework, Sect. 4 presents the conclusion and future enhancement. Finally, references are included in chronological order.

2 Related Work

The data from various resources is growing exponentially with every second. However, the heterogeneity of data makes it difficult to organize, interpret, co-relate, and analyze to make it useful. The most fundamental challenge for the big data applications is to explore the large volumes of data and extract useful information or knowledge for future actions, as per Leskovec et al. [5]. Conventionally, Management Information Systems (MIS), Decision Support Systems (DSS) were used for decision making in any business. But these systems were having a limited access to data and hence limited analytics capabilities. Gorry et al. [6] conceived the term "decision support systems" and built a framework for improving management information systems using Anthony's [7] categories of managerial activity and Simon's [8] taxonomy of decision types. Keen et al. [9] proposed a scope by narrowing down the definition to semi-structured managerial decisions, that is relevant today. There is plenty of complex unstructured data floating everywhere and to analyze and understand this data, there is a strong need of algorithms and tools to perform analytics on big data. Various industries, used the behavioral model proposed by Simon [10] for contemporary decision making systems but with the explosion of data in modern era, big data is the relevant data provider for the same. As per Thomas H. Davenport, it is important to remember that the primary value from big data comes not from the data in its raw form, but from the processing and analysis, followed by the insights, products, and services which emerge from analysis [11]. Artificial intelligence, machine learning, semantic web, data mining are the ideal way of addressing a majority of challenges for big data. Data mining can be defined as the discovery of "models" for data [12]. Machine learning and data mining, combined with big data can give great results with far less dependability on human.

Dr. Michael Wu, chief scientist of San Francisco-based Lithium Technologies [13], in his blog series, (March 2013) mentioned, that the first step of data reduction in big data, is descriptive analytics which can turn the big data into smaller, more useful nuggets of information. By utilizing variety of statistical modeling, data mining, and machine learning techniques on recent and historical data, predictive analytics forecast the future. By recommending one or more courses of action, and presenting the likely outcome of each decision, prescriptive analytics goes beyond descriptive and predictive models, as it does not predict one possible future, but "multiple futures" based on the decision-maker's actions. Diagnostics, on the other hand analyses the past performances to determine the state. Thus many type of analysis can be performed on the data depending on the requirement, more the data more better the analysis.

Every organization need data from multiple sources which can then be mixed and matched for centralized data analytics and business intelligence. With the evolution of big data analytics, Big Data as a Service (BDaaS) is on the horizon. It is a combination of Data as a Service (DaaS), Platform as a Service (PaaS), and Software as a Service (SaaS). Pramita Ghosh in her paper writes that with Hadoop technology, data is more accessible and controlled, and has a combination of on-premise and cloud technologies, making it possible to store data both on-premise and on cloud platforms [14]. Gartner predicts rapid evolution of modern BI and analytics market and the BDaaS will be offered via web and social sites. Information systems will allow sensor-driven product tracking. The information sources will include government data sources, social data, and connected IoT data to get a competitive edge for the organizations. As per the report of EMC Solutions Group [15], in the new data technology world, Big Data as a Service will survive and prosper due to the available cloud platforms. Those organizations that will use the best combination of technologies and tools for scaled data analytics at a reasonable cost, will ultimately win the race. The service providers, providing only hardware infrastructure services, will have to combine powerful big data analytics services in an integrated package.

3 Proposed Work

Big data analytics is actually a mix of skills, technologies, applications and processes that enable the analysis of an immense volume, variety and velocity of data across a wide range of networks to support decision making and action taking [16]. In every field, success depend on the right decision taken at right time from available and implementable alternatives. If the alternatives available are more and are easily accessible, the possibility to have a right decision increases multi-fold. Today, there are numerous challenges faced by the society at either personal or professional level. Some have solutions and some may not have it at present; but in majority cases, due to lack of awareness the problems remain unresolved. Therefore, there is a need of a platform to connect the people who are facing problems with those who have some idea or solution which may solve their problems. In this paper, a framework is proposed, which will suggest a set of solutions to resolve the challenges faced by the people in any corner of the world. The proposed framework intend to explore and exploit the power of advanced technologies like big data analytics, text mining, natural language processing. The initial design of the framework suggests that there would be primarily three types of the users of the system

(i) Requester: The one who is facing some problem and requesting for solutions.
(ii) Knowledge provider: The expert who is willing to share his experience in the domain or may have a ready solution for the requester. The organizations which are willing to share data can also register to be a part of the system as a Knowledge Provider. They can add their products details, which might be useful to provide solutions.

The Requester and the Knowledge Provider may belong to one of the two categories:

   a. Naïve users: Naïve users are the users who are not much familiar with the technology and may have language barriers.
   b. Sophisticated users: These users are technical or semi-technical and can use english for communication.

(iii) Big Data Administrator (BDA): The one who would be administrating the big data getting collected from the users and also from the available free resources which may include Government data like census, government policies, etc. The BDA will have right to add keywords in Keyword Bank and also approve the request to add new keywords in the keyword bank. The keyword bank is a table of keywords which would be used by the both types of users to input their data.

Fig. 1 6-Tier structure of the proposed framework

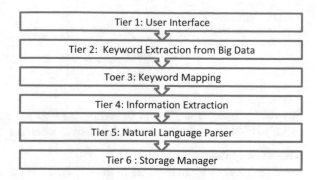

To avoid the misuse of the data which is shared, only the statistical information can be used to generate the solution of the problem of the registered Requester. The proposed 6-Tier structure as shown in Fig. 1 incorporates many complex algorithms to provide smart solutions to the requester. The structure includes concepts like keyword extraction, keyword mapping, information extraction, natural language parser and storage manager.

Tier 1: User Interface

The tier 1 is the user interface for all the types of users. Separate interface is designed for different type of users. The users can be broadly categorized as naïve requester, naïve knowledge provider, sophisticated requester, sophisticated service provider or big data administrator. It is a mandatory requirement for each user to register themselves before using the system. The big data administrator, initially will create a keyword bank with a list of commonly used keywords from various domains or verticals. Since it is not possible to create such a huge bank, the users are also allowed to append the new keywords. This keyword bank is accessible to the users. The naïve or sophisticated requester submits the problem statement through their respective interface. Requester can also submit few important keywords which describes the challenge to a great extent. Similarly, the knowledge provider enters success stories, products details, implemented ideas or implementable ideas to solve the challenges along with the relevant keywords. These keywords are subsequently added to the "Keyword Bank" by BDA, if not already available. The user interface also tries to predict the keywords, making it easy for the user to list the keywords. To make the system user friendly, the framework allows the user to choose the language as per convenience, at the time of registration. Finally, tier-1 stores the registration details on the cloud storage after proper verification of the data submitted by the users.

Tier 2: Keyword Extraction from Big Data

A suitable "text extraction algorithm" is used to extract the keywords from the textual big data and possible keywords are be added to the "Keyword Bank". The label of the video and pictures are also considered as keywords and extracted to be added to "Keyword Bank". All the extracted keywords along with their references are then added to the Keyword Bank. Table 1 demonstrates the structure of

Table 1 Structure of keyword bank table

Keywords	Link to file
Kwrd1	Link to Service Providers knowledge
Kwrd2	Link to the Big Data reference
Kwrd3	...

"Keyword Bank" table. The table mainly has two attributes i.e. keywords and link to file. The link to file attribute is basically a reference to the keyword.

Tier 3: Mapping

Once the Keywords are received in the required form from tier 2, using appropriate text mining algorithm, the keywords would be matched.

Let us assume the keywords entered by the Requester are $R\_Kwrd1, R\_Kwrd2, \ldots,$ R_Kwrdn, and the keywords entered by various knowledge Providers are $KP1\_Kwrd1, KP1\_Kwrd2, \ldots, KP2\_Kwrd10, KP2\_Kwrdn$, then using selected text mapping algorithm the keywords would be matched/mapped with the keywords from the keyword bank and set is formed in the following manner:

$$\{R\_Kwrd1, R\_Kwrd2, \ldots, R\_Kwrdn\} = \{KP\_Kwrd1, KP\_Kwrd2, \ldots, KP\_Kwrdn\}$$

Text mining would be performed again to extract the keywords from the big data (BD_Kwrd) from cloud storage. These keywords also would be match with the requested keywords, adding the elements to the final set of mapped keywords as follows:

$$\{R\_Kwrd1, R\_Kwrd2, \ldots, R\_Kwrdn\} = \{KP1\_Kwrd1, KP1\_Kwrd102, \ldots, KP8\ Kwrd20,$$
$$BD1\_Kwrd1, \text{ and } BD18\_Kwrd200\}$$

Tier 4: Information Extraction

Once the mapped keyword sets are ready, the respective information of the knowledge provider-owner and big data is also extracted. Thus the set to provide smart solution to requester's problem is formed by extracting the registration number details of the mapped knowledge providers and matched big data file details as follows:

$$\text{Requester Problem} = \{KP\_Reg3, KP\_Reg100, KP\_Reg30, \ldots,$$
$$BD18\_Kwrd200, BD101\_Kwrd2\}$$

This set would help the requester to get the details of the knowledge provider through the registration number and the big data files through the keyword reference. Subsequently, using the above extracted keywords, the concept map is formed for the submitted problem statement using the graph data structure which will give a conceptual idea of the solution as shown in Fig. 2. Thereafter, prescriptive analysis tool are used to predict the smart solution(s) for the given problem.

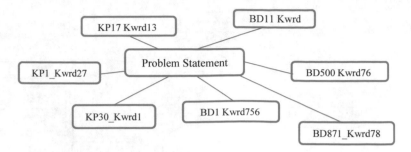

Fig. 2 Concept map/graph for the problem statement

The concept map which will keep on building up over the time till the requester is active/registered to the service. The map such developed, would have the permanent linkages developed with their contributors. In case, requester keyword do not get any match with the knowledge keywords, then the data would be saved in the cloud as would always be done and in future when the keywords finds the match, the requester would be informed on his dashboard.

Tier 5: Natural Language Parser
Natural language processing algorithms are used in case of naïve requester to convert the problem statement in English language and the solution to the desired/ selected language.

Tier 6: Storage Manager
The concept map and the solution is sent to the dashboard of the user and also stored on the cloud storage to make it available for future use.

The flowchart for the suggested algorithm for the proposed 6-tier framework is shown in Fig. 3:

Even predictive data will be cross checked with the actual facts to understand the correctness of the predicted solutions. For this it requires two additional components: actionable data and a feedback system that tracks the solution provided by the system based on the knowledge provided. Figure 4 presents the 6-Tier framework for smart solution as a service.

4 Conclusion

At present, it is just the beginning of Big Data transformation; those who do not seriously consider becoming "intelligent", will left behind. There are plenty of big data opportunities and only sky is the limit. The data which is getting collected should be analyzed innovatively using proper technology, thinking out of box, only then will able to get the real value out of the big data and also the results which may not have imagined earlier.

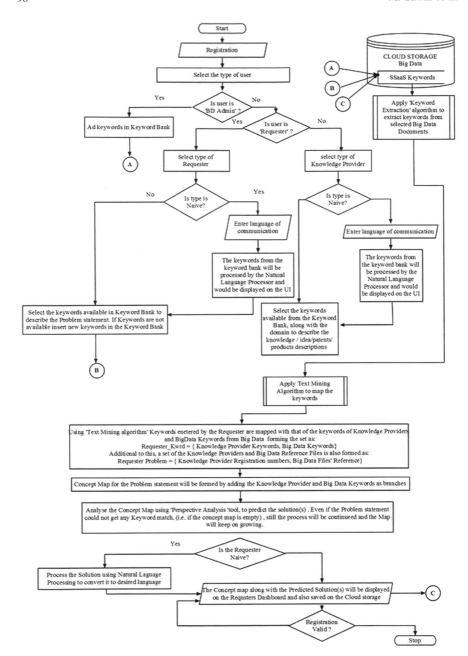

Fig. 3 Flow chart for the suggested algorithm

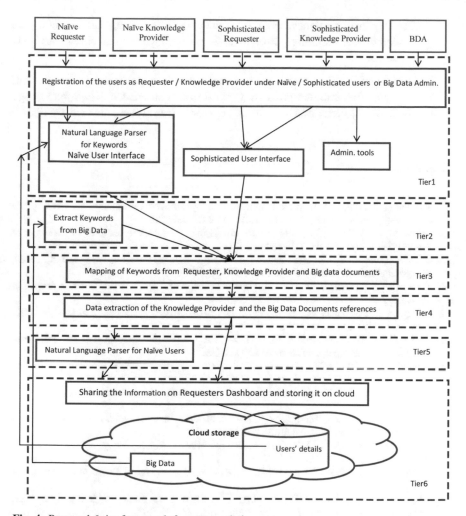

Fig. 4 Proposed 6-tier framework for smart solution as a service

The framework design is an excellent example exhibiting that the technical and human resources can be combined for exploiting big data's value. The framework will able to help transform the data into insights, knowledge and decisions, in turn making the individual or organization self-reliant. The whole process of providing smart solution as a service can be summarized into two important steps:

Step 1: To break down the knowledge and the problem statements into nuggets of information in the form of keywords. Besides, to extracted the keywords from available big data and to prepare the keyword bank.

Step 2: To generate a concept map from the mapped keywords and to predict the appropriate solution(s)/future action(s) using the prescriptive analysis tools.

To implement the framework, open source software, commodity servers, and massively parallel-distributed processing platforms is be needed. The future scope of our work would include extraction of keywords from image and video clips.

References

1. Kleinberg, J. (1999). Authoritative sources in a hyperlinked environment. *Journal of the ACM, 46*(5), 604–632. https://doi.org/10.1145/324133.324140.
2. Mihalcea, R., & Tarau, P. (2004). TextRank: Bringing order into texts. In *Proceedings of EMNLP 2004* (pp. 404–411), Barcelona, Spain. Association for Computational Linguistics.
3. Brin, S. (1998). Page l: The anatomy of a large-scale hypertextual web search engine. *Computer Networks and ISDN Systems, 30*(1–7), 107–117.
4. Ercan, G., & Cicekli, I. (2007). Using lexical chains for keyword extraction. *Information Processing and Management, 43*(6), 1705–1714.
5. Leskovec, J., Rajaraman, A., & Ullman, J. D. (2014). *Mining of massive datasets* (2nd ed.). New York, NY, USA: Cambridge University Press.
6. Gorry, G. A., & Scott Morton, M. S. (1971). A framework for management information systems. *Sloan Management Review, 13*(1), 1–22.
7. Anthony, R. N. (1965). *Planning and control systems: A framework for analysis*. Cambridge, MA: Harvard University Graduate School of Business Administration.
8. Simon, H. A. (1977). *The new science of management decision* (rev. ed.). Englewood Cliffs, NJ: Prentice-Hall.
9. Keen, P. G. W., & Scott Morton, M. S. (1978). *Decision support systems: An organizational perspective*. Reading, MA: Addison-Wesley.
10. Simon, H. A. (1955). A behavioral model of rational choice. *The Quarterly Journal of Economics, 69*(1), 99–118.
11. Davenport, T. H., & Dyché, J. (2013). *Big data in big companies*. International Institute of Analytics.
12. Wu, X., Zhu, X., Wu, G. Q., & Ding, W. (2014). Data mining with big data. *IEEE Transactions on Knowledge and Data Engineering*, (1), 97–107.
13. Wu, M. (2014). Big Data Analytics: Descriptive vs. Predictive vs. Prescriptive. http://bigdata-madesimple.com.
14. Ghosh, P. (2017). Big Data as a Service: What Can it Do for Your Enterprise? http://www.dataversity.net.
15. EMC Solutions Group, White paper Big Data as a Service A Market and Technology Perspective (2012).
16. Hajli, N., Wang, Y., Tajvidi, M., & Hajli, S. (2017). People, technologies, and organizations interactions in a social commerce era. *IEEE Transactions on Engineering Management*. https://doi.org/10.1109/TEM.2017.2711042.

Part II
Big Data Management

How Marketing Decisions are Taken with the Help of Big Data

Niraj A. Rathi and Aditya S. Betala

Abstract In today's fast-moving era, all decisions are made on the basis of the data available. Companies are always running behind an important piece of information, which would help them to cater to the larger market and add additional pie of profit to their stats. Although there is an ample of data available, it is very much important to scrutinize and choose the appropriate data, which would help the company to have a cutting edge over its competitors. Big data management is a boon to marketing as data is collected in huge numbers but providing significant data at the right time, at a right juncture and at the right place is very much crucial to the marketers in order to make decisions, which would decide the fortune of the companies. This paper discusses the importance of big data in marketing decisions by framing long-term strategies and short-term tactics in the business scenario to have a sustainable competitive advantage. This paper is entirely based on secondary data research, and accordingly the required analysis is done (Prof. John A. Deighton Harvard Business Review in Big data in marketing, 2016) [19].

Keywords Big data · Brand clustering · Marketing decisions

1 Introduction

A few decades earlier, organizations had meant to cultivate client–customer benefit to oversee firms along with clients highlighted as customer relationship management. In order to upgrade once connections and benefits, associations have conveyed firm-wide CRM in the wake of gathering and examining the outline occurs socio-statistic information and exchange of their customers [1].

N. A. Rathi (✉) · A. S. Betala
Universal Business School, Mumbai, India
e-mail: niraj25rathi@gmail.com

A. S. Betala
Cardiff-Met University, Cardiff, UK
e-mail: adityabetala78@gmail.com

© Springer Nature Singapore Pte Ltd. 2019
V. E. Balas et al. (eds.), *Data Management, Analytics and Innovation*,
Advances in Intelligent Systems and Computing 839,
https://doi.org/10.1007/978-981-13-1274-8_8

The greater part of corporate advertisers relies upon a review to distinguish once buyer's inclinations, items assessments and buying expectations. An overview strategy can lead the marketing strategists to startling positioning, as the repliers are very liberal to provide great focuses on the assessments. An optional way to deal with a review technique is gathering individuals' straightforward perspectives on the Internet [1].

Many individuals record and offer their day-by-day feelings on news, items and brands in SNS. Thus, web-based social networking, for example, Facebook, Twitter and LinkedIn, assume a critical part of collecting, transmitting and imparting insights of individuals. Several organizations are anxious in order to distinguish the aggregate assessments about organization's items. As clients probably share their thoughts relating to organization's items honestly on the Internet—particularly via web-based networking media—assembling and mining sentiments have turned into a basic factor for marketers who are endeavouring to distinguish client inclinations [1].

2 Big Data and Marketing

From various perspectives, the big data revolution gives a parallel to the scanner data revolution. Big data is frequently characterized by volume, speed and assortment. Firms and particular data providers now track and keep up to a great degree of substantial databases on buyers' shopping and buying conduct (volume). This information is frequently accessible consistently (velocity) empowering marketing models that tweak marketing instruments to buyers as shoppers look for data. Big data comes in many formats beyond the simple numerical data with which we have managed for numerous years (variety). This information incorporates numerical information, content, sound and video documents, which are progressively interconnected. To exploit huge information, marketing science should hold onto embrace disciplines, for example, data science, machine learning, text processing, audio processing and video processing [2].

3 What Is Big Data?

Big data alludes to the consistently expanding volume, speed, assortment and data complexity. In order to promote associations, it is the basic result of the advanced marketing scene, and conceived surrounding of the Internet. The expression 'big data' does not simply allude to information only but additionally alludes to difficulties, abilities and competences related to putting away and investigating such gigantic informational collections to help neck and neck of basic leadership that is more exact and convenient else already tried—big data—drives basic analytical thinking [3].

4 Why Do Big Data Matters in Marketing Decision-Making?

Having enormous information does not consequently prompt in proper marketing—yet, there is potential. Considering big data as one's mystery fixing, one's crude material and one's basic component. It is not simply the information that is so imperative or maybe the big data's experiences, the choices one makes and moves people. Through integrating big data with marketing strategies, it gives an additional edge in planning and executing the tactics and strategies [3].

1. **Customer engagement**: Big data can convey inputs to the company not only exactly who our clients are but also where they are, what they need, how they need to be reached and when [4].
2. **Customer maintenance and reliability**: It can enable the company to find what impacts its client dedication and what holds them returning over and over [5].
3. **Marketing streamlining/execution**: Along with the help of big data, one could decide ideal spent over different channels of marketing, and ceaselessly improve marketing programs through testing, estimation and examination [3].

The Three Major Kinds of Big Data Which are Very Much Crucial in Marketing

1. **Client**: The client segment is the most curious and typical for big data as it involves various kinds of marketing tools, which helps for pilot testing and it also keeps the track of the client's pre- and post-purchase behaviours [6].
2. **Functional**: It ordinarily incorporates target measurements which measures the nature of progressions involved in marketing by identifying its functioning, asset designation, administration of resources, budgetary controls, etc. [7].
3. **Monetary**: Typically housed in an association's money-related frameworks, this classification will incorporate income and different target information sorts, which evaluates the budgetary well-being of a firm [3].

Encounters in Big Data Management in Marketing
The difficulties identified with the compelling utilization of big data can be particularly overwhelming for promoting. It is on account of many investigative/Research frameworks do not adjust to marketing firm's information, procedures and choices [8].

- **Knowing what information to accumulate**: Huge chunk of Information is available everywhere, Companies have heavy client base, functional and budgetary information to fight with. However, an excess of anything is no good, and it must be perfect enough [9].
- **Knowing which systematic devices to utilize**: The quantity of big data grows and the period accessible aimed at settling on choices, and following up on them is contracting. Logical instruments could enable one to total and investigate the

information, and additionally, distribute significant bits of knowledge and choices fittingly all through the firms—yet, which are those need to be understood thoroughly [10].

- **Knowing how to go from information to knowledge to impact**: When one has the information, how would he transform it into knowledge? Also, how would he utilize that knowledge to have a beneficial outcome on one's marketing programs [3]?

10 Conducts How Big Data is Transforming Marketing and Sales

1. **Separating valuing techniques for the client product level and upgrading pricing utilizing big data are fetching additional reachable**. McKinsey got the result that a successful firm's 75% income originates from its typical items and the remaining 30% comes from the great many evaluating choices the organizations influence each year to neglect to convey the best cost. With a 1% cost increment and converting it to around 8.7% expansion hip working benefits, accept that there is no reduction in quantity and estimating takes huge advantage latent aimed at enhancing productivity [11] (Table 1).

Table 1 Patterns in the analysis highlight opportunities for differentiated pricing at a customer–product level, based on willingness to pay [12]

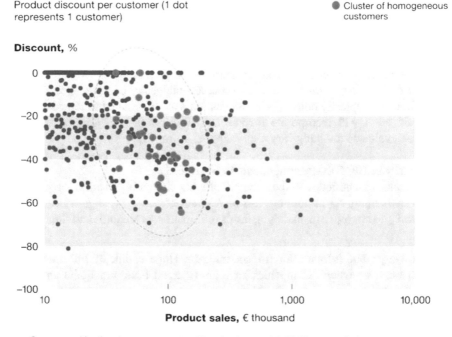

Source: multinational energy company (disguised example); McKinsey analysis

Source Using Big Data To Make Better Pricing Decisions.

2. **Big data is changing by what method organizations achieve more note-worthy client's responsiveness and increase more prominent client's bits of knowledge**. Forrester research institute estimated that nearly 44% of business to customer sellers are utilizing big data analytics in the direction of enhanced receptiveness and nearly 36% are effectively utilizing investigation in additional information pulling out to increase more noteworthy bits of knowledge to design more relationship-driven methodologies [13] (Table 2).

   Source Marketing's Big Leap Forward Overcomes the Urgent Challenge to Improve Customer Experience and Marketing Performance.

3. **Buyer analytics 48%, Operative analytics 21%, Scam and Compliance 12%, New Product and Service Innovation (10%), and Enterprise Data Warehouse Optimization (10%) are among the most popular big data use cases in sales and marketing**. A current report by Datameer discovered that client investigation dominates big data use in sales and marketing departments, supporting the four key systems of expanding client securing, diminishing client beat, expanding income per client and enhancing existing products [14] (Table 3).

   Source Big Data: A Competitive Weapon for the Enterprise.

4. **Reinforced via big data besides the aforementioned associated innovations, it is currently conceivable to implant knowledge into contextual marketing**. The marketing stage stack in many organizations is developing quickly in view of advancing client, sales, service and channel which need not met with existing frameworks today. Accordingly, many marketing stacks are not totally

Table 2 What are the 3 most critical factors for the success of a Marketing Programme?

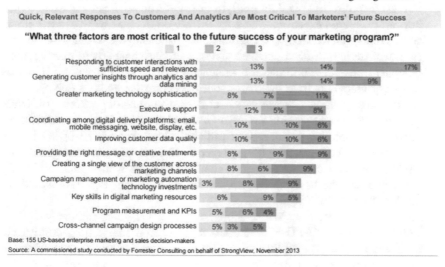

Table 3 What all big data uses in Business?

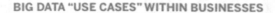

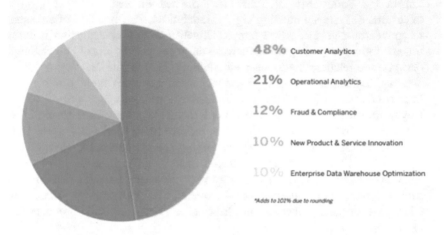

coordinated at the information and process levels. Big data analytics gives the establishment of making versatile systems of insight to help ease this issue [15]. The following graphic is from the Forrester study made available: Combine Systems of Insight and Engagement for Contextual Marketing Tools and Technology: The Enterprise Marketing Technology Playbook (Table 4).

5. **Forrester found that big data analytics expands advertisers' capacity to get beyond campaign execution and concentrate on the best way to make client connections more effective**. By utilizing big data analytics to characterize and control client improvement, marketers increment the capability of making more prominent client faithfulness and enhancing client lifetime value [16]. The following graphic is from the SAS-sponsored Forrester study (Table 5).

6. **Optimizing selling techniques and go-to-market plans utilizing geo-analytics are beginning to occur in the biopharma business**. McKinsey found that biopharma organizations ordinarily burn through 20–30% of their incomes on selling, general and managerial. If these organizations could precisely adjust their offering and go-to-market strategies with regions and territories that had the best sales potential, the go-to-market expenses would be quickly decreased [13] (Table 6).

Source Making Big Data Work: Biopharma, McKinsey and Company.

7. **Totally, 58% of C.M.O.'s articulate of SEO and marketing, email and mobile marketing is the place where big data is having the biggest effect on marketing programs today**. 54% trust that the big data and analytics will be fundamental to marketing strategy over the long haul [11] (Table 7).

Table 4 Enterprise marketing technology components support systems of insight and engagement

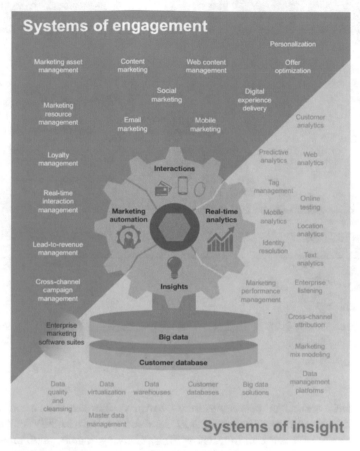

Source Big Data and the CMO: What is Changing for Marketing Leadership?

8. **Marketplace pioneers hip 10 industries according to Forbes acumen followed in its current review stand increasing more prominent client engagement and client loyalty using big data and cutting-edge analytics**. In this research, it was instituted that crosswise over ten ventures, department-explicit big data and analytics proficiency remained adequate towards getting methodologies for practicability and fruitfulness. Firm extensive mastery besides gigantic values variation stayed proficient later experimental runs lineups conveyed positive outcomes [13].

Source Forbes Insights, The Rise of The New Marketing Organization.

Table 5 Drive customer value across the life cycle

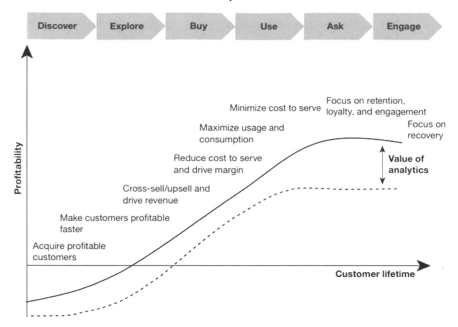

Source- How Analytics Drives Customer Life-Cycle Management Vision: The Customer Analytics Playbook.

9. **Big data stands empowering firms towards increase more noteworthy experiences and significant insight hooked on individually of the vital chauffeurs of once commercial setting**. Creating income and plummeting expenses and decreasing working capital remain in three centre territories, where big data stays conveying trade esteem currently. An undertaking esteem chauffeurs gauge effectively once overseen utilizing big data and unconventional analytics. The accompanying guide towards esteem represents these facts.
   Source Big Data Stats from Deloitte.

10. **Customer value analytics in light of big data stands constructing its feasible driving vendors towards conveying predictable omnichannel client experience in overall stations**. It is developed by the way of practical arrangement of big data constructed skills that quicken the sales series, although holding besides ascending the customized idea of client relationships. Basically, it is currently a practical arrangement of advancements for organizing incredible omnichannel client encounters over a selling system [13].
   Source Capgemini Presentation, From Customer Insights to Action Ruurd Dam, November 2015.

Table 6 Big data delivers high impact targeting for marketing and sales

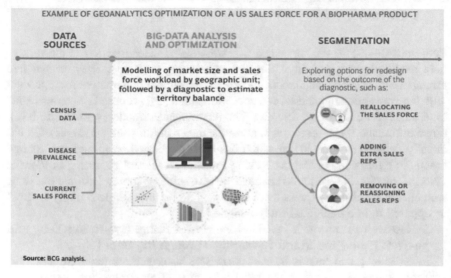

Table 7 Which are the Streams in which Big Data is having maximum Impact?

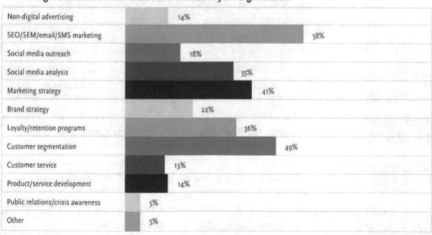

5 Brand Clustering Through Big Data Management— A Review

Punj and Moon [17] recommended a psychological categorization framework to help marketing managers in settling on positioning decisions. They created five arrangements of recommendations interfacing positioning alternative for a product with the preparation of band-level data. Chintagunta [18] proposed a heterogeneous logit model for branding positioning. The recommended model is utilized to break down information with respect to analyze the data regarding the purchases of liquid laundry detergents. Very less research has been performed on grouping corporate brands in light of the online networking information. Brand clustering in view of SNS information is required to conquer the restriction of survey information, since web clients are able to express their actual feeling on the web particularly on SNS as opposed to in a traditional study method [1].

A Research Framework Provided by Ha-Na Kang, Hye-Ryeon Yong and Hyun-Seok Hwang for Brand Clustering Through Big Data [1]

In order to group brands by means of SNS data, authors have proposed the subsequent process shown in the picture by way of an exploration outline. The elementary expectations of the investigation carried out by researchers are as follows: Subsequently discovering the obtainable groups, one needs to recognize the

Table 8 7 Levels for Clustering of Brand with the help of Big Data

Step 1	Choosing brand names
Step 2	Creating Co-mention matrix
Step 3	Calculating Distance (similarity) between brand names
Step 4	Projecting brand names onto both 2D and 3D positioning Map
Step 5	Grouping brand names using cluster analysis
Step 6	Identifying characteristics of clusters
Step 7	Building Brand Positioning Strategy

features of an individual group aimed at an improved sympathetic of the groups. The last stage will remain structuring a brand positioning approach towards the installation of publicizing activities.

(a) The added successive two brands stay specified and the nearer two brands are seen via buyers.
(b) The separation amongst two brand names remains in reverse extent of recurrence (Table 8).

6 Conclusion

Big data and marketing are supportive to each other as both work hand in hand and if companies make a robust use of this, the companies can reach new heights following the three major steps, which the companies should follow in order to move after big data towards improved marketing. Big data remains a major arrangement in marketing. In any case, several things each marketer should remember to aid guarantee are that the big data resolves prompt huge achievement [19].

1. **Using big data towards excavation aimed at more profound understanding**: Big data manages the chance to give further and then more profound hooked on the information, shedding back levels to uncover wealthier bits of knowledge. The knowledge which it picks up from underlying investigation can be investigated further, with wealthier and more profound bits of knowledge developing each time. This level of knowledge can enable companies to create particular plans and schedules for get-up-and-go development.
2. **Getting experiences as of big data towards the individuals who can utilize it**: There is not at all doubt that C.M.O requires the significant piece of knowledge that big data probably gives, nonetheless, thus fix store administrators. What great is about knowing whether it remains inside the limits of a boardroom? once it gets under the control of the individuals who can follow up on it.
3. **Trying not to attempt to spare the world in any event not at first**: Going up against big data can now and again appear to be overpowering, so one should begin by concentrating on a couple of key targets. What results might one want to make strides? When one concludes that he can recognize what information he would need to help with the related investigation. When a company has finished that activity, it must proceed onwards to its next goals [3].

Bibliography

1. Kang, H.-N., Yong, H.-R., & Hwang, H.-S. (2016). Brand clustering based on social big data: A case study. *International Journal of Software Engineering and Its Applications, 10*(4), 27–36. ISSN: 1738-9984 IJSEIA.

2. Chintagunta, P., Hanssens, D. M., & Hauser, J. R. (2016). Marketing science and big data. *Marketing Science, 35*(3), 1–2. ISSN 0732-2399. ©*2016 INFORMS.*
3. SAS. (2016). Big Data, Bigger Marketing. https://www.sas.com/en_us/insights/big-data/big-data-marketing.html.
4. Kim, D. H., & Kim, D. (2016). A feasibility study on the research infrastructure projects for the high-speed big data processing devices using AHP. *International Journal of Software Engineering and Its Applications, 10*(4), 37–46.
5. Sunny Sharma, P. S. (2016). A review toward powers of big data Sunny Sharm. *International Research Journal of Engineering and Technology (IRJET), 03*(04). e-ISSN: 2395 -0056, Apr-ISO 9001:2008 Certified Journal.
6. Shkapsky, A., Yang, M., Interlandi, M., Chiu, H., Condie, T., & Zaniolo, C. (2016). *Big data analytics with datalog queries on spark.* ACM: University of California, Los Angeles. ISBN 978-1-4503-3531-7/16/06.
7. Boland, M. V. (2015). Big data, big challenges. *The American Academy of Ophthalmology.* ISSN 0161-6420/15.
8. Svilar, M., Chakraborty, A., & Kanioura, A. (2017). Big data analytics in marketing. *The Institute for Operations Research and the Management Sciences INFORMS.*
9. Bradlow, E. T., Gangwar, M., & Kopalle, P. (2015). The role of big data and predictive analytics in retailing. *Journal of Retailing.*
10. Niranjanmurthy, B. P., Dr. (2016). The study of big data analytics in e-commerce. *International Journal of Advanced Research in Computer and Communication Engineering.* ISSN (Online) 2278-1021.
11. Columbus, L. (2016). Ten ways big data is revolutionizing marketing and sales. *Forbes.*
12. Baker, W., Kiewell, D., & Winkler, G. (2014). Using big data to make better pricing decisions. *McKinsey Analysis.*
13. Forbes. (2016). Ten Ways Big Data is Revolutionizing Marketing and Sales. https://www.forbes.com/sites/louiscolumbus/2016/05/09/ten-ways-big-data-is-revolutionizing-marketing-and-sales/#3d57e95721cf.
14. Dholakia, R. R., & Dholakia, N. (2013). Scholarly research in marketing: Trends and challenges in the era of big data. University of Rhode Island.
15. Riahi, A. & Riahi, S. (2015). The big data revolution, issues and applications. *International Journal of Advanced Research in Computer Science and Software Engineering, 5*(8), August. ISSN: 2277 128.
16. Arthur, L. (2013). *Big data marketing.* Hoboken, New Jersey: Wiley. ISBN 978-1-118-73389-9.
17. Punj, G., & Moon, J. (2002). Positioning options for achieving brand association: A psychological categorization framework. *Journal of Business Research, 55*(4), 275–283.
18. Chintagunta, P. K. (1994). Heterogeneous logit model implications for brand positioning. *Journal of Marketing Research, 31*(2), 304–311.
19. Harvard Business Review. (2016). Big Data in Marketing. *HBR* http://www.hbs.edu/coursecatalog/1955.html.

Big Data Forensic Analytics

Deepak Mane and Kiran Shibe

Abstract Forensics is an important topic for civil, academics, and software professionals who deal with very complex investigations. Today there is increases usage of big data technologies such as hadoop in all sectors like BFSI, oil energy, etc. for analytics. Forensic investigators have required new approaches to how analysis conducted also need to understand how to work with big data solutions/ technologies. The number of organizations who implemented big data solutions had spent lots of time for forensic investigation by using of traditional forensic approach. These big data clusters contain critical information that provides information about organization's strategies and operations. Since hadoop has distributed architecture, in-memory data storage and huge volume of data, forensic on hadoop offers new challenges to investigators. Big data forensic is a new area within forensic which focuses on the forensics of big data systems. Since big data are scalable, high volume of data, which provides practical limitation for forensic investigation by using traditional approaches. Big data Forensic analytics provides new approaches to collect data, for analyzing collected data.

Keywords Forensic · Big data · Analytics · Hadoop · HDFS
HIVE

D. Mane (✉)
AAA Consulting Group, Analytics and Insights, Tata Consultancy Services,
Hadapsar, Pune 411028, India
e-mail: deepak.mane@tcs.com

K. Shibe
Banking and Finance Group, Tata Consultancy Services Hadapsar,
Pune 411028, India
e-mail: kiran.shibe@tcs.com

© Springer Nature Singapore Pte Ltd. 2019
V. E. Balas et al. (eds.), *Data Management, Analytics and Innovation*,
Advances in Intelligent Systems and Computing 839,
https://doi.org/10.1007/978-981-13-1274-8_9

1   Introduction

In current era big data attracted to more peoples/researchers as attention as it changes the way of computation and analytics services to customers for examples Hadoop platform provides massively parallel computation. Today, leading banking customers like Bank of America, Deutsche Bank had implemented enterprise big data solution platform which provides intelligent analytics services for customer's application with single click at low cost.

Recently, one of the large banking and insurance organization has suffered a data affecting millions of customers on its international health insurance plan. Similarly, one of large search companies, which provides users with an online guide to restaurants, cafes and clubs, reported that data from more than 1 million users had been stolen, which includes credentials information such as credit card numbers, address, etc.

This incident not only brings loss of business but also causes some major impacts on trust of companies. A major impact on trust on technologies about data storage, access, etc., above two incidents spend lots of man-hours for investigation using traditional forensic approach. Big data forensic is considered as new area of forensic and service model in which forensic investigator undertakes activities of big data system for customers using new investigation methodologies.

There are many research/technical white papers published which they are discussing about forensic process, technologies, models, design, and management. Big data forensic analytics are still new subjects in forensic domain. Hence forensic investigators encountered many issues and challenges in Big data forensic.

Typical questions are listed below.

1. What is Forensic analysis in big Data? and What are different better forensic process and scope, requirements, and features?
2. What are attributes of big data forensic analysis?
3. What types of big data forensic analysis?
4. What are the major difference between traditional forensic analysis and big data forensic analysis?
5. What are the special requirements and distinct features of big data forensic analysis?
6. What are the special issues, challenges, and needs for big data forensic analysis?
7. What are key benefits of big data forensic analysis?
8. What are big data forensic approaches and risk and mitigations in big data forensic?
9. What are the current practices, tools and its features and major players for big data forensic analysis?

2 Digital Forensic—Definition

There are many ways to define Digital Forensic but, Ken Ratko defined digital forensics in different ways:

> The application of computer science and investigative procedures for a legal purpose involving the analysis of digital evidence after proper search authority, chain of custody, validation with mathematics, use of validated tools, repeatability, reporting, and possible expert presentation. (Ratko 2007)

In digital forensic domain, following steps are involved

- Identification
- Requirement
- Analytics
- Design
- Presentation of digital evidence.

3 Types of Digital Forensic

There are following types of digital forensic, please refer below diagram

- Network Forensic
- Computer Forensic
- Mobile Forensic
- Live Forensic
- Big Data Forensic
- Database Forensic (Fig. 1).

4 What Is Big Data Forensic

Big Data Forensic is the study of big data which includes about data creation, storage and investigation methodology. It is a branch of digital forensic where we need to collect/study/store different types of evidence to prove it's a crime.

Big data forensic may include may different tasks such as data recovery, data lineage, and data governance. It might cover the tracking information about all jobs, user activities within a large hadoop clusters. Forensic investigators can also use various methods to purse data forensic such as data decryption, advance search engines, reverse engineering, high level data analysis, data collection, and backup methodologies.

Fig. 1 Types of digital forensic

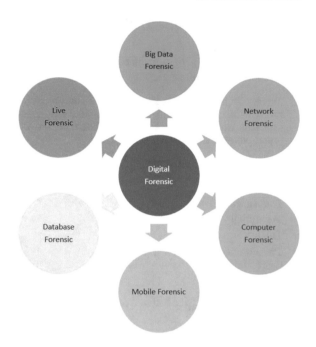

Based on data, big data forensic analysis divides data into two types

- Persistent Data—Data is permanently stored on data, so it is easier to find
- Volatile Data—Data is stored on namenode. That is, data stored into memory or cache. It stores metadata about user activities, operations, etc.

Forensic investigators often always focus on different ways to retrieve volatile data from namenode. Sometimes it is very difficult to get current state of volatile data from memory/namenode.

5 Key Benefits—Big Data Forensic

All levels of big data forensic can be carried out in big data environments, and indeed, some types of forensic benefit greatly from a forensic analysis in the big data environment. In an early pilot with a We identified a number of benefits to big data forensic in hadoop environments.

- **Platform for Big Data Forensic Analytics**
  This is platform for analysis of different types of data like Volatile or static data which captured through process. This platform provides capability to search and analyze a mountain of data quickly and efficiently with help of data mining and analytics tools. It also helps to convert unstructured data into structured data.

- **Driving Standardization**
  Driving standardizations in big data forensic domain is a longer journey which will give huge impact on hadoop cluster infrastructure. This will be major impact on f catalyze IT modernization and improve internal IT services maturity, quick way of resolution new big data forensic issues. Also helps in Application security portfolio rationalization.
- **Geographic Transparency and Traceability**
  Company or organization data is always sensitive including geographical locations. Data movement across borders which includes national and international rules and regulation such as the EU Data Protection Directive. Big data forensic analytics process is always transparent about geographic locations which include information about data and service storage locations, VPN Connections, access mode of data, etc.
- **Simplicity**
  Simplicity offers in terms of training, experimental environment for bug fixing. Which also can platform for quickly investigation or simulation for end-to-end scenarios which helps to generate similar kind of logs.

6   Types of Big Data Forensics

This paragraph explains different types of big data Forensic based on two major components

- Hadoop Cluster Components
- Hadoop Ecosystem Components (Fig. 2).

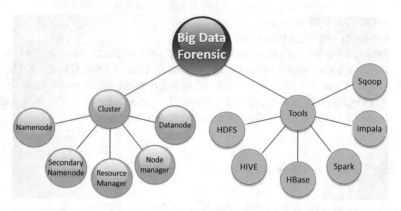

Fig. 2 Types of big data forensics

7 Operational Challenges to Big Data Forensic

Despite the bright upside, digital forensic has limitations, too. Organizations must contend with a different set of challenges in their quest to reap big data forensics

- **Collection methods**
  Big data systems are complex, distributed systems with business requirements and business critical information. As such, they may not be able to be taken offline for a forensic investigation. In big data investigations, hundreds or thousands of storage hard drives may be involved, and data is lost when the big data system is brought offline. Also, the system may need to stay online due to business requirements.
- **Collection Verification**
  Both MD5 and SHA-1 are disk-access intensive. Verifying collections by computing an MD5 or SHA-1 hash comprises a large percentage of the time dedicated to collecting and verifying source evidence. Spending the time to calculate the MD5 and SHA-1 for a big data collection may not be feasible when many terabytes of data are collected. The alternative is to rely on control totals, collection logs, and other descriptive information to verify the collection.
- **Lack of standards**
  Presently, there are no universal/standard solutions to investigate forensic analytics in big data clusters. Even organization do not have any resources who have experience in big data forensics'.
  Big data clusters have their own architecture, operating models, and offer very little interoperability. This poses a big challenge for companies when they want to perform forensics.
- **Infrastructure**
  Big data clusters contains different types of configurations, technology, servers and storage, networking. Forensic investigator has a difficult to create real-time scenarios because of unavailability, complexity.
- **Security in the Big Data clusters**
  Security in the big data clusters is a complex process there are multiple types of encryption techniques implemented in hadoop cluster level (Disk, HDFS, and Encryption zone) level. Digital forensic investigators spend more time to convert encrypted into non-encrypted format. The main cause for concern is that the data may be stored in a different datanode with huge volume.

8 Big Data Forensic Process

Big data forensic process is consisting of activities and steps within a circular and redundant hierarchy. Big data forensic process can originate from any activity/steps and can subsequently lead to any other phases. Big data forensic process must establish support from administrative, application, security, and infrastructure steam to effectively support the activities and tasks.

Fig. 3 Big data forensic processes

Figure 3 illustrates different phases that make up big data forensic process model.

a. **Preparation**

Preparation is essential first step in digital forensic science for the activities and steps performance in all other phases of the workflow, if the preparation activities and steps are deficient in any way. All steps are comprehensive and reviewed regularly for accurateness (Fig. 4).

b. **Discovery**

As the second phase of the investigative workflow, discovery is made up of the activities and steps performed to identify, collect, and preserve evidence. These activities and steps are critical in maintaining the meaningfulness, relevancy, and admissibility of evidence for the remainder of its life cycle (Fig. 5).

c. **Analytics**

As the third phase of big data forensic, analytics involves the activities and steps performed by the investigator to examine and analyze digital evidence. Apply machine learning algorithms by identifying patterns, build recommendation engines these activities and steps are used by investigators to examine duplicated evidence in a forensically sound manner to identify meaningful data and subsequently reduce volumes based on the contextual and content relevance.

There are four types of analytics in big data forensic (Fig. 6).

d. **Presentation**

As the fourth and last phase of the investigative workflow, presentation involves the activities and steps performed to produce evidence-based reports of the investigation. These activities and steps provide investigators with a channel of

Fig. 4 Preparation

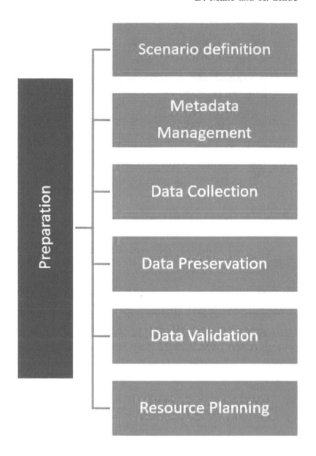

demonstrating that processes, techniques, tools, equipment, and interactions maintained the authenticity, reliability, and trustworthiness of digital evidence throughout the investigative workflow.

9   Case Study—Detecting Transaction Fraud

a.   Problem Statement

A telecom client and its distributor company uncovered a false invoicing fraud. The client suspected other instances of false invoicing fraud over a period of 5 years. For the time period in questions, procurements had billions of transactions from more than 6000 vendors. These transactions exhibited a huge range of PO values: from a few dollars to hundreds of millions we were not informed of which transactions the client had identified as fraudulent.

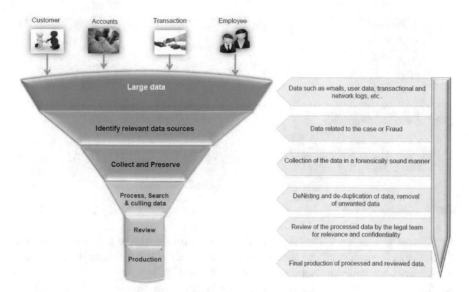

Fig. 5 Discovery

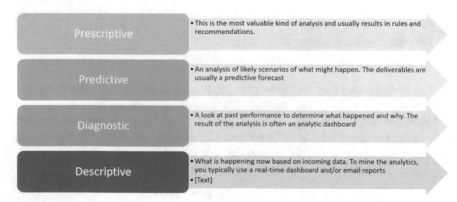

Fig. 6 Analytics

b. **Current Challenges with traditional forensic Analytics methods**

Typically, we would solve this type of problem with a traditional red-flag approach, on limited set of data, i.e., decide whether any transactions broke pre-agreed rules but we cannot solve based on following reasons:

- Traditional forensic methods have following limitations:
  - They tend to be rule-based
  - Exceptions are only treated in isolation
  - They assume that the fraud pattern is known

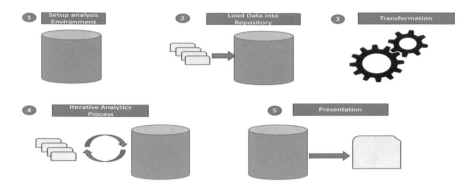

Fig. 7 Analysis Process

- By using traditional forensic analytics approach in multiple scenarios there are multiple indicators but no clear rules that definitely show that fraud has occurred
- All data are structured data having size of 500 GB.
- All data are stored in distributed way on hadoop cluster
- Quality Review
- Complex process
- Reconciliation
- Large number of reviewers
- Automated visualizations

c. **Analysis Process**

The investigator relies on large-scale database systems to load, transform, and analyze the data to reach his findings. The investigator sets up the analysis environment and prepares the data for analysis before beginning any analysis. The data transformation or preparation is the process of converting and standardizing the data from Hadoop applications into a form that can be readily analyzed.

Figure 7 shows about analysis process.

d. **Technology used in Big Data Forensic process**

Following are the details of the tools and components proposed in the solution.

No.	Technology	Description
1.	Apache Hive	Apache Hive is a data warehouse system for Hadoop that facilitates easy data summarization, ad hoc queries, and the analysis of large datasets
2.	Apache Hadoop/HDFS	Hadoop Distributed File System (HDFS) is the primary storage system used by Hadoop applications
3.	Bulk Extractor	Bulk Extractor is a keyword search and file carving tool that can extract text, graphics, and other information from forensic images

(continued)

(continued)

No.	Technology	Description
4.	Autopsy	Autopsy is a freeware forensic tool that provides a number of useful functions, including keyword searching and file and data carving. Autopsy is a graphical version of the Sleuth Kit
5.	Apache Spark	Apache Spark is a fast and general engine for big data processing, with built-in modules for streaming, SQL, machine learning and graph processing

e. **Data/Evidence Collections Methods**

Following methods/approaches used to collect data from Hadoop cluster

  (i) Backup-based collection: This collects a newly created or archived application backup
  (ii) Query-based collection: This collects all data or a subset of the data via queries
  (iii) Script-based collection: This collects all data or a subset of the data via scripts or another application (for example, via Pig)
  (iv) Software-based collection: This collects data through an application that connects to the source application

f. **Evidence Collection from HIVE**

In current state all transactional data stored into HIVE, it is a platform for analyzing data. It uses a familiar SQL querying language. Hive has several important components that are critical to understand for investigations:

- Hive Data Storage: The type and location of data stored and accessed by Hive, which includes HDFS,
- Megastore: The database that contains Hive data metadata (not in HDFS)
- HiveQL: The Hive query language, which is a SQL-like language
- Databases and Tables: The logical containers of Hive data
- Hive Shell: The shell interpreter for HiveQL
- Hive Clients: The mechanisms for connecting a Hive server, such as Hive Thrift clients, Java Database Connectivity (JDBC) clients, and ODBC clients

g. **Evidence Collection from HIVE Backup**

A Hive application collection differs from collecting Hive data through HDFS. With a Hive application collection, the data is collected either as a backup from HDFS with the Hive megastore, or the Hive service is used to interface with the data for collection (Fig. 8).

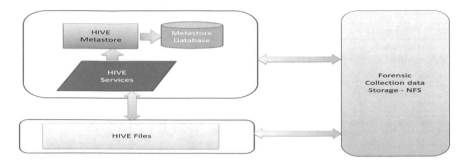

Fig. 8 Evidence collection from HIVE backup

h. **Evidence Collection using python scripts**

Python script creates a connection to Hive and exports the results to the supplied output file. Information about the table is exported to the supplied log file, with the collection date, number of records, and query also being written to the log.

```
import pyhs2
import date time
output File = open ('/home/extract/outputFile.txt', 'w')
log File = open ('/home/extract/logFile.txt', 'w')
logFile.write(str(datetime.date.today()) + '\n')
with pyhs2.connect(host = 'localhost',

port = 10000,
authMechanism = "PLAIN",
user = 'root',
password = 'pwdVal',
database = 'default') as conn:

with conn.cursor() as cur:
logFile.write(cur.getDatabases() + '\n')
cur.execute("select * from table")
logFile.write('select * from table\n')
logFile.write('Number of rows: ' + str(cur.rowcount) + '\n')
logFile.write(cur.getSchema() + '\n')
#Fetch table results
for i in cur.fetch():
outputFile.write(i + '\n')
outputFile.close()
logFile.close()
```

i. Evidence Collection using Bash scripts

If all tables need to be collected, the process can be automated through a HiveQL script and a Bash script.

```bash
#!/bin/bash
hive -e "show tables;" > hiveTables.txt
for line in $(cat hiveTables.txt);
do
hive -hiveconf tablename=$line -f tableExport.hql >
${line}.txt
done
```

j. Analysis Preparation

Once we collected all data evidence from Hadoop Cluster, the first step is to attach a copy of the evidence to the environment in a read-only manner. Because the amount of forensic data is large in a big data investigation, the hard drives containing the evidence should be attached to a sufficiently large storage device in the read-only mode. The big data analysis environment should be attached to a **network-attached storage** (**NAS**).

k. Data-Analysis Preparation

After preparing analysis preparation platform, we need to carry out below primary analysis techniques

- Keyword searching
- File and data carving
- Metadata analysis, such as file modification timeline analysis
- Cluster reconstruction

Above analysis techniques serve can be performed as an individual or combined in various orders cluster reconstruction.

l. Metadata analysis

Analyzing metadata is important or relevant for a big data investigation. Because HDFS data is distributed and the data is created, modified, and accessed using shared or system accounts, the metadata is a valuable as it is in typical forensic investigations.

Autopsy's timeline feature generates graphical, interactive timelines based on the evidence's **Modified, Accessed, and Created** (**MAC**) metadata times. To create a timeline, click **Tools Timeline**. Autopsy generates a timeline based on the loaded evidence.

The following metadata are analyzed using Autopsy tool:

- Filename
- File Extension
- File Type

- Deleted/Not Deleted
- Last Accessed/Created/Modified
- Size
- Hash
- Permissions
- File Path

m. **Tag Cloud**

One of the most widely used visual techniques is a Tag Cloud. This is a good example of expressing complex data that can be understood intuitively. A Tag Cloud is the visual representation of communication relating to transactional data entries. It is represented by a combination of words in varied fonts, sizes or colors. This format is useful for quickly determining the important terms to identify key fraud issues.

n. **Sentiment Analysis**

Known as behavioral analysis, this refers to the application of text analytics to identify and extract subjective information including the attitudes of writers, their affective state and the intended emotional quotient. It determines whether expressed opinions in a document are positive, negative or neutral. The "fraud triangle" can be applied to categorize events into rationalization, opportunity, and pressure to identify sentiments.

o. **Grouping and Clustering Analytics**

Cluster analysis is a powerful statistical technique for analyzing large sets of data. Clustering can be achieved by running a number of algorithms to group data, assess the distribution, and identify outliers. A cluster is a grouping of like subsets of data for the purpose of classifying them. Cluster analysis typically shows multiple clusters from a data set, and these clusters can be devised and structured in a number of ways.

Numerous cluster analysis techniques exist in the field of data mining. The two primary algorithms used for data clustering are K-means and **expectation-maximization** (**EM**). The approaches an investigator can take to perform clustering are

1. Select a distance measure.
2. Select a clustering algorithm.
3. Define the distance between two clusters.
4. Determine the number of clusters based on the data.
5. Validate the analysis.

The most common distance measure is the Euclidean measure, which computes the distance using spatial coordinates. Here, the Euclidean measure can be applied to calculate distance using the values from a numeric field.

Fig. 9 Sample screenshot
using *K*-means algorithm

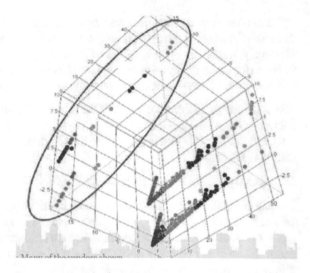

Additional fields can be added to the distance calculation, if required. The formula for adding fields is

$$\text{sqrt}((\text{Record A}_{\text{field1}} - \text{Record B}_{\text{field1}})^2 + (\text{Record A}_{\text{field2}} - \text{Record B}_{\text{field2}})^2)$$

The *K*-means clustering algorithm is a well-known clustering algorithm and is computed as follows:

1. Define the number of clusters, *k*.
2. Assign each data point to its closest cluster center.
3. Recomputed the cluster centers.
4. Repeat these steps until there are minimal or no changes to the cluster centers.

By using *K*-means clustering algorithms we identified distributed client. Please see Fig. 9.

Findings

- Uncovered 42 "outlier" vendors for further investigation
- Two of these vendors were confirmed as the anonymized frauds

p. **Time Series Analytics**

Events in the data can be plotted to establish a chronology, highlight key pattern changes, or establish what the normal patterns are in the data. Time series analysis computes specific metrics using a sequence of data points based on a defined date interval. The date interval can be chosen by the investigator, and the time period can either be the entire date range of the data or a selected subset of the dates.

The first step is to select the date range and date interval. In the Transaction data, there are 1 million individual dates spanning a 5-year period. The investigator can plot every single day, plot a subset of days, or plot the data using an aggregated interval (for example, by months or years). Given the large number of days, the data can best be reviewed when aggregated by month and year:

SELECT MONTH(date) + "-" + YEAR(date), SUM (forensic volume)
FROM [001_FORENSIC]
GROUP BY MONTH(date) + "-" + YEAR(date)
ORDER BY 1

q. **Anomaly Detection based on Rule Engine**

Data anomalies are a major issue. Anomalies can be natural occurrences due to data being incorrectly entered or imported into a system, or they can indicate or be proof of fraud or other wrongdoing. An investigator can analyze a data set for anomalies to either pinpoint where evidence of wrongdoing exists or to indicate or rule out the possibility of wrongdoing.

Rule-based analysis is an effective method for isolating specific types of anomalies or key records. This method requires knowing the rules that the data should adhere to and then executing the rules against the data to identify the records that violate the rules.

The rules can take many forms and include multiple criteria, such as

- Date ranges
- Acceptable values
- Numeric value ranges
- Acceptable combinations of values across fields
- Data values confirmed against known events.

10 Conclusion

Since big data is distributed system consisting of programming frameworks, tools, and software. Forensic analysis in big data is complex process. We need to adopt new methodology for conducting/investigation. Above forensic analytics process in big data helps in deep process/methodology in forensic also it provides different analytics methodologies/algorithms for different stages. This also provides end-to-end process information for big data forensic analytics.

References

1. EY Publications. http://www.ey.com/Publication/vwLUAssets/EY_-_Forensic_Data_Analy tics_(FDA)/$FILE/EY-forensic-data-dnalytics.pdf.
2. Big data as a forensic Challenges. https://link.springer.com/chapter/10.1007%2F978-3-658-03371-2_17.
3. Big Data challenges, approaches. https://www.researchgate.net/publication/281257770_ Digital_Forensics_in_the_Age_of_Big_Data_Challenges_Àpproaches_and_Opportunities.
4. Analytics and Forensic Technology. https://www2.deloitte.com/ch/en/pages/forensics/solut ions/analytic-and-forensic-technology.html.
5. Big data Fraud Magazine. https://www.google.co.in/url?sa=t&rct=j&q=&esrc=s&source= web&cd=53&cad=rja&uact=8&ved=0ahUKEwj8_NbF-57WAhUGp48KHSRHAqo4MhAW-CDMwAg&url=http%3A%2F%2F, www.fraud-magazine.com%2Farticle.aspx%3Fid%3D4 294983057&usg=AFQjCNHfGZHtdusS31V_UDv GcdphkPp_0A.
6. Fraud Analysis. https://iaonline.theiia.org/2016/Pages/Proactive-Fraud-Analysis.aspx.
7. Fraud Analytics using big data. http://dsicovery.com/forensics/forensic-analysis/.

An Approach for Temporal Ordering of Medical Case Reports

Rajdeep Sarkar, Bisal Nayal and Aparna Joshi

Abstract Temporal ordering is important in deducing time sequence of medical events in biomedical text and has significant application in summarization, narrative generation, and information extraction tasks. We attempt temporal ordering of events in medical case reports. Our approach is deterministic in extracting explicit temporal expressions and probabilistic in extracting implicit boundaries. We introduce event context as a set of features to learn a CRF model. We achieve F1 score of 0.78 in detecting boundaries. We apply rule-based normalization and ordering of identified temporal expressions.

Keywords Machine learning · NLP · Temporal expressions · CRF
Medical case reports · Maximum entropy · SVM · Naïve base classifier

1 Introduction

Temporal ordering is of important significance in any information extraction task from biomedical text where the goal is to identify medical events in a time sequenced basis. For instance, medical case reports contain case-specific treatment information linearly presented as a series of events in natural language text. The ability to sequence medical events in the correct temporal order allows, among other applications, an accurate determination of causal relations, and meaningful summarization.

The text in medical case reports may contain expressions that indicate explicit temporal breaks. Consider Table 1. Sentences s2 and s6 clearly indicate events in the past by way of explicit expressions ('*one week before*', and, '*a month ago*', respectively). Sentences s3 and s7 do not contain such explicit expressions. However, based on the narrative, it becomes evident that s3 implicitly indicates a temporal break-away from s2, whereas, s7 indicates continuation with respect to s6.

R. Sarkar (✉) · B. Nayal · A. Joshi
Analytics and Insights, Tata Consultancy Services, Pune, India
e-mail: rajdeep.sarkar@tcs.com

© Springer Nature Singapore Pte Ltd. 2019
V. E. Balas et al. (eds.), *Data Management, Analytics and Innovation*,
Advances in Intelligent Systems and Computing 839,
https://doi.org/10.1007/978-981-13-1274-8_10

131

Table 1 Temporal markers in medical case reports

s1	A 40-year-old lady presented with fever, headache, arthralgia, and myalgia, after returning from the Philippines
s2	She recalled mosquito bite *one week before*
s3	Investigations showed lymphopenia, increased hepatic enzymes, C-Reactive-Protein (CRP), normal renal function (Table 1), and prominent spleen (11.6 cm) on ultra-sonogram
...	...
s6	The pain started about *a month ago* and it was getting progressively worse
s7	It was constant, 8 out of 10 in intensity and throbbing sensation involving the right upper and lower quadrants

Determination of such temporal expressions, explicit as well as implicit, becomes vital for the task of temporal ordering.

The goal of this paper is to identify such temporal expressions in medical case reports and attempt to order medical events based on the identified expressions. Medical case reports follow a typical narration technique. One such typicality is the ordering of clinically significant events in the text. Case report narratives present events related to case presentation, investigations, diagnosis, treatment-interventions, case outcome, follow-up reporting, etc. in an ordered sequence. An understanding of this contextual ordering often provides invaluable cues to determine implicit temporal breaks. For instance, in Table 1, s1 and s3 indicate contextual information related to case-presentation and investigation, respectively, and therefore, suggest temporal continuity (s1 → s3). We introduce 'event context' as a feature to enable this aspect of detection of implicit temporal breaks. Event context assigns a contextual label to an event that says whether the event is related to a drug administration, an investigation, a case presentation, etc.

Another typicality with medical case reports is that, in most cases, the entire narrative is presented in past tense. This, as a result, limits the use of tense-based (part of speech) features for determining temporal ordering.

A third important aspect is that the narratives of medical case reports follow a pattern that is inherently time sequenced. In other words, in the absence of an explicit temporal expression, we can assume that all events are narrated in the order of their occurrence. Consider Table 2. Event s2 describes an event in the relative future ('postoperatively'). Because s3 does not have an explicit temporal expression, and because the narrative is inherently temporally sequenced, s3 can be assumed to be in continuation with s2. Similar temporal continuity can be observed with s6 → s7. As such, we note that temporal ordering of events does not depend only on the contextual and syntactic features but also on the temporal aspect of the previous event. In other words, the temporal label assigned to an event t_i depends on temporal label assigned to event t_{i-1}.

A fourth aspect typical to medical case reports is that in most cases, there is an absence of any date mentions, an important expression for determination of temporal sequencing. Unlike in other biomedical narratives such as discharge

Table 2 Labeling events in medical case reports

s1	Based on these findings, the patient underwent a near-total small bowel resection	Current
s2	*Postoperatively*, the patient was started on enoxaparin for anticoagulation due to the findings of thrombosis in the mesenteric vein	Future
s3	Intravenous uids and antibiotics were continued	Continuation
s4	The patient initially responded favorably and was started on liquid diet	Continuation
s5	Improvement continued and antibiotics were discontinued on *postoperative* Day 6	Future
s6	*Fourteen days after surgery*, however, the patient developed acute abdominal pain	Future
s7	CT at this time showed pneumato-intestinalis of remaining part of proximal small gut along with pneumobilia and air in portal vein, along with massive splenic and liver infarct	Continuation

Table 3 Temporal labels assigned to medical events

Event	1	2	3	4	5	6	7	8
Label	c	cont	p	cont	c	f	cont	cont
Temporal expression	x	x	'1 year ago'	x	Implicit	'POD 6'	x	x

summaries, explicit temporal expressions in case reports are therefore primarily relative (e.g., *'one year ago'*). We note that, for this reason, use of publicly available date detection tools may not be appealing for our situation.

Our approach for temporal ordering is to first identify explicit and implicit temporal breaks in case report narratives and then use a deterministic normalization and ordering technique. We assign labels to each event to mark temporal breaks. The labels assigned are

- c: current event
- cont: continuation with the event before
- p: past event
- f: future event

Table 3 shows labels assigned to a series of events.

For the detection of explicit temporal expressions, we use word cues (e.g. 'ago', 'before') with a rule-based extraction mechanism. Next, we detect a contextual aspect of events by learning a set of 7 'event contexts'. These contexts assign specific aspects such as 'drug application', 'symptom detection', 'investigations', etc., to each event and allow for the detection of implicit breaks. Temporal label assignment to a sequence of events is done using a probabilistic sequence classifier. Probability of a label is given by

$$P(y|x) = \exp \sum (O(x,y) + T(x,y))$$

$$\text{Or}, P(y|x) = \frac{1}{Z(d)} * \exp \sum (\alpha_i | t_{i-1}, t_i + \beta_i | t_i, d)$$

where

$O(x, y)$ is the observation probability.
$T(x, y)$ is the transition probability.
α is the weight vector assigned to transition functions.
β is the weight vector assigned to observation functions.
$Z(d)$ denotes the partition function that normalizes over all possible sequences.

With each identification and labeling of temporal breaks, we also extract the explicit temporal expressions and derive normalization of these expressions. These are then used to perform rule-based ordering of events.

To the best of our knowledge, we are the first to attempt temporal ordering of medical case reports. We derive our own annotation scheme. Further, our contribution with this paper is the use of 7 'event contexts' that help identify implicit temporal markers using a sequence-based model.

Since our scope is limited to identification and ordering of temporal expressions, we consider events as given and do not attempt to detect them. For the sake of simplicity, therefore, we assume each sentence to describe an event, although, we note that multiple events can be detected from within a sentence and temporal expressions can be assigned to each event separately.

2   Relevant Studies

Temporal ordering of natural language text has been extensively studied in the recent past. TimeBank [1] corpus was developed as a first significant step towards such studies to cover temporal ordering in news articles. This annotation was, however, found to be inadequate for biomedical text. In the biomedical domain, Bramsen et al. [2] studied temporal ordering of a manually annotated set of discharge summaries. The task of temporal ordering was broken into (i) identification of sections of text bounded by temporal breakages, and, (ii) ordering sections of text in a time sequenced manner. They proposed 'topical continuity' as a feature to detect temporal continuation of events. In the 2012 i2b2 Challenge, the annotation guidelines used for the creation of TimeBank corpus were modified to explicitly address temporal ordering of biomedical text. Discharge summaries were used to create the corpus for the challenge. The challenging task was divided into (i) identification of medical events, (ii) identification of temporal expressions, and (iii) identification of temporal relations between events and temporal expressions. Noteworthy approaches for detection of temporal expressions included using modified version of rule-based extraction systems such as HeidelTime [3–5].

For temporal relations, participants used hybrid approaches comprising of rule-based and ML mechanisms consisting of Support Vector Machines (SVMs) [6], maximum entropy [7] classifiers, Bayesian and CRF sequence models. Raghavan et al. [8] studied temporal ordering of medical events found in clinical narratives. They defined coarse time bins relative of the date of admission of patient and found that use of sequence taggers gave better results as opposed to using non-sequence based classifiers. One of the features they used is the 'section-based feature' typical to clinical narratives and discharge summaries.

It is noted that discharge summaries, owing to the presence of explicit dates, are quite different from case reports where most often the temporal information are either relative (e.g. '8 weeks earlier') or implicit. Furthermore, whereas explicitly demarcated sections such as lab tests, patient outcome, etc. exist in discharge summaries, such is not the case with case reports. Case reports, therefore, call for a different approach for detection of temporal boundaries and subsequent ordering.

3 Data

We use 150 medical case reports comprising 3347 sentences to build our corpus. First, the narratives are tokenized into sentences. Next, two independent annotators are assigned the task to label the sentences as one of the 4 temporal markers: c—current, p—past event, f—future event, or, cont—continuation of the event before. The annotators are handed the following set of rules.

- A sentence is marked p (past) or f (future) in case there exists an explicit temporal expression (such as, '1 year before', '2 weeks later', respectively).
- The first sentence of a narrative is always marked between c (current), p (past), or f (future), but never a cont (continuation).
- Sentences expressing temporal continuity with the previous sentence is marked as cont (continuation). We note the need for annotator's discretion to detect such continuity.
- A sentence following a p or f label can be marked as a c (current) if the annotator agrees that an implicit cue requires such a labeling (refer Table 1, events s2, s3).
- In case explicit dates are mentioned, the annotator is encouraged to use the same in assigning labels appropriately. As date mentions in case reports are sparse, label assignment based on dates alone is usually not feasible.

On an average, 670 markers are detected with 30% of them being implicit markers. The inter-annotator agreement is calculated as 0.85 on the Kappa coefficient scale. We use 120 case reports for training and 30 for testing.

4 Methodology

4.1 Detecting Explicit Temporal Expressions

Detection of explicit temporal expressions is very significant for the task of assigning p (past) and f (future) labels. We note that explicit temporal expressions are sparse in our documents, and date mentions are sparser. Identification of relative expressions such as 'one day ago' is therefore vital for our ordering task. We experiment with temporal tagging applications for our extraction [9]. While we observe very high accuracy of such systems in detecting dates such as 'January, 2013', extraction of domain-specific expressions such as 'postoperatively', 'upon follow-up', etc., are the problems. We also note that the systems extract only the date information from text and leave out past/future cues ('one month' is extracted from 'one month ago'). Based on these bindings, we attempt to create our own rule-based mechanism for detection of temporal expressions. We identify explicit temporal markers as either direct phrases (such as, 'history', 'yesterday', 'the next day', etc.) or as associative phrases (e.g. 'ago' in '3 weeks ago', 'later' in '3 days later'). A set of phrases used for such markers is given in Table 4. We create rules to extract temporal markers using associative words. Such rules look for expressions comprising of date mentions (such as '3 years') in the proximity of associative words. For the detection of explicit dates, we rely on a native date parsing utility.

We estimate the accuracy of our explicit temporal expression detection mechanism on all 150 documents as 0.95. We observe that limitations to our detection mechanism are mostly introduced by domain-specific biomedical terms (e.g. PODn $\rightarrow$ Post operative Day n).

4.2 Detecting Event Contexts

Event contexts assign a contextual aspect to an event and help in the detection of implicit temporal breaks. We categorize a clinical event as being one of seven contexts as illustrated in Table 5. As our experimentation is done on the sentence level, we learn event contexts for whole sentences instead of atomic events. It is noted that although an event can be assigned only 1 context, a sentence can be

Table 4 List of words/phrases for extracting explicit temporal markers

Event	Associative marker phrases	Directive marker phrases
Past	Past, ago, before, back, last, earlier, prior, since, previous(ly)	Yesterday, yesteryear, history, time ago
Future	Later, after, following, next, by	Next time, after discharge, after surgery, postoperatively, subsequently

Table 5 Event contexts

Context name	Context aspect	Example
Patient details	Describes details regarding a patient, namely, patient gender, ethnicity, date of birth, smoking habit, etc.	Patient was a smoker
Symptoms	Symptoms experienced by the patient	He reported pain and paraesthesia
Investigations	Describes lab investigations performed	A CT scan was performed
Diagnosis	Formal diagnosis of illness	The patient was diagnosed with MS
Drug intervention	Application of drug as a treatment intervention	He was started on LEV
Treatment intervention	Nondrug treatment procedures	A surgery of the lower abdomen was performed
Case outcome	Outcome of the treatment	Patient was discharged on POD23

assigned more than 1 contexts depending on the number of events described in the sentence (e.g., Sentence with 2 events: 'Patient was *diagnosed with MS and started on methotrexate.*'). Seven separate classifiers are trained on 828 sentences each manually annotated with seven labels. Our feature set comprises of Bag of Words from each sentence. We perform the following transformations on each word:

- Lemmatize the words using nltk [10] parser.
- Emphasize clinically important terms by substituting lemma forms. For example, words ending with '-therapy' are substituted as '*therapy*'.
- Apply ontology matching and substitute: Generic drug names such as 'methotrexate' are substituted as '*drug*'.
- Create a tf-idf representation of the words to feed into a classifier.

We use ontologies to detect drug names and reactions. Generic drug names are matched against a dictionary created from UMLS [11]. For reaction (symptom) ontology, we use MedDRA [12].

Our test data for testing event contexts comprises of 126 sentences. We experiment with two sets of features:

(i) Features comprising of a vocabulary consisting of a handpicked collection of 120 words.
(ii) Features comprising of all words from the sentences in the training set.

We observe better accuracy with feature set comprising of all words in the sentence rather than with feature set comprising of (handpicked) 120 words. We further experiment between using a Naive Bayes classifier [13] and a Support Vector Machine [14]. The SVM outperforms the NB classifier as illustrated in Fig. 1. The F1 score is calculated based on the following considerations:

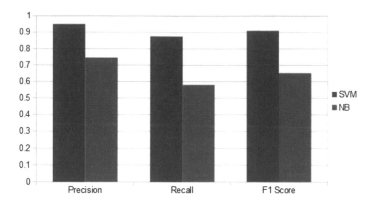

Fig. 1 Sentence context results using SVM, Naive Bayes

- If all labels in a sentence are correct, only then we assign the labeling as true positive (tp)
- Even if a single label is incorrect, we assign the labeling as fp (false positive)
- All missed labels are assigned fn (false negative)

Overall F1 score is calculated as

$$\text{Precision} = \frac{\text{tp}}{\text{tp} + \text{fp}} = 0.95$$

$$\text{Recall} = \frac{\text{tp}}{\text{tp} + \text{fn}} = 0.875$$

$$\text{F1Score} = \frac{2 * \text{Precision} * \text{Recall}}{\text{Precision} + \text{Recall}} = 0.91$$

4.3 Detecting Temporal Breaks

Our approach for temporal break detection is based on sequence classification technique. We use CRF++ implementation [15] of Conditional Random Fields [16]. Our feature set is described as follows.

1. 8 event context classes based on 7 event context labels (8th class, namely 'Other' is assigned 1 if all other classes are labeled as 0). 2 sets of features were derived from the event context classes: (i) a set of 8 labels for the current sentence, (ii) a set of 16 labels consisting of labels from previous 2 sentences. These features are considered to provide cues to detect implicit temporal breaks.
2. Temporal labels derived from explicit temporal expressions. We derive labels such as 'p', 'f' for expressions 'eight months earlier', '2 days later', respectively. We provide the CRF with labels from current sentence as well from previous 2 and next 2 sentences

3. Words from explicit temporal expressions in the current sentence (e.g. '*8 months later*' transformed into '**num**', '**time**', '*later*')

The CRF is trained on 2660 sentences from 120 case reports. The resulting CRF model is tested on 30 case reports (687 sentences). We achieve a precision of 0.81, a recall of 0.75 and an F1 score of 0.78 in detecting temporal breaks.

4.4 Ordering Events

Because of the absence of dates in medical case reports, the task of temporal ordering relies on the ability to normalize and sequence relative date expressions. Consider a narrative (Fig. 2) consisting of 9 sentences out of which 4 temporal breaks are detected (marked in gray). The first task is to identify groups of sentences that have temporal continuity. This task can be accomplished by starting with any temporal break (1, 2, 4, and 7) and extending up to the next break (or, the end of the narrative). We identify 4 such event groups: group with sentence 1 (belonging to 'c' → current), group with sentences 2 and 3 (belonging to 'p' → past with explicit expression 'One year back'), group with sentences 4, 5 and 6 (belonging to 'p' → past with explicit expression '2 months ago'), and group with sentences 7, 8 and 9 (belonging to 'f' → future with explicit expression '2 months later').

Next, we take groups belonging to past events and attempt to order them. For this ordering task, we first separate the numeric part from the date part ('2' and 'months' from '2 months') and then attempt ordering. The entire logic is rule-based and relies on dictionaries of numbers and dates. A similar ordering is done for the groups belonging to future events.

Event groups belonging to current are placed one after the other. We note that because of lack of explicit temporal expression in such groups, we can assume that they are inherently aligned in a sequence.

The final task of ordering is accomplished by adjacently placing the ordered past event groups, current event groups and the future event groups, in that order. For our example, the final ordering obtained is shown as follows: 2 → 3 → 4 → 5 → 6 → 1 → 7 → 8 → 9.

Sentence #	1	2	3	4	5	6	7	8	9
Labels applied	c	p	cont	p	cont	cont	f	cont	cont
Temporal Expressions	x	One year back	x	2 months ago	x	x	2 months later	x	x

Fig. 2 Temporal ordering of events

5 Discussions

In this paper, we attempt temporal ordering of medical case reports. We annotate 150 publicly available case reports, train a CRF model and achieve an F1 score of 0.78 in detecting temporal breaks in narrative. We subsequently perform rule-based normalization and ordering of medical events.

Our mechanism to detect explicit temporal expressions uses domain-specific cues (such as 'postoperatively') along with natural language cues (such as 'later', 'ago'). We note that a more exhaustive list of such phrases will help make the detection more robust.

We introduced 'event context' as a feature to enable the detection of implicit temporal breaks in case report narratives. This feature is specific to the biomedical domain and its effectiveness has only been tested on medical case reports. However, we believe this feature can be used in other biomedical narratives as well.

Finally, our event ordering approach takes events as given and instead attempts to order sentences. We note that this approach can easily be applied to the task of ordering of atomic events.

References

1. Pustejovsky, J., Hanks, P., Sauri, R., See, A., Gaizauskas, R., Setzer, A., et al. (2003). The Timebank corpus. In *Corpus linguistics* (p. 40). Lancaster, UK.
2. Bramsen, P., Deshpande, P., Lee, Y. K., & Barzilay, R. (2006). Inducing temporal graphs. In *Proceedings of the 2006 Conference on Empirical Methods in Natural Language Processing* (pp. 189–198). Association for Computational Linguistics.
3. Cheng, Y., Anick, P., Hong, P., & Xue, N. (2013). Temporal relation discovery between events and temporal expressions identified in clinical narrative. *Journal of Biomedical Informatics, 46,* S48–S53.
4. Jindal, P., & Roth, D. (2013). Extraction of events and temporal expressions from clinical narratives. *Journal of Biomedical Informatics, 46,* S13–S19.
5. Xu, Y., Wang, Y., Liu, T., Tsujii, J., & Chang, E. I.-C. (2013). An end-to-end system to identify temporal relation in discharge summaries: 2012 i2b2 challenge. *Journal of the American Medical Informatics Association, 20*(5):849–858.
6. Nikfarjam, A., Emadzadeh, E., & Gonzalez, G. (2013). Towards generating a patient's timeline: Extracting temporal relationships from clinical notes. *Journal of Biomedical Informatics, 46,* S40–S47.
7. Chang, Y.-C., Dai, H.-J., Wu, J. C.-Y., Chen, J.-M., Tsai, R. T.-H., & Hsu, W.-L. (2013). Tempting system: A hybrid method of rule and machine learning for temporal relation extraction in patient discharge summaries. *Journal of Biomedical Informatics, 46,* S54–S62.
8. Raghavan, P., Fosler-Lussier, E., & Lai, A. M. (2012). Temporal classification of medical events. In *Proceedings of the 2012 Workshop on Biomedical Natural Language Processing* (pp. 29–37). Association for Computational Linguistics.
9. Strötgen, J., & Gertz, M. (2010). Heideltime: High quality rule-based extraction and normalization of temporal expressions. In *Proceedings of the 5th International Workshop on Semantic Evaluation* (pp. 321–324). Association for Computational Linguistics.
10. Bird, S. (2006). Nltk: The natural language toolkit. In *Proceedings of the COLING/ACL on Interactive Presentation Sessions* (pp. 69–72). Association for Computational Linguistics.

11. Bodenreider, O. (2004). The unified medical language system (UMLS): Integrating biomedical terminology. *Nucleic Acids Research, 32*(suppl 1), D267–D270.
12. Brown, E. G., Wood, L., & Wood, S. (1999). The medical dictionary for regulatory activities (MedDRA). *Drug Safety, 20*(2), 109–117.
13. Murphy, K. P. (2006). Naive Bayes classifiers. *University of British Columbia.*
14. Cortes, C., & Vapnik, V. (1995). Support-vector networks. *Machine Learning, 20*(3), 273–297.
15. Kudo, T. (2005). Crf++: Yet another CRF toolkit. Software available at http://crfpp. sourceforge.net.
16. Lafferty, J., McCallum, A., & Pereira, F. C. (2001). Conditional random fields: Probabilistic models for segmenting and labeling sequence data.

Big Data Analytics: A Trading Strategy of NSE Stocks Using Bollinger Bands Analysis

Gokul Parambalath, E. Mahesh, P. Balasubramanian
and P. N. Kumar

Abstract The availability of huge distributed computing power using frameworks like Hadoop and Spark has facilitated algorithmic trading employing technical analysis of Big Data. We used the conventional Bollinger Bands set at two standard deviations based on a band of moving average over 20 minute-by-minute price values. The Nifty 50, a portfolio of blue chip companies, is a stock index of National Stock Exchange (NSE) of India reflecting the overall market sentiment. In this work, we analyze the intraday trading strategy employing the concept of Bollinger Bands to identify stocks that generates maximum profit. We have also examined the profits generated over one trading year. The tick-by-tick stock market data has been sourced from the NSE and was purchased by Amrita School of Business. The tick-by-tick data being typically Big Data was converted to a minute data on a distributed Spark platform prior to the analysis.

Keywords Big data · Bollinger bands · Chartists · Intraday trading
NSE · Spark

G. Parambalath · E. Mahesh
Department of Economics, Christ University, Bengaluru, India
e-mail: gokul96@gmail.com

E. Mahesh
e-mail: mahesh.e@christuniversity.in

P. Balasubramanian
Amrita School of Business, Amrita University, Coimbatore, India
e-mail: bala@amrita.edu

P. N. Kumar (✉)
Department of Computer Science and Engineering, Amrita University,
Coimbatore, India
e-mail: pn_kumar@cb.amrita.edu

© Springer Nature Singapore Pte Ltd. 2019   143
V. E. Balas et al. (eds.), *Data Management, Analytics and Innovation*,
Advances in Intelligent Systems and Computing 839,
https://doi.org/10.1007/978-981-13-1274-8_11

1   Introduction

The National Stock Exchange of India Limited (NSE) situated at Mumbai is one of the two major stock exchanges of India. The sectors of the Indian economy is represented by Nifty 50, the main stock index of NSE symbolizes the performance of Indian stock market. With the advent of high-speed computers, it has now been made possible to execute huge transactions at high frequency and speed (Big Data), which thus far have not been found possible manually for human traders. When stocks are bought and sold intraday, it is called "Day trading". Fundamental analysis and technical analysis are the two diverse strategies employed to analyze stocks for the purpose of investment in securities. The intrinsic value of a stock is estimated in fundamental analysis, whereas, the technical analysts (chartists), track only the movements of stock prices, without considering the value assignable to a company. Considering the 50 listed companies of Nifty 50, this work explores the profitability strategizing day trading employing the technical trading concept of Bollinger Bands. The buy-and-sell points are identified by the movement of the traded prices while they cross the upper and lower Bollinger bands. The tick-by-tick data (Big Data) was converted to a minute data on a distributed Spark platform, and subsequently analyzed by implementing in the open source R program. The methodology demonstrated here can be extended to implement algorithmic trading as well.

This article is organized as follows. Section 2 presents the literature review. Section 3 states the Research Problem followed by Sect. 4 on Methodology. Sections 5 and 6 gives the Implementation and Results, respectively. Section 7 gives our Conclusion and Sect. 8 suggests scope for Future Work.

2   Literature Review

2.1   National Stock Exchange, India

The Nifty 50 is the central index on the NSE, comprising about 70% of the market capitalization of the stocks listed as on March 31. It represents well-diversified blue chip companies and encompassing 13 sectors of the economy. In India along with the Bombay Stock Exchange (BSE) index, it is used for benchmarking fund portfolios, derivatives, and index funds, Kabasinskas and Macys [1]. Day trading defines the act of transactions in a stock intraday, with profits being made on investing large amount of capital taking advantage of minor price movements. Liquidity and volatility are the important parameters which are pertinent while taking a decision on the day trades. Liquidity allows entering and exiting a stock at a good price and volatility is an indicator of the daily price ranges expected. Higher volatility can be exploited to garner greater profit or loss, Shleifer [2].

A fundamental approach evaluates a company's value with the financial statements, analyzing the balance sheet, the cash flow statement, and income statement. A "buy" decision is made if the intrinsic value is more than the current share price, i.e., if a company is found worth the sum of its discounted cash flows and is assessed to be worth all of its future profits added together, duly discounted to cater for the time value of money [3]. Technical analysts determine the trend of price movements, with an attempt to understand the emotions at play in the market. The demand versus supply position in a market is evaluated, future trend refers to the anticipated price movement [4]. A trade is executed if favourable past prices and volume are important parameters which enables analysis of the trend in stocks. Technical analysts identify patterns without considering the intrinsic value of a stock. The past trading pattern of the stock and its movements are the main considerations. The technical analysts believe that history tends to repeat itself, and price moves in trends with the market discounting all factors.

2.2 Technical Analysts

Ignoring the fundamental factors, the technical analysts consider only the price movements. The future price movement is expected to follow a trend that may have been established. In contrast to the technical analysis, Fama ([5], pp. 34–105) and Fama and Miller [6], presented the Efficient-Market Hypothesis (EMH), which states that it is impossible to predict future prices based on past history alone. They argued that the market price reflects the price of the stock absolutely. EMH, therefore, highlights that any attempt to find undervalued companies is a futile exercise, and thereby counters the technical analysts. EMH theory, however, has been challenged by Baltussen [7], who stated that the stock price levels are a manifestation of traders' behavior, and Mandelbrot [8–10] demonstrated that the daily stock price returns do not follow a Gaussian distribution. These views effectively questioned the EMH. For a survey on the field of financial forecasting, the article by Nair and Mohandas [11, 12] can be referred. They have covered most relevant articles published over the period 1933 up to 2013.

2.3 Trends

Identification of presence of trends is the cornerstone in technical analysis. However, it is difficult to see trends in stock price movements. Therefore, more often than not, it is the movement of the highs and lows that constitutes a trend. An uptrend is a series of higher highs and higher lows, versus a downtrend, is one of lower lows and lower highs. There are three types of trends: *Up trends*—each subsequent peak and trough is higher; *Downtrends*—the peaks and troughs are getting lower; and *Sideways/Horizontal Trends*—there is negligible movement up or down in the peaks and troughs, i.e., absence of well-defined trends.

There are three trend classifications along the trend directions. Long term: A trend lasting longer than a year with several intermediate trends. Weekly/daily charts over a 5 year period are used by chartists for analysis. Intermediate trend: Period is between one and 3 months and occurs when a major trend is in the upward direction with a downward price movement followed by an uptrend. An analysis of daily charts assists in its analysis. The short-term trends comprise of both major and intermediate trends. A trend line represents resistance levels faced by stocks in their movements. A channel can be defined as the addition of two parallel trend lines (highs or lows). This depicts a region with strong areas of support and resistance, implying that trade between the two levels of support/resistance would be fruitful. A channel displays the trend as well. A sharp move can be expected in the direction of a break as and when one occurs. Support is the price level through which a stock/market holds versus resistance. This is the price level which is rarely penetrated.

2.4 High-Frequency Trading

Using algorithms, it is possible to carry out high-frequency trading based on mathematical models. Nair and Mohandas [11, 12] have presented a Genetic Algorithm-based technical indicator decision tree and a Support Vector Machine-based intelligent recommender system. The proposed method can learn patterns from the movements for a 1-day-ahead trading decision. As and when the trading conditions are met, buy/sell is enabled, eliminating emotional biases. Moreover, a trader s not required to monitor live prices and graphs. The common trading strategies used in algorithmic trading are: (a) Trend following strategies, (b) Arbitrage opportunities, (c) Strategies based on mathematical models, (d) Trading Range (Mean Reversion), and (e) Volume-Weighted Average Price (VWAP) and Time-Weighted Average Price (TWAP), Baltussen [7]. Kumar et al. [13] have illustrated prediction strategy using feed forward neural networks.

Bollinger Bands (Fig. 1) are trading bands plotted in and around the price structure to form an envelope of two standard deviations above and below a 20-day Simple Moving Average (SMA). The significance of 20-day Simple Moving Average is to gauge the intermediate-term trend. The standard deviation and the Simple Moving Average are calculated using the same trade data. The SMA is an intermediate-term trend, having achieved wide acceptance [3, 4]. Notable among the study applying Bands, is the work of Lai and Tseng [14]. According to behavioral finance, the decision-making of investors is modified and altered by their feelings with greed and fear driving investors towards irrational behavior that affects their portfolio allocation.

Therefore, value averaging concept has been used as the main strategy, with Bollinger band as a tool to assess volatility for trade (buy/sell). Few authors have published strategies which outperform Bollinger bands. The work of Lento et al. [15]

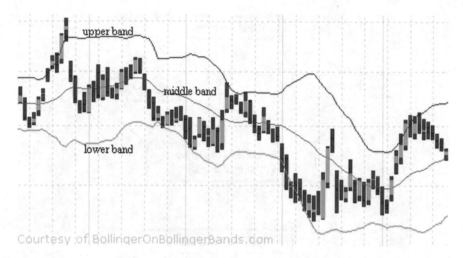

Fig. 1 Bollinger Bands (*Source* BollingerBands.com)

has shown that the Bollinger Bands are consistently unable to earn profits surpassing the buy-and-hold trading strategy after adjusting for transaction costs. Similarly, another study by Leung and Chong [16] compared the profitability of Moving Average Envelopes versus Bollinger bands. Despite the fact that Bollinger Bands can capture sudden price fluctuations (which Moving Average envelopes cannot), their study demonstrated that results employing Bollinger Bands do not surpass the methodology employing Moving Average Envelopes.

3   Research Problem

In our work, we first surmount the challenge of handling the Big Data stock on a Spark platform, and subsequently evaluate the investment methods based on technical analysis (Bollinger Bands methods) for the investment in the NSE [17]. The main objectives of this paper, therefore, is to adopt Bollinger Bands to study the NSE market and determine the investment period (i.e., long term versus short term), where Bollinger bands is more efficient. In this work, the transaction costs are not considered while evaluating the expected profits.

4   Methodology

The proposed system employs the methodology of Bollinger Bands for technical analysis of Indian stocks listed in Nifty50 of NSE. It consists of the following steps:

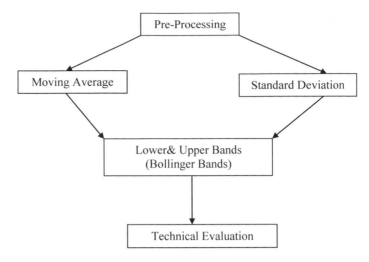

Fig. 2 Architecture: construction of Bollinger Bands

- Preprocessing.
- Construction of Bollinger Bands.
- Technical Evaluation.

4.1 Proposed System Architecture

The architecture of the proposed system is given in Fig. 2.

5 Implementation

5.1 Minute-Wise Abstraction of Big Data in Spark

The tick-by-tick stock market data has been sourced from the NSE. The intraday files depict a recalculation of the market index on the occurrence of a trade in an index component. The files are organized with Trade ID number; Symbol; Series; Time Stamp (hh:mm:ss); Price; and Quantity traded. To perform analysis more effectively, the data needs to be available at a lower resolution. Thus minute-wise amalgamation was carried out. The huge tick-by-tick data (Big Data) has been converted to minute-wise data in Spark platform at Amrita Vishwa Vidyapeetham. The minute-wise dataset consists of a trade ID for each transaction followed by the symbol and series of the shares traded. The other two important attributes are the price at which the trade is taking place and the quantity of shares traded.

It comprises of transactions for all companies, which had to be preprocessed in order to do prediction of the stock price for a particular company. Company-wise trade dataset was taken out from the minute-wise amalgamated dataset.

5.2 Preprocessing the Dataset in R

This data is converted into a comma separated format (.csv), and loaded into R using the loader command. The NSE dataset comprises of trades in respect of all companies. In order to perform the analysis, company-wise trade information is needed. The analysis is done in R programming.

5.3 Construction of Bollinger Bands

The steps implemented are as follows:

- 20 day Moving Average of the prices is calculated using the Financial Tool TTR. 20 day Moving Standard Deviation(s) is calculated.

$$s = \sqrt{\frac{\sum (x - \bar{x})^2}{n - 1}} \tag{1}$$

- Band separation is taken as twice the Standard Deviation.

$$UB(p) = MA + 2s \tag{2}$$

$$LB(p) = MA - 2s \tag{3}$$

where

UB—Upper band
LB—Lower band
P—Stock price
MA—20 day Moving Average

- The crossover points occur where the price cuts the Lower/Upper Bollinger Bands.
- The prices lower than Bollinger Bands and above the average are considered for "BUY" and those above the Bollinger Bands are earmarked for "SELL".

- An amount of Rs. 1,00,000/ is taken as the investible fund for all "BUY" signals.
- The aggregated difference of the sum invested and the amount received consequent to the "SELL" orders is taken.
- The profit realized is calculated.
- The average profit is arrived at and the volatility of the stocks is calculated.

5.4 Programming in R

Two modules have been developed. The R program, for calculating the *Intraday profits* and *one-year profits*, generated using the Bollinger Band algorithm is given at Appendix.

6 Analysis of Results

Employing the theory of Bollinger Bands, the profits generated (extracts of the results obtained) intraday and also over the trading year are given at Tables 1, and 2, respectively. From the results obtained, it is deduced that the shares possessing highest profit percentage can be filtered and trading can be focused on these stocks so as to maximize returns. The last column gives an idea of the volatility of the stocks. This would enable investors to concentrate on stocks conforming to their risk appetite.

Table 1 Intraday analysis of NSE stocks employing Bollinger Bands theory (part results)

Stocks (NIFTY)	Quantity traded	Shares sold (INR)	Shares bought (INR)	Profit (INR)	Profit percentage	Volatility
1	1156	1,713,476	1,700,000	13,476	0.79	3.05
2	12,710	3,110,611	3,100,000	10,611	0.34	0.64
3	8931	2,006,363	2,000,000	6363	0.32	0.45
4	5661	3,315,124	3,300,000	15,124	0.46	1.30
5	1456	2,805,113	2,800,000	5113	0.18	2.94
6	559	1,306,235	1,300,000	6235	0.48	3.03
7	2397	2,106,906	2,100,000	6906	0.33	2.17
...	...	...	...	...	...	...
50	3366	3,413,169	3,400,000	13,169	0.39	1.21

Table 2 One-year analysis of NSE stocks employing Bollinger Bands theory (part results)

Stocks (NIFTY)	Quantity traded	Shares sold (INR)	Shares bought (INR)	Profit (INR)	Profit percentage	Volatility
1	463,209	692,960,291	692,400,000	560,291	0.08	4.12
2	2,193,984	647,856,930	647,100,000	756,930	0.12	1.22
3	3,009,009	686,123,381	685,500,000	623,381	0.09	0.72
4	886,474	640,521,609	639,800,000	721,609	0.11	2.09
5	1,553,397	796,947,466	796,300,000	647,466	0.08	2.28
6	244,313	549,600,186	548,800,000	800,186	0.15	5.69
7	3,497,277	669,083,172	668,400,000	688,172	0.10	2.24
...	...	...	...	...	...	...
50	516,090	641,450,388	641,000,000	450,388	0.07	3.59

7 Conclusions

In this work, we examine an intraday trading strategy of NSE stocks performing technical analysis employing the concept of Bollinger Bands. We also highlight the suitability of the method for carrying out algorithmic trading. We examine the profits generated over trading for one full year based on the theory of Bollinger Bands. Bollinger Bands enables one to read price charts and offers a visual presentation of the price movements over two standard deviations. 96.5% of the data can be expected to fall within the Bollinger Bands. Overbought and oversold scenarios can be demonstrated statistically based upon a moving average of the lows and highs.

This enables an astute investor to identify price patterns, trend direction, and reversals, enabling low-risk trades. Employing the theory of Bollinger Bands, the profits generated intraday and also over the entire trading year has been analyzed. It has been demonstrated that it is possible to deduce the shares possessing highest profit percentages. Such shares can be filtered and trading can be focused to these stocks so as to maximize returns. The information on volatility enables investors to concentrate on stocks conforming to their risk appetite.

8 Future Work

While Moving Average Envelopes are more suitable for short-term investments, the strategy employing Bollinger Bands is profitable for long-term investments [16]. However, it is possible to factor in sudden price fluctuations in Bollinger Bands, which is not possible while using Moving Average Envelopes. This characteristic of Bollinger Bands can be exploited for various combinations of strategies aimed at optimizing profits for varying trading horizons and strategies. Development of a tool for implementing such combinations of trades will be helpful for investors to implement different trading strategies of choice.

Acknowledgements The authors would like to thank Amrita School of Business for providing the tick-by-tick trade data and the Centre of Excellence in Computational Engineering & Networking (CEN), Amrita University for support of Spark platform to handle the Big Data of stock prices.

Appendix

R Program Implementation of Trading Strategy Employing Bollinger Bands

Intraday Profits

The R program for calculating the intraday profits generated using the Bollinger Band algorithm and volatility are given below:

```
main_intraday.R
setwd( "C:\\Users\\Desktop\\bb\\yearfullsorted")
stock_list<-dir();
for(i in 1:49)
{rm(stock);
work<-stock_list[i]
setwd( "C:\\Users\\Desktop\\bb\\yearfullsorted");
stock1<-read.csv(work)
stock<-stock1[1:350,];
names(stock)[1]<-"stock";
names(stock)[2]<-"date";names(stock)[3]<-"time"
names(stock)[4]<-"price";
names(stock)[5]<-"qty";
setwd( "C:\\Users\\Desktop\\bb\\work")
source("bollinger.R");
setwd( "C:\\Users\\Desktop\\bb\\yearfullsorted")
}
```

5.4.2 One-Year Profits

The R program for calculating the 1-year profits generated using the Bollinger Band algorithm and the volatility calculated are given below:

```
main-full-year.R
setwd( "C:\\Users\\Desktop\\bb\\yearfullsorted")
stock_list<-dir();
```

```
for(i in 1:49)
{
rm(stock)
work<-stock_list[i];
setwd("C:\\Users\\Desktop\\bb\\yearfullsorted")
stock<-read.csv(work);names(stock)[1]<-"stock";
names(stock)[2]<-"date";names(stock)[3]<-"time";
names(stock)[4]<-"price";names(stock)[5]<-"qty";    setwd("C:\\Users\\Desktop\
\bb\\work")
source("bollinger.R");
setwd("C:\\Users\\Desktop\\bb\\yearfullsorted")
}
bollinger.R
library("TTR");
stock[6]=round(SMA(stock[4],n=20,na.rm=TRUE),digits=2)
names(stock)[6]<-"SMA";k=19;p=nrow(stock)-k;
for (i in 1:p)
{r=k+i;
stock[r,7]=round(sd(stock$SMA[i:r]),digits=2)};
names(stock)[7]<-"SD"
stock[,8]=round((stock[,6]+2*stock[,7]), digits=2);
stock[,9]=round((stock[,6]-2*stock[,7]), digits=2);
names(stock)[8]<-"UB";names(stock)[9]<-"LB"
volatility<- round(mean(stock$SD,na.rm="true")*4, digits=2)
avg_qty<-round(mean(stock$qty, digits=2));
tot_rows=nrow(stock);qty_bought<-0; qq<-0;
sold_amt<-0.0; buy_amt=0.0;sold<-0.0;
invest_fund=100000; for(i in 40:tot_rows)
{if ((stock$price[i]<stock$LB[i]) && (stock$qty[i] > (avg_qty)))
{qty_bought=qty_bought + round((invest_fund/stock$price[i]),digits=0);
bb<-round(invest_fund/stock$price[i], digits=0);
qq<- qq+bb; buy_amt<-buy_amt+invest_fund;}
if ((stock$price[i]>stock$UB[i])&& (stock$qty[i] >avg_qty))
{sold<-(stock$price[i]*qty_bought);sold_amt<-sold_amt+sold;
qty_bought=0;}}
invest_amt_bal=(qty_bought*max(stock$price))
tot_sale=sold_amt+invest_amt_bal; intraday_profit=tot_sale-buy_amt;
profit<-round(intraday_profit/buy_amt*100, digits=2);
result<-c(qq,tot_sale,buy_amt,intraday_profit,profit,volatility)
print(result);
```

References

1. Kabasinskas, A., & Macys, U. (2010). Calibration of Bollinger Bands parameters for trading strategy development in the Baltic stock market. *Inzinerine Ekonomika-Engineering Economics, 21*(3), 244–254.
2. Shleifer, A. (2000). *Inefficient markets: An introduction to behavioral finance.* Oxford University Press.
3. Bollinger, J. (1992)."Using Bollinger", Bands. *Stocks & Commodities, 10/2*, 47–51.
4. Bollinger, J. (2001). *Bollinger on Bollinger Bands.* New York: McGraw-Hill.
5. Fama, E. (1965). The behavior of stock market prices. *Journal of Business, 38*, 34–105.
6. Fama, E., & Miller, M. H. (1972). *The theory of finance.* New York: Holt, Rinehart and Winston.
7. Baltussen, G. (2009). *Behavioral finance: An introduction.* http://ssrn.com/abstract=1488110.
8. Mandelbrot, B. B. (1963). The variation of certain speculative prices. *Journal of Business, 36*, 394–419.
9. Mandelbrot, B. B. (1963). New methods in statistical economics. *Journal of Political Economy, 71*, 421–440.
10. Mandelbrot, B. B. (1964). The variation of certain speculative prices. In P. Cootner (Ed.), *The random character of stock prices.* Cambridge: MIT Press.
11. Nair, B. B., & Mohandas, V.P. (2015). Artificial intelligence applications in financial forecasting—a survey and some empirical results. *Intelligent Decision Technologies, 9*, 99–140. https://doi.org/10.3233/idt-140211 (IOS Press).
12. Nair, B. B., & Mohandas, V. P. (2015). An intelligent recommender system for stock trading. *Intelligent Decision Technologies, 9*, 211–220. https://doi.org/10.3233/idt-140220 (IOS Press).
13. Kumar, P. N., Rahul Seshadri, G., Hariharan, A., Mohandas, V. P., & Balasubramanian, P. (2011). Financial market prediction using feed forward neural network. In *International Conference on Technology Systems and Management.* Communications in Computer and Information Science, CCIS (Vol. 145, pp. 77–84).
14. Lai, II. C., Tseng, T. C., & Huang, S. C. (2016). Combining value averaging and Bollinger Band for an ETF trading strategy. *Applied Economics, 48*(37), 3550–3557.
15. Lento, C., Gradojevic,* N., & Wright, C. S. (2007). Investment information content in Bollinger Bands. *Applied Financial Economics Letters, 3*, 263–267.
16. Leung, J. M. J, & Chong, T. T. L. (2003). An empirical comparison of moving average envelopes and Bollinger Bands. *Applied Economics Letters, 10*, 339–341.
17. Harris, L. (2003). *Trading and exchanges: Market microstructure for practitioners.* Oxford University Press.

Handling Concept Drift in Data Streams by Using Drift Detection Methods

Malini M. Patil

Abstract The growth and development of the information and communication technology of the present era resulted in huge amount data generation. It is found that the rate of data distribution is very high. The data which is generated with varying distributions is referred to as data stream. Few examples to quote, data generated with regard to applications related to mobile networks, sensor networks, network traffic monitoring and network traffic management, etc. It is found that, the data generation process often change with respect to data distribution for any kind of concept, i.e. application which is referred to as concept drift. Handling concept drift is a challenging task. It is impossible to develop a model as it will be inconsistent in nature because of continuous change. The present work emphasises on handling the concept drifts, using different drift detection methods using Massive Online Analysis Framework. The important feature of the present study is varying size of a data stream (50,000–250,000). Totally the Concept Drift is handled using 11 drift detection methods using 2 stream generators abrupt and gradual under this frame work respectively.

Keywords Concept drift · Data streams · Data distribution · Drift detection
Massive online analysis framework

1 Introduction

In the recent past the data generation process is found to be very huge, with varying data distributions. Few real-time examples which are listed under this feature are weather forecasting data applications related to mobile networks, sensor networks, network traffic monitoring, and network traffic management, social network data,

M. M. Patil (✉)
Department of Information Science and Engineering,
J S S Academy of Technical Education, Bengaluru, India
e-mail: drmalinimpatil@gmail.com

© Springer Nature Singapore Pte Ltd. 2019
V. E. Balas et al. (eds.), *Data Management, Analytics and Innovation*,
Advances in Intelligent Systems and Computing 839,
https://doi.org/10.1007/978-981-13-1274-8_12

etc. Data which is generated with varying distributions is called as a data stream or a streaming data. Data Stream is far different from traditional databases.

Some of the features of data streams are identified as: (i) Data streams are huge in size. (ii) Data streams are ubiquitous in nature. (iii) Data streams require fast response. (iv) It is not possible to access the data streams randomly. (v) They require limited memory for storage. (vi) Require highly sophisticated techniques for mining. The challenges of data streams are: (i) Data stream processing algorithms does not permit multiple scans when compared to traditional data mining algorithms. (ii) Faster mining methods are to be employed with respect to speed of incoming data for faster response. (iii) Handling the change in data distribution is the major challenge for mining data streams. From the literature it is identified that the data streams can be studied under two headings (1) *Static streams* (2) *Evolving streams*. History data or regular bulk arrivals are termed as Static streams e.g. queries on data warehouses. Real time data which gets updated constantly are termed as evolving data streams [1] e.g. stock market data, and sensor data.

1.1   Concept Drift

Weather forecasting data [2, 3] which is quoted as one of the important real-time example to understand the feature of the data stream is mainly used to explain the concept drift. As it is known that the forecast variation is seasonal and the data distribution changes are likely to occur continuously. Another example is customer buying pattern in an inventory system. It is clearly observed that the customer buying preferences change quite often with respect to time. From both the examples it is clear that the cause of the change in weather forecast and change in customer preferences is hidden. Changes are hidden and unpredictable. The study reveals that an effective learner should be designed so that the learner can detect and able to track the hidden and unpredictable changes. One of the important problems identified in handling the concept drift is differentiation between noise and concept drift. Normally it is advised to design the ideal concept drift handling system which can have a method to adapt to concept drift by differentiating with noise.

a. **Types of concept drifts**
   The different types of concept drifts identified in the literature [4] are listed as follows:

   a. **Abrupt concept drift**: This type of drift refers to sudden or instant changes that occur. e.g., Changes on seasonal demand on sales.
   b. **Incremental concept drift**: This type of drift occurs when the values of variables are changed slowly over a period of time. e.g. slow increase in prices.
   c. **Gradual concept drift**: This type of drift occurs when variables change their class distribution slowly over time. e.g. spam data.

d. **Recurring concept drift**: This type of drift represents temporary changes that occur in data streams. They can be reverted back to their original state after some time. e.g. Trends in market.

b. **Drift detection**

Drift detection is a technique used to determine concept drift between two or more time periods. *Drift detector* is an algorithm that accepts input as stream of instances of varying sizes. The output of an algorithm is identification of concept drift, i.e. the detection of change in the distribution of the data. Drift detection or Change detection is a challenging task which consists of detecting the true changes. Study reveals that different drift-detection techniques are available. But the present work focuses mainly Handling Concept Drift in data streams by using Different Drift Detection Methods in Massive Online Analysis Framework.

2 Literature Survey

From the study of literature survey it is found that STAGGER [4] is one of the important bench mark data set which was able to handle concept drift. Later it was found to be the most popular bench mark data set. The study also reveals that many learning algorithms were used in handling concept drifts. They are rule-based learning, decision trees, Naïve Bayes, Radial Basis Functions instance-based learning. The interesting point to be high lighted here is that the lazy learning techniques are also found to be appropriate to handle concept drift which is presented very efficiently in [5]. The authors have proposed a case-based system for spam filtering using dynamic approach. In [6] the authors present the new ensemble learning methods. In [7] the authors elaborated on the study of concept drift in continuos domains. In [8] the authors present a very novel approach based on the learning method which is incremental. The method is based on a distributed concept description which is composed of a set of weighted, symbolic characterizations. The method utilizes previously acquired concept definitions in subsequent learning by adding an attribute for each learned concept to instance descriptions. Another important aspect of detecting concept drift is proposed in [9] in which the authors have developed a classification model of data stream. The adaptive window based approach proposed by authors of the paper [10] can detect different types of drift, is based on processing data chunk by chunk and measuring differences between two consecutive batches, as drift indicator. The experimental results show that the proposed method is capable to detect drifts and can approximately find concept drift locations. On the other side the platform used for running experiments was a break through contribution to the research developed by [11–16]. It is Massive Online Analysis framework. The authors have used the frame work extensively for the mining of data streams using MOA frame work. Perceptron learning model on evolving streams [17], study of recommender systems [18], performance analysis of hoeffding trees [19], frequent itemset mining on data streams [20], regression modelling using IBL streams [21] and mining data streams with concept drift [22] are some of the important works of the authors.

3   Drift Detection Methods

The MOA release 2014.04 provides 11 different drift detection methods. This section provides a brief note on all the methods.

a. *Adwin Change detector (DD1)*: ADWIN [10, 16] stands for Adaptive Windowing and uses sliding windows. Sliding window is one of the key data processing models in data streams. The size of the sliding window is application or machine dependent. Adwin automatically readjusts the window size with respect to change in data distribution and the window size is recomputed online.

b. *CusumDM (DD2)*: CusumDM stands for Cumulative Sum detection method [23–25] CUSUM is published in Biometrika in the year 1954, E. S. Page. It is a sequential analysis technique. In drift detection it is used for monitoring change detection.

c. *DDM (DD3)*: DDM stands for Drift Detection Method. It uses a Binomial Distribution [26] to detect the changes. Binomial distribution gives the general form of the probability for the random variable that represents the number of errors in a sample of '*n*' examples. DDM handles the classification errors produced by the learning model during prediction.

d. *EDDM (DD4)*: EDDM stands for Early Drift Detection Method (EDDM) [27]. This method is developed as an improvement over DDM to detect the concept drift. The basic idea is to find the distance between classification errors to detect change. It can detect the change without increasing the rate of false positives. It is also capable of detecting the slow gradual changes. The study reveals that it will improve the predictions.

e. *EWMAChartDM (DD5)*: EWMACHARTDM [28] stands for Exponentially Weighted Moving Average Chart Detection Method. It is a new modular approach for detecting concept drift. The method is designed in such a way that it is able to monitor the misclassification rate of a streaming classifier. It can also detect the change without increasing the rate of false positives during prediction.

f. *GMA DM (DD6)*: GMADM stands for Geometrical Moving Average Detection Method [29]. It is based on the concept of assigning weights to the observations for detecting changes in data streams. The method of assigning weights is based on geometric progression. The latest first observation is assigned with the greatest weight. The previous weights assigned to the observations were found to decrease in geometric progression.

g. *HDDM-A-Test (DD7) and HDDM-W-Test (DD8)*: HDDM-A-Test stands for The Hoeffding-based Drift Detection Method. It involves moving averages to detect abrupt changes (HDDM) [30]. HDDM-W-Test stands for The Hoeffding-based Drift Detection Method which is also a window based approach as mentioned earlier. It mainly detects gradual changes using weighted moving averages.

h. *PageHinkley DM (DD9)*: PageHinkley test [25, 31] is the sequential analysis technique typically used for monitoring concept drift detection. It allows effi-cient detection of changes in the normal behaviour of a process which is established by a model.

i. *SeqChange1DriftDetector (DD10) and SeqChange2DriftDetector (DD11)*: Sequential change-point detection [16, 31] is concerned with the design and analysis of techniques for online detection of a concept drift to a tolerable limit on the risk of a false detection. It is also referred to as quickest change detection method.

4 Methods and Models

This section presents the methodology used in this paper to compare the above mentioned drift detectors which are briefly explained in the previous section. The section covers the brief introduction of Massive Online Analysis framework (MOA), the process of evaluation in MOA, performance evaluators of MOA, Data sources in MOA.

4.1 *Massive Online Analysis (MOA) Framework*

Massive online analysis frame work [15] is an open source software environment to handle massive evolving data which is potentially infinite. It is also supported with features used for implementing algorithms and running experiments for evolving data streams. The present work uses MOA-14 for running the algorithms and conducting the experiments. MOA-14 is designed in such a way that it can meet the challenges of handling both the online and offline data streams. It can also handle real world data sets. It consists of offline and online algorithms for classification and clustering. It also consists of tools for evaluation. It is able to meet the important challenges of data streams. Study reveals that many of the works found in the literature establish the challenges of the data streams. MOA mainly permits the evaluation of data stream learning algorithms on massive data streams under explicit memory limits.

4.2 *Methodology*

The sequence of steps involved in configuring the MOA framework for detecting the concept drift using 11 Drift Detection Methods is as shown in Fig. 1.

The main steps involved in the drift detection method are listed as follows: (a) Selection of the generator. (b) Selection of the performance evaluator.

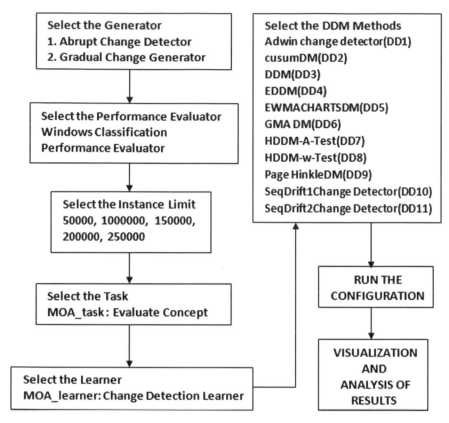

Fig. 1 Methodology used in drift detection method

(c) Selection of the instance limit. (d) Selection of the task. (e) Selection of the learner. (f) Selection of the drift detection method. (g) Visualization of the results. The other details of the steps are shown in Fig. 1. MOA is embedded with two types of concept drifts namely *abrupt* and *gradual*. The present work uses both the types. The performance evaluators available in MOA are *windows classification performance evaluator* and *basic classification performance evaluator*. The present work uses the windows method. Selection of instance is an auto generation method embedded in MOA. For the evaluation purpose the instance limit selected is 50,000–250,000. In the moa_task selection process selects *evaluate_concept drift* as the task which is the main purpose of the present work. Naïve Bayes is the learner used for handling the concept drift. The Whole selection process mentioned above is kept same for both the types of concept drifts and only the drift detection method is varied every time before running the experiment. The time taken to run the configuration is tabulated accordingly.

5 Experiments and Results

The configuration of the drift detection method using MOA framework is as shown in Fig. 2.

Experiments are carried out on the 11 drift detection methods mentioned in Sect. 3 by using abrupt and gradual drift generators. Naïve Bayes learner is used for the evaluation purpose. The results are tabulated in Tables 1, and 2 respectively. The graphical representation of the results is shown in Figs. 3, and 4 respectively.

The glance at Table 1 reveals that the behaviour of all the drift detectors is very interesting in case of gradual generator method. Following are the observations made in the analysis.

- For an instance size of 5000, CUSUMDM (DD2) performs well. DD2 takes minimum execution time (1.93 s) where as SEQUENTIAL CHANGE 1 DETECTOR (DD10) takes maximum execution time (5.04 s).
- For an instance size of 100,000, 150,000, 200,000 SEQUENTIAL CHANGE 2 DETECTOR (DD11) takes maximum execution time (14.76, 50.48, 45.29 s) and EDDM (DD4) takes minimum execution time (4.12, 5.32, 8.05 s).
- For an instance size of 250,000, SEQUENTIAL CHANGE 2 DETECTOR (DD11) takes minimum execution time (9.3 s) SEQUENTIAL CHANGE 1 DETECTOR (DD11) takes maximum execution time (21.96 s).
- EDDM (DD4) performs better with respect to all the drift detectors for varying instance sizes.
- The graphical representation of behaviour of all the drift detectors in case of gradual generator method is as shown in the Fig. 3 and is self-explanatory.

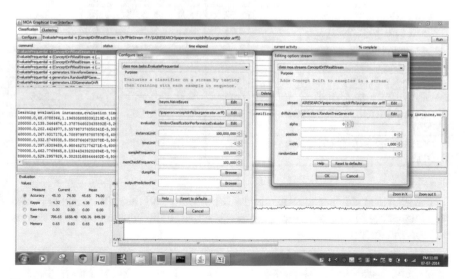

Fig. 2 Configuration window in MOA

Table 1 Time (in sec) recorded for gradual generator method

Drift detector	Instance size				
	50,000	100,000	150,000	200,000	250,000
ADWIN Change detector (DD1)	2.39	5.66	9.09	9.59	14.76
CUSUMDM (DD2)	1.93	5.13	6.58	10.90	11.9
DDM (DD3)	3.00	6.43	5.35	8.30	11.23
EDDM (DD4)	2.37	4.12	5.32	8.05	10.83
EWMAChartDM (DD5)	2.71	4.93	6.65	9.63	12.18
GMA DM (DD6)	2.73	4.96	6.55	9.70	12.60
HDDM-A-Test (DD7)	2.78	5.05	10.92	10.48	11.89
HDDM-W-Test (DD8)	2.96	5.94	12.95	11.01	12.96
PAGEHINKLEY DM (DD9)	2.89	5.34	10.87	10.59	13.37
SEQCHANGE1DRIFTDETECTOR (DD10)	5.04	9.92	21.61	16.61	21.96
SEQCHANGE2DRIFTDETECTOR (DD11)	4.37	14.76	50.48	45.29	9.30

Table 2 Time (in sec) recorded for abrupt generator method

Drift detector	Instance size				
	50,000	100,000	150,000	200,000	250,000
ADWIN Change detector (DD1)	2.48	5.23	6.86	10.92	11.31
CUSUMDM (DD2)	2.32	5.16	8.03	10.22	13.79
DDM (DD3)	1.92	4.93	6.46	8.6	13.54
EDDM (DD4)	1.67	4.21	5.05	8	8.42
EWMAChartDM (DD5)	2.04	4.96	5.44	9.94	13.48
GMA DM (DD6)	2.40	4.96	6.65	9.63	12.68
HDDM-A-Test (DD7)	0.05	5.18	6.85	10.14	15.33
HDDM-W-Test (DD8)	0.05	5.46	8.13	9.98	13.10
PAGEHINKLEY DM (DD9)	0.05	5.41	6.74	10.31	10.89
SEQCHANGE1DRIFTDETECTOR (DD10)	0.05	11.17	21.62	28.59	14.38
SEQCHANGE2DRIFTDETECTOR (DD11)	0.05	5.37	7.00	10.42	9.97

The glance at Table 2 reveals that the behaviour of all the drift detectors is also very interesting in case of abrupt generator method. Following are the observations made in the analysis.

- For an instance size of 5000, the drift detectors DD7 to DD11 performs well and are consistent with the minimum execution time (0.05 s). Adwin Change detector (DD1) takes maximum execution time (2.48 s).

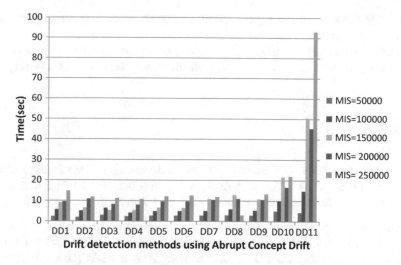

Fig. 3 Graph of drift detection methods using abrupt concept drift for varying instance sizes

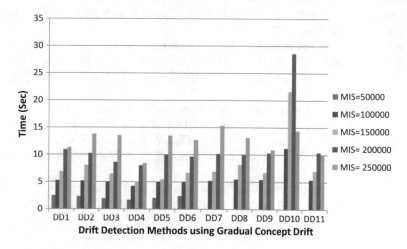

Fig. 4 Graph of drift detection methods using gradual concept drift for varying instance sizes

- For instance sizes 100,000, 150,000, 200,000, 250,000 it is very interesting to note that EDDM (DD4) performs very well with minimum execution time (4.21, 5.05, 8 and 8.42 s) respectively.
- For an instance sizes of 100,000, 150,000 and 200,000 SEQUENTIAL CHANGE 1 DETECTOR (DD10) takes maximum time (11.17, 21.62 and 28.59 s) respectively.
- For an instance size of 250,000 HDDM (DD7) take maximum execution time (15.33 s) and EDDM (DD4) takes minimum time (8.42 s).

- EDDM (DD4) performs better with respect to all the drift detectors for varying instance sizes.
- The graphical representation of behaviour of all the drift detectors in case of abrupt generator method is as shown in the Fig. 4 and is self-explanatory.

6 Conclusions

The present work is mainly emphasis on the understanding of the concept drift in data streams by using different drift detection methods for gradual and abrupt generators in MOA framework which is concluded with the following observations.

- The present analysis uses drift detectors namely Adwin Change Detector (DD1), CUSUMDM (DD2), Drift Detection Method (DD3), EDDM (DD4), EWMAChartDM (DD5), Geometric Moving Average DM (DD6), HDDM-A-Test (DD7) and HDDM-W-Test (DD8), Page HinkleyDM (DD9), Sequential Change 1 Detector (DD10), Sequential Change 2 Detector (DD11).
- The maximum instance size used in the analysis varies from 500,000, 100,000, 150,000, 200,000, and 250,000.
- Naïve Bayes learner is used for the evaluation purpose.
- The performance of EDDM (DD4) is found to be good with respect to abrupt and gradual generators especially in case of instance sizes 100,000, 150,000 and 200,000.
- Another important observation is for an instance size of 500000 the drift detectors HDDM-A-Test (DD7) and HDDM-W-Test (DD8), Page HinkleyDM (DD9), Sequential Change 1 Detector (DD10), Sequential Change 2 Detector (DD11) perform well and are consistent with the minimum execution time (0.05 s).
- The results of the present investigation are unique in handling concept drift in MOA frame work and also they provide an excellent platform for future investigation.

References

1. Aggarwal, C. (2007). *Data streams: Models and algorithms.* Series: Advances in database systems (Vol. 31, XVIII, p. 354). Berlin, Heidelberg: Springer (ebook).
2. Gaber, M. M., Zaslavsky, A., & Krishnamurthy, S.(2005). *Data streams: Models and methods* (Vol. 31, pp. 39–59). Berlin, Heidelberg: Springer.
3. Gao, J., Fan, W., Han, J., & Yu, P. S. (2007). General framework for mining concept-drifting data streams with skewed distributions. In *SIAM International Conference on Data Mining* (pp. 3–14), Minneapolis.
4. Brzesinski, D. (2010). *Mining data streams with concept drift (Ph.D. thesis).*

5. Cunningham, P., Nowlan, N., Delany, S. J., & Haahr, M. (2003). A case-based approach to spam filtering that can track concept drift. In *The Proceedings of ICCBR-2003 Workshop on Long-Lived CBR System.*

6. Kolter, J. Z., Maloof, M. A. (2003). Dynamic weighted majority: A new ensemble method for tracking concept drift. In *3rd IEEE International Conference on Data Mining ICDM-2003* (pp. 123–130). IEEE CS Press.

7. Kubat, M., & Widmer, G. (1994). *Adapting to drift in continuous domains* (Technical Report). Vienna: Austrian Research Institute for Artificial Intelligence.

8. Schlimmer, J. C., & Granger, R. H. (1986). Incremental learning from noisy data. *Journal of Machine Learning, 1*(3), 317–354.

9. Xiaofeng, L., & Weiwei, G. (2014). SERSC study on a classification model of data stream based on concept drift. *International Journal of Multimedia and Ubiquitous Engineering, 9* (5), 363–372. http://dx.doi.org/10.14257/ijmue.2014.9.5.37. ISSN: 1975-0080.

10. Bifet, A., & Gavalda, R. *Learning from time changing data with adaptive windowing.* Univeritat Poliyecnica De Catalunya.

11. Bifet, A., Holmes, G., Kirby, R., & Pfahringer, B. (2010). MOA: Massive online analysis. *Journal Machine Learning Research,* 1601–1604.

12. Bifet, A., Holmes, G., Kirby, R., & Pfahringer, B. (2011). MOA: Massive online analysis. *Journal of Machine Learning Research,* 1601–1604.

13. Bifet, A., Kirkby, R., Kranen, P., & Reutemann, P. (2009). *Massive online analysis, technical manual.* Hamilton, New Zealand: University of Waikato.

14. Bifet, A., & Kirkby, R. (2009). *Data stream mining: A practical approach* (Technical report). New Zealand: The University of Waikato.

15. https://moa.cms.waikato.ac.nz/

16. Bifet, A., & Gavaldà, R. (2009). Adaptive learning from evolving data streams. *Advances in Intelligent Data Analysis,* 8, 249–260. Berlin, Heidelberg: Springer.

17. Srimani, P. K., & Patil, M. M. (2012). Simple perceptron model (SPM) on evolving streams in MDM. *International Journal of Neural Networks, 2*(1), 20–24. E-ISSN 2249-2771.

18. Patil, M. M. (2015). A comprehensive study of recommender systems. *International Journal of Advancements in Engineering Research, 10*(86), 332–337.

19. Srimani, P. K., & Patil, M. M. (2015). Performance analysis of hoeffding trees in MDM using MOA framework. *International Journal of Data Mining, Modeling and Management, 7*(4), 293–313. http://dx.doi.org/10.1504/IJDMMM.2015.073865.

20. Srimani, P. K., & Patil, M. M. (2015). Frequent item set mining using INC_MINE in massive on line analysis framework. *Science Direct Journal of Elsevier Publication, 45,* 133–142.

21. Srimani, P. K., & Patil, M. M. (2014). Regression modeling using IBLSTREAMS. *Indian Journal of Science and Technology, 7*(6), 864–870. Print ISSN 0974-6846, Online ISSN 0974-5645.

22. Srimani, P. K., & Patil, M. M. (2016). Mining data streams with concept drift in massive online analysis framework. *WSEAS transactions on computers, 15*(#14), 133–139. Article is in the press. ISSN/E-ISSN 1109-2750/2224-2872.

23. Barnard, G. A. (1959). Control charts and stochastic processes. Journal of the Royal Statistical Society. *B (Methodological), 21*(2), 239–271. JSTOR 2983801.

24. Grigg, Farewell, V. T., Spiegelhalter, D.J., et al. (2003). The use of risk-adjusted CUSUM and RSPRT charts for monitoring in medical contexts. *Statistical Methods in Medical Research, 12*(2): 147–170. https://doi.org/10.1177/096228020301200205.pmid12665208.

25. Page, E. S. (1954). Continuous inspection schemes. *Biometrika, 41,* 100–115.

26. Gama, J., & Bifet, A. (2014). *A survey on concept drift adaptation.* Portugal: University of Porto.

27. Gavaldµa, R., & Morales-Bueno, R.(2011). *Early drift detection method manuel.* Spain: University of Malaga.

28. Rossa, G. J., Adamsa, N. M., Tasoulisa, D. K., & Handaa, D. J. (2012). *Exponentially weighted moving average charts for detecting concept drift*. Department of Mathematics, Imperial College, London SW7 2AZ, UK.
29. Roberts, S. W. (2012). In M. Brama & M. Petridis (Eds.), Control chart tests based on geometric moving averages. *Research Development in intelligent systems, XXIX*, 97–101. https://doi.org/10.1007/978-1-4471-4739-8-6 .
30. Frías-Blanco, I., Campo-Ávila, J. Ramos, G., Bueno, R., Díaz, A., & Caballero Mota, Y. (2015). Online and non-parametric drift detection methods based on Hoeffding's bounds. *IEEE Transactions on Knowledge and Data Engineering, 27*(3), 810–823.
31. Moustakides, G. V. (2008). Sequential change detection revisited. *The Annals of Statistics 2008, 36*(2), 787–807. https://doi.org/10.1214/009053607000000938.

Policy-Based Access Control Scheme for Securing Hadoop Ecosystem

Madhvaraj M. Shetty, D. H. Manjaiah and Ezz El-Din Hemdan

Abstract Hadoop is a framework for distributed, fault-tolerant storage and processing of Big Data. Apache Hadoop ecosystem enables users to harness the potential of Big Data fully with the help of other open-source tools such as Hive, HBase, Pig, and Storm. It was hugely adopted in various areas such as training, online media, government, and web-based social networking to deal with the huge development of data in their respective domains. However, the core architecture totally relies on a trusted cluster and lacks native techniques for securing sensitive data of users stored and processed on the system. Designing and implementing access control system is an essential step to secure Hadoop ecosystem components by defining a user policy with different roles as well as offering centralized administration for policy management. In this chapter, we have proposed a multi-layer policy-based access control scheme for Hadoop ecosystem, based on concepts of role-based access control system. The proposed model provides a way to prevent unauthorized access to cluster resources and imposes access control for data owners.

Keywords Access control · Role-based access control · Authorization
Hadoop ecosystem components · Big data

M. M. Shetty · D. H. Manjaiah (✉) · E. E.-D. Hemdan
Department of Computer Science, Mangalore University, Mangalore,
Karnataka, India
e-mail: manju@mangaloreuniversity.ac.in

M. M. Shetty
e-mail: madhvarajj@gmail.com

E. E.-D. Hemdan
e-mail: ezzvip@yahoo.com

© Springer Nature Singapore Pte Ltd. 2019   167
V. E. Balas et al. (eds.), *Data Management, Analytics and Innovation*,
Advances in Intelligent Systems and Computing 839,
https://doi.org/10.1007/978-981-13-1274-8_13

1   Introduction

Over the decade, the size of data has significantly increased. This made analyzing large datasets complicated by using traditional tools and applications. Recently, such large datasets have been named as "Big Data", which describes a situation where the speed at which data comes in, different data formats, and its size exceeds the capacity of existing storage and hardware. Analysis becomes more and more complex due to the difficulties we face in extracting specific useful information from large datasets. Apache Hadoop [1] is one of the powerful programming frameworks to store and process enormous quantity of data produced in varied formats. Because of its cost-effective, scalable, distributed storage for storing and analyzing structured, unstructured, and semi-structured data, it has established itself as an essential open-source framework for Big Data. Rich analytical capabilities and resilient storage furnished by Hadoop framework makes it a primary choice for a Big Data in government sectors as well as in industries, which utilizes Hadoop on a bigger level. The Hadoop platform includes two main elements: Hadoop Distributed File System (HDFS) and MapReduce as discussed in Table 1.

The remaining of this chapter is structured as follows. Section 2 provides a background about the access control mechanisms available and briefs about existing Hadoop security. Section 3 discusses role-based access control concepts followed by the current Hadoop access control security mechanism in Sect. 4. Next, we introduce the proposed role-based access control model and its formal description in Sect. 5. Security analysis and features are discussed in Sect. 6, and Sect. 7 concludes the paper.

Table 1   The main elements of Hadoop ecosystem

HDFS	The HDFS is a distributed file system used to store large datasets with streaming data access, running clusters on commodity hardware [20]. The file system is extremely fault tolerant that splits data and stores on multiple nodes throughout the cluster, with a number of replicas. It gives an extremely consistent, reliable, efficient, and profitable approach to store bulk data. The HDFS stores the data that is split in multiple nodes called DataNodes. The namespace, metadata of each file and its blocks are stored in a master node called NameNode
MapReduce	MapReduce is the component responsible for the distributed and parallel processing of data stored throughout the HDFS; it splits each main problem into multiple individual subtasks that are executed in parallel in Hadoop. This can be well explained by taking an example like if a machine takes 30 min to process a particular task, and if that task is broken into various subtasks in 30 various machines that can be computed in 1 min. Mainly, it consists of two fundamental functions, namely, Mapper and Reducer. The Mapper responsible for processing input splits in parallel through the help of multiple map tasks and submits shuffled, sorted outputs to the Reducers that, in turn, process and group them using reduce tasks for each key available

2   Related Work

As Hadoop is being generally utilized as a part of government and private sector, however, its security has been a noteworthy concern. Such extensive acceptability of Hadoop environment comes with the duty to make it secure against cyberattacks. However, distributed nature and the scale of the platform make it more difficult to protect the framework resources. Data saved in Hadoop multi-tenant data lake often consists of sensitive data from numerous vital assets, together with credit card numbers, medical details, Social Security Numbers (SSNs), and information from banking and intelligence corporations, which should only be accessed by authorized applications and users. It requires the cluster to be secured against digital threats since unapproved access to these information resources can seriously affect its uprightness and secrecy. Threats including noxious, client-killing YARN applications, disguising Hadoop administrations like DataNode, YARN, NameNode, and so on and attacks such as denial of resources can have serious implications. Additionally, an inside client can disguise by running malevolent code to impersonate Hadoop core services.

2.1   Hadoop Security

With the growing need of demand for data architecture, today many big companies such as Yahoo, Facebook, Microsoft, and IBM along with other educational and government institutions have generally accepted Hadoop as a primary key for their data framework. Oracle and Amazon wrapped Hadoop as one of their key businesses by raising it to the "cloud" and offering "Platform as a Service" [2]. The day by day growth of its acceptance over various sectors and use for sensitive data management, the data protection has become quite a matter of importance and prevalence. However, the Hadoop core architecture was designed for cluster environment which is trusted. All of its operations are primarily based on assumptions that clusters would comprise of trusted machines utilized by trusted users.

A security scheme was later introduced by Yahoo for addressing these concerns by Kerberos protocol [3]. However, it lacks requirements of active directory and access control mechanisms [4–6] and is unable to provide a comprehensive solution for authorization. Similar to POSIX file systems, the HDFS file system enables authorization based on Access Control Lists (ACL). But, as the numbers of users as well as data files have increased, it has become very difficult to manage the access control lists and its permissions. Also, by supplying an appropriate data block id, Hadoop clients may directly read or write data bypassing the NameNode [5, 6].

Table 2 Common access control models

Common access control models are Discretionary Access Control (DAC), Mandatory Access Control (MAC), and Role-Based Access Control (RBAC)	
1.	In MAC, the permissions and assignments are typically defined by the security policy administrator in an organization. MAC policies are defined based on the sensitivity of the object such as "Confidential, Unclassified, Secret and Top Secret." Therefore, this model is insufficient especially in the context of a large distributed system like Big Data, Cloud computing
2.	DAC is another form of access control in which the flexibility of rules allocation are given to the users themselves without the intervention of a system administrator, usually the owners of the information. As presented in [21], it permits system to allow or to reject other user's access to the objects under their charge. This benefit leads to a security drawback that the policies do not provide a real assurance on the flow of information in the system
3.	RBAC is a progression in the field of access control [11]. In this model, access decisions are based on the roles of individual users. This includes specifying user's responsibilities, duties, qualifications, the actions that are allowed to be performed within the context of an organization. So, it helps in the administration of access rights and users access control

2.2 Access Control Mechanism

Private data in any organization must be secured appropriately against the external security threats as well as internal threats. A large number of threats originate from the internal sources in an organization [7, 8]. An access control mechanism is a security parameter that is used to enforce security against internal threats. It [9, 10] is defined as the restriction of access to a resource for a selective group of users. This limitation is accomplished by defining a set of rules to access the authorized resource. Here, the system grants or revokes the privilege to perform some activity or to access some data, based on the rule which is predefined. So, access control mediates every request to resources and determines whether the request should be denied or granted. The common access control models are discussed in Table 2.

3 Role-Based Access Control

Among various access control models, Role-Based Access Control (RBAC) [11–13] is the most widely used and becomes a research focus over the decades. Here, a set of permissions are associated with a role, and each of the roles has a different set of predefined permissions. Then, the roles are assigned to users which, in turn, associate users with sets of permissions. It has increased flexibility compared with other models concerning providing access to multiple users at once, rather than allocating permission to each user separately. Here, we can easily audit all access permissions associated with a user by checking the permissions related to the specific role and is easier to determine the associated risk exposure. Initially, permissions are defined based on the actions to be performed and then these permissions are assigned to roles. So, instead of assigning privileges to users directly, access rights are associated with roles. The role acts as a bridge between users and their privileges. Users are assigned to appropriate roles, thereby acquiring the corresponding privileges of that role [14, 15]. By this method, it enables simple and safe management of access control system.

System administrator creates appropriate roles according to the actions performed on the context and grants permission to each role. Then, a particular user is assigned to the relevant role according to the security requirement [16]. It is majorly famous because of high security, robustness, easy administration for managing roles as well as for various other permissions. Meanwhile, designing role and structuring access control system for a large-scale organization is a difficult and more complex task to complete [17].

4 Hadoop Access Control Mechanisms

Hadoop utilizes multi-layer authorization framework using ACLs to authorize clients to access data, infrastructure resources, and other services available in the cluster. The first layer of access control mechanisms is provided by service layer authorization that is used to verify and authorize a user much before data underlying the services are accessed. By default, this is disabled in Hadoop; it can be enabled by configuring core-site.xml file by setting the below property in all the nodes of cluster.

hadoop.security.authorization = true

The HDFS file system supports POSIX-style permissions on files stored. The access control for various Hadoop services must be defined in Hadoop-policy.xml file. ACLs authorization is disabled by default in HDFS, which can be enabled by setting the below property in configuration file hdfs-site.xml for HDFS NameNode.

dfs.namenode.acls.enabled = true

Apache Sentry [18] and Apache Ranger [19] are two important, proficient authorization frameworks used to execute fine-grained access control in Hadoop ecosystem services. The two frameworks enable centralized administration to store and manage security policies. Ranger assigns permissions to users and groups whereas Sentry supports role-based authorization.

5 The Proposed Model

In this part, we present the multi-layer role-based access control model to authorize
users to Hadoop ecosystem, where the first layer of defense checks whether the user is
authorized to access the Hadoop ecosystem daemons. This prevents unauthorized
access early in the access request lifecycle because it is performed before data and
service objects permissions are assessed. Another layer used for fine-grained access
which handles permissions that are specified in the policy manager with the help of
Access Control Lists (ACLs) for individual datasets that are stored in different nodes
throughout the Hadoop cluster. Figure 1 shows main components of our proposed
access control system, web server, and policy manager.

As HDFS handles Hadoop data storage layer, it is likely to have access to data
stored in HDFS through certain data engines, such as Hive or Pig, and no direct
access to HDFS. For example, a sales manager may get access to data using Pig
query, but may not be allowed to access corresponding data files in the file system
through MapReduce jobs. So, in this proposed model, the user will communicate
with Hadoop using web server with valid credentials to log in. After successful
login, the user can see the list of components for which he has permission to access,
and through these components the user is able to access data and process the data
stored in Hadoop HDFS. All the roles and permissions are defined in policy
manager, which handles all the complexities of maintaining various roles with its
permissions list with the simple dashboard implemented along with this model.
Admin of the system has full control over the users and roles permissions through
which he can add new roles as well as modify existing roles in the database and
block a specific IP or a particular user with one click from the admin dashboard.
Each role will be differentiated with components that user can access as well as the
individual datasets that are stored in HDFS, which guarantees fine-grained access
control. Figure 2 depicts the relationships between individual users, roles/groups,
and system resources as well as data files stored in HDFS. These relations collectively determine whether a particular user will be allowed to access a particular
resource or piece of data.

Fig. 1 Framework of the
proposed model

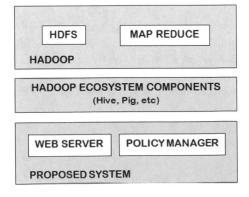

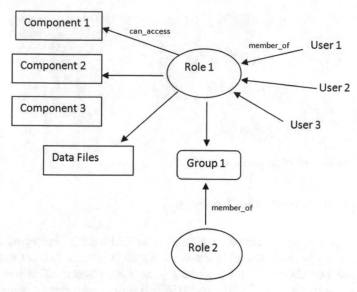

Fig. 2 Relationship between users, roles, groups, and resources

$Roles = \{Role_1, Role_2, Role_3, ..., Role_m\}$
$Resources = \{Resource_1, Resource_2, Resource_3, ..., Resource_n\}$
$Role_x = \{RoleID, RoleName, \{Resource_1, Resource_2, Resource_3, ..., Resource_m\}\}$
$User_y = \{UserDetails, Role_z\}$

User can access resources only if he has assigned role where that role is authorized to that specific set of resources. As shown in Fig. 3, in our proposed model, the permission of the role will be verified by the web server centrally before submitting any job to Hadoop. The system throws error while a user without permission tries to access any component/dataset from HDFS, and that error will be logged in a log database for monitoring.

We considered Pig, one of the components of the Hadoop ecosystem for the use case in this access control system. Here, the user who has the permission to access the Pig will be able to see Pig option in his dashboard after successful login. Also, gets space to enter Pig queries to submit to the Hadoop which will be processed by MapReduce. A user with relevant permission to access ecosystem component can submit jobs directly via web application which will run on Hadoop by utilizing input from the HDFS and get the result on the screen as well as stored in HDFS file system.

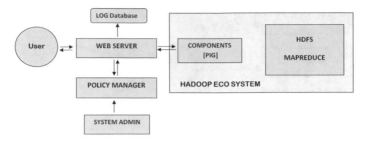

Fig. 3 Block diagram of the proposed model

6 Implementation and Analysis

We have used Apache Tomcat as a web server and MySQL server as a policy manager. The proposed method is tested on Hadoop cluster with three nodes, one master, and two slaves. It used Apache Pig as a component that is used to run queries on the dataset stored in Hadoop HDFS. The user with certain permission to access Pig and dataset will able to run the queries successfully. Also, the system throws permission-denied error on accessing the component or dataset without proper privileges in the role assigned to that user.

The proposed model makes security management easier for the Hadoop ecosystem because of the following facts:

- *Centralized administration*: The model follows centralized administration, which is responsible for maintaining policies all at one place for different services as well as data objects, and all access requests go through a central authority that grants or denies the request.
- *Flexibility*: The admin can manage permissions of various roles by some clicks directly from the dashboard. The administrative task can be granting or revoking permission from the specified role within the system.
- *Easy Management*: Users can be easily reassigned from one role to another. Roles can be granted new permissions as new components and resources are incorporated. The roles can be grouped to define master role, assigning this role to any user gives all individual permissions of the all roles to that specific user.
- *Auditing*: All the actions of users are stored in log database. The collected activity logs allow administrators to monitor who is using the resource and how it is being used. Also, logs that are gathered can be used to identify suspicious or unusual activity.
- *Confidentiality and integrity*: Access control supports both the confidentiality and the integrity properties of a secure system. Often, this approach requires more administrative maintenance, but it provides more security than more permissive strategies.

7 Conclusion

As Hadoop continues to be widely used in government and private sectors, its security has been a major concern. Such wide acceptability comes with the huge responsibility to make it secure against cyberattacks. However, the scale and distributed nature of the platform makes it harder to protect the infrastructure assets. Data stored in Hadoop multi-tenant data lake often includes sensitive information from various critical sources, which should only be accessed by authorized applications and users. In this chapter, we proposed an access control paradigm using the role-based access control concepts. It mainly aimed at allowing the data owners/admins to control and audit access to their data as well as to make security management easier for the administrator. There are still many challenges for future work like how to improve role-based access control model for multiple applications under one ecosystem, real-time activity monitoring of all users, dynamic permission management, integrating with native POSIX permissions, securing web communication channels, etc.

References

1. Apache Hadoop. http://hadoop.apache.org/.
2. Lal, K., & Mahanti, N. C. (2010). A novel data mining algorithm for semantic web based data cloud. *International Journal of Computer Science and Security (IJCSS)*, 4(2), 160.
3. O'Malley, O., Zhang, K., Radia, S., et al. Hadoop security design. https://issues.apache.org/jira/secure/attachment/12428537/security-design.pdf.
4. Becherer, A. Hadoop security design just add kerberos? Really? http://media.blackhat.com/bh-us-10/whitepapers/Becherer/BlackHat-USA-2010-Becherer-Andrew-Hadoop-Security-wp.pdf.
5. Hadoop poses a big data security risk: 10 reasons why. http://www.eweek.com/security/slideshows/hadoop-poses-a-big-datasecurity-risk-10-reasons-why/.
6. Brus, A. The odd couple: Hadoop and data security. http://www.zdnet.com/the-odd-couple-hadoop-and-datasecurity-7000018468/.
7. Habib, M. A. (2011). *Secure RBAC with dynamic, efficient, & usable DSD*. Ph.D. Institute for Information Processing and Microprocessor Technology, Johannes Kepler University (JKU).
8. Aftab, M. U., Nisar, A., Asif, M., Ashraf, A., & Gill, B. (2013). RBAC architectural design issues in institutions collaborative environment. *International Journal of Computer Science Issues, 10*, 216–221.
9. Gupta, M., & Sandhu, R. (2016). The GURAG administrative model for user and group attribute assignment. In *Proceedings of NSS* (pp. 318–332). Springer.
10. Sandhu, R. S., Coyne, E. J., Feinstein, H. L., & Youman, C. E. (1996). Role-based access control models. *IEEE Computer, 29*(2), 38–47.
11. Incits, A. (2004). Incits 359-2004, role based access control. American National Standard for Information Technology.
12. Sahafizadeh, E., & Parsa, S. (2010). Survey on access control models. In *Future Computer and Communication (ICFCC), 2010 2nd International Conference on 2010* (Vol. 1, pp. V1-1–V1-3).
13. Li, F.-H., Su, M., Shi, G.-Z., & Ma, J.-F. (2012). Research status and development trends of access control model. *Tien Tzu Hsueh Pao/Acta Electronica Sinica, 40*(4), 805–813.

14. Habib, M. A., & Praher, C. (2009). Object based dynamic separation of duty in RBAC. In *International Conference for Internet Technology and Secured Transactions 2009* (pp. 1–5).
15. Ferraiolo, D., Kuhn, R., & Chandramouli, R. (2003). *Role based access control: Artech House*.
16. Biswas, A. K., & Nandy, S. K. (2015). *Nano communication networks* (Vol. 7, pp. 46–64).
17. Kuhn, R., Coyne, E. J., & Weil, T. R. (2010). Adding attributes to role-based access control. *Computer, 43,* 79–81.
18. Apache Sentry. http://sentry.apache.org/.
19. Apache Ranger. http://ranger.apache.org/.
20. White, T. (2012). *Hadoop: The definitive guide*. O'Reilly.
21. McCollum, C. J., Messing, J. R., & Notargiacomo, L. (1990). *Beyond the pale of MAC and DAC-defining new forms of access control*, Oakland, CA, May 7–9, 1990.

Defect Detection in Oil and Gas Pipeline: A Machine Learning Application

Gitanjali Chhotaray and Anupam Kulshreshtha

Abstract Being an extremely powerful Artificial Intelligence tool, Machine learning algorithms enable the modern analytics applications to spot patterns in petabytes of data and provide robust models. These models, trained with the historical data and relationship, help identifying complicated defect patterns hidden in the large datasets and in turn, facilitate high prediction accuracy. The application of machine learning in defect identification process of oil and gas pipelines makes it better and simpler without missing any of the actual defects. Defects/leakage in the Oil & Gas pipeline may result severe losses to people's lives or public safety. Identifying such defect in oil and gas pipeline are quite tedious and time-consuming process. This paper presents an automatic defect detection and classification methodology for the safety of the pipeline. Pipeline conditions are captured by a sensor based device called PIG (pipeline inspection Gauge). Identifying defects, from the huge amount of sensor data generated, is a tricky and extremely complex activity. Since the activity involve huge amount of computation, we opted for parallel processing of this data, providing results in considerable lesser time. The classification task using machine learning algorithms successfully remove over 90% of false alarms. This paper provides a unique approach to handle huge sensor data, process it successfully and build machine learning based algorithm for defect detection.

Keywords Oil & gas · Machine learning · Pipeline defect · Ovality

G. Chhotaray (✉) · A. Kulshreshtha
Analytics and Insights, Tata Consultancy Services, Bangalore 560066, Karnataka, India
e-mail: gitanjali.chhotaray@tcs.com

A. Kulshreshtha
e-mail: anupam.kulshreshtha@tcs.com

© Springer Nature Singapore Pte Ltd. 2019   177
V. E. Balas et al. (eds.), *Data Management, Analytics and Innovation*,
Advances in Intelligent Systems and Computing 839,
https://doi.org/10.1007/978-981-13-1274-8_14

1   Introduction

The computational analysis of machine learning algorithms helps in creating a generic model which can work well on new unexplored examples with minimal chance of wrong classification.

Pipelines, one of the most economical modes of transport for oil and gas, suffer from metal loss, shape change or holes, etc., due to usage conditions such as aging, thermal expansion and contraction, human inflicted damages or natural reasons. Timely and accurate monitoring information about such defects is necessary for the maintenance of pipelines and prevention of accidents. As missing any of such defects is extremely dangerous, traditional monitoring processes are quite expensive, effort intensive, and time-consuming.

As Oil and Gas Pipelines are long and connecting across countries, it is very difficult to inspect the pipeline physically and identifying the defects. As a standard procedure, the inspection activity is carried out using a sensor based tool called PIG (pipeline inspection gauge) which is inserted in the pipeline at the time of inspection. The tool records pipe diameter at each 0.02 mm (or less distance) on pipeline. The tool generates huge amount of data for each inspection, however the actual defects are very less in number. The proportion mismatch makes the defect identification extremely tedious and complex.

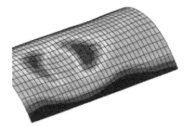

Pipeline Dent

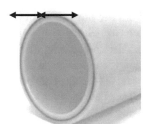

Pipe line Ovality

1.1   Business Challenges

The complexity of modern system and its maintenance is highly dependent on sensor technology which helps to act upon unexpected events. Many systems, still operated manually, requires large number of engineers and analysts to control the system and identifying the defects on the system.

The current system generates vast measurement data using sensors, manual data interpretation is time-consuming and quite challenging.

Another challenge was there are huge false alarms generated from the existing legacy software system. A team of analysts use an image-based software where they go through each and every defect generated from existing legacy system and segregate the actual defects from false defects. The whole process of detecting the defect was tedious and quite expensive too.

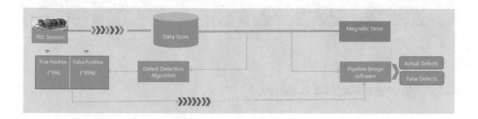

2 Machine Learning (ML) for Reduction of False Alarms

Machine learning algorithms are able to detect the patterns from the huge data without being explicitly instructed. It uncovers the hidden insights through continuous learning and trends in the data. As sensor data are huge and quite imbalanced like events are very less as compared to non-events, an ML approach is used to make the data balanced and train the model on balanced data to improve the true positives.

Processing sensor data intelligently and use of robust machine learning approach can make the task simpler and can save time, money by reducing huge false defects, not missing any actual defects which can enhance the safety, productivity, and life of the assets.

As the current system generates huge sensor data through PIG and spotting patterns inside huge data is really difficult. Only machine learning algorithms will work better with huge data and can provide robust model in terms of accurate prediction in less time [1–3].

3 The Process Framework

3.1 Analytic Solution Steps

1. Aggregation mechanism
2. Data Balancing
3. Application of Machine learning algorithms
4. Model Validation and Accuracy.

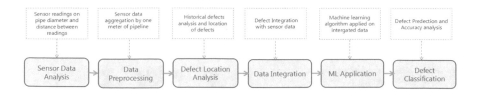

3.2   Aggregation Mechanism

As sensors exponentially increase the volume the data, it is required to process these data and organized in a suitable manner in order to act on it. Sensor readings from different pipelines are collected separately, processed based on the nature of defects and then combined into a single data unit for in depth analysis like descriptive, diagnostic, predictive, and prescriptive analytics. Here different pipelines means pipe lines with different diameter like 42, 60 and 48 in. pipes, etc. Number of sensors use in the pig varies according to inch. For example, PIG with 42 sensors will be used for 42 in. pipe and PIG having 60 sensors will be used for 60 in. pipes.

Data from 26 different inspections are used for analysis. Sensor readings were collected within 0.02 mm distance on the pipeline. In these huge data few defects were present. Hence aggregation of the data is required to maintain the balance between events (defects) and non-events. Data within one meter of pipeline were aggregated into a single record/observation. For example, in 100 m pipe we should get 100 rows of data.

max(sensor 1), max(sensor2) …, max(sensor n) # In Case of Dent
min(sensor 1), min(sensor2) …, min(sensor n) # In Case of Offtake

After aggregating the data separately on different runs, we combine all separate data sets into a single data set. Then we create the dependent variable in the data set, which is the target of prediction.

We look at the historical reports and find out the defect location and identify the same location in the data set and create the events.

3.3   Data Balancing

Imbalance data refers to a problem of rare events or unequal distribution of target variable which can produce biased models. The challenge of learning from imbalanced data is that defect class cannot draw equal attention to the learning algorithm compared to the non-defect class, which often leads to very specific classification rules or missing rules for the event class without much generalization ability for future prediction and machine learning algorithms may produce unsatisfactory classifiers which will be biased towards majority class. An over-sampling and under-sampling approach is followed in order to derive a balance data set from the existing imbalance data set by adding few events to the data or removing some of the non-events from the data. As the recognition of event class is more important, because miss-classifying an example from the event class is usually more costly.

After aggregating the data and combining it into a single data set, we look at the event percentage in the data set, which, in our experiment, was hardly 0.01% of total data. There should be some significant percentage of events in your data to run any prediction algorithm, otherwise result will be biased.

We split the data randomly into two sets, training, and validation. Training set contains 70% of the data and testing set contains 30% of the data, but it is not mandatory to keep this ratio always. The ratio can be changed, based on the data availability. Both data sets are quite imbalanced. As we train the model in training data set, we need to make it balanced. It is not required for the test data set. We use OVUN method to increase the number of events and reduce the non-events.

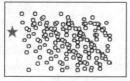   ★ - Event ○ - Non-event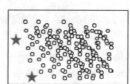

Imbalance data set   Balanced data set

3.4 Application of Machine Learning Algorithms

The main objective of the machine learning algorithm is to improve the performance of the prediction by selecting the best parameter and increase the number of iterations as per user's choice.

In the model building step we use more than 10 machine learning algorithms to select the best model. We experimented with Support Vector Machine, Gradient Boosting Machine, Random Forest, Decision tree, Linear Discriminate Analysis, Artificial Neural Network, etc. with 10 fold cross validation on the training data set. Support vector machine(**SVM**) and gradient boosting Machine(**GBM**) with 10-fold cross-validation result with high accuracy and no over-fitting. We fine tuned the model by running several iterations and changing the parameter value.

Support Vector Machine uses the hyper planes and look for the maximum margin hyper plane(MMH) to separate the classes. The data point labeled as defects are having either maximum values or minimum values compare to other data points. Those data points are clearly separated by the MMH.

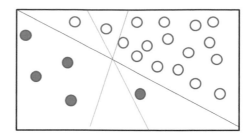

In the above figure open circles are non defects and filled circles are defects. We can see there are three hyper planes and the diagonal line separates the filled circles and open circles very clearly, where as other two hyper planes miss-classified one defect. We can say the diagonal line is the MMH.

3.5   Model Validation and Accuracy

Model is validated against the validation sample for its goodness of fit and checked for the accuracy. There is a huge reduction of false positive observed after applying machine learning techniques and none of the actual defects are missed which indicates 100% true positive rate.

Below figures are confusion matrix for three prevalent defects, i.e., Dent, Offtake and Ovalities.

Confusion Matrix—Dent
All the true dents are detected as dents by our approach, and less than 1% of non-dents are detected as dents. Able to reduce false alarms by >95%.

Actual	Predicted	
	No defect	Defect
No defect	8403	75
Defect	0	87

Confusion Matrix—Offtake
All the true offtakes are detected as offtakes by our approach, and less than 1% of no-offtakes are detected as offtakes. Able to reduce false alarms by >95%.

Actual	Predicted	
	No defect	Defect
No defect	5466	7
Defect	0	24

Confusion Matrix—Ovality

All the true ovalities are detected as ovalities by our approach, and less than 1% of no-ovality are detected as ovalities. Able to reduce false alarms by >95%.

Actual	Predicted	
	No defect	Defect
No defect	72,347	36
Defect	0	7

4   Operational Efficiency

Huge sensor data processing in Hadoop clusters and machine learning algorithms deployed across big data management system, predicts the defect most accurately and minimize the time and cost as well. Sensor data are aggregated within one meter of pipeline depending on the type of the defect. This task is paralleled across nodes to reduce the processing time. Machine learning algorithms are deployed on a separate node which is applied to processed data to build the model and validated as well.

New information from sensors are collected and processed, then stored in data server where machine learning models and rules are applied instantly on the processed sensor information to check for anomalies and observe the values of impacting parameters. Controlling the parameters or corrective actions can be taken to avoid the failure or sudden disasters.

5   Conclusion

In this paper we propose sensor data analysis and machine learning algorithms for detecting defects on oil and gas pipelines which significantly reduce the false defect generation and while not missing out on any of the actual defects. Parallel processing of sensor data in the distributed environment system speed up the data processing task.

We use machine learning algorithms on the processed sensor data to predict the undesired events or defects in the unit. Accuracy of prediction improves a lot by using of machine learning and reduce huge false alarms.

The industry recognizes this problem to be one of the most complex and tedious ones, having extremely high security and safety related concerns. Our approach will ensure the industry safety standards while significantly reduce the cost of analysis, thus enhancing the system efficiency.

References

1. Bishop, C. M. (2006). *Pattern recognition and machine learning, information science and statistics*. Heidelberg: Springer.
2. Michie, D., Spiegelhalter, D. J., & Taylor, C. C. (1994). *Machine learning, neural and statistical classification*. Chichester: Ellis Horwood Limited.
3. Marsland, S. Machine Learning: An Algorithmic Perspective.

Measuring Individual Performance with the Help of Big Data—IPES Framework

Manisha Kulkarni, Anushka Kulkarni and Abhay Kulkarni

Abstract Big Data has already started to infuse the field of performance measurement and management. With data being referred to as the 'New oil of the twenty-first century' and organisations are striving for utilising this to their advantage, the notion of Big Data has become a symbol for today's decision makers. Despite the many positive aspects associated with Big Data, there is a need for a sound, critical perspective. The discussions are based on the 3 Vs regarded as the core definition of Big Data (Volume, Variety and Velocity) and illustrated by new concept. The volumes, variety and velocity of both structured and unstructured data generated by the Internet, social media, and the 'Internet of Things' (IoT) are unfolding massive innovation opportunities. There is a need to develop capabilities to understand and interpret Big Data to take advantage of the opportunities it provides. This paper has taken bold step to create a basic framework to measure individual performance on totally new perspective which is building over the famous concept created by Dr. Norton & Dr. Kaplan, 'Balanced Scorecard'. This new concept and framework talks about mapping 'Emotional Quotients'—i.e. Intelligence, Physical, Emotional and most important is Spiritual (IPES). This IPES framework which this paper is trying to establish so that in future a IT framework, Software and artificial intelligence & machine learning with Neural network can be used to measure and analyse these IPES quotients to benefit individuals in organisations. Measurement of Physical quotient is relatively possible as it has certain

M. Kulkarni (✉) · A. Kulkarni
IICMR, HS-2, Sector 27A, Behind Tukaram Garden, Pradhikaran,
Nigdi, Pune 411044, Maharashtra, India
e-mail: kulkarni.iicmr@gmail.com

A. Kulkarni
e-mail: abhaykulkarni2@gmail.com

A. Kulkarni
Cummins College of Engineering for Women, Pune 411052, India
e-mail: anucool1798@gmail.com

A. Kulkarni
IICMR, Adishree, 2/99, Pawana Nagar, Behind Tejas Health Centre,
Chinchwad, Pune 411033, Maharashtra, India

© Springer Nature Singapore Pte Ltd. 2019
V. E. Balas et al. (eds.), *Data Management, Analytics and Innovation*,
Advances in Intelligent Systems and Computing 839,
https://doi.org/10.1007/978-981-13-1274-8_15

185

dimensions and conditions but other three factors are relatively tough to measure and put yardsticks around. Thus, authors are exploring potential of an individual by mapping/measuring as is IPES quotients and will like to measure the performance in the organisation by using IT tools and techniques.

Keywords Artificial intelligence · Balanced Scorecard · BSC · Emotional Quotients · Intelligence · Physical · Spiritual · IPES framework IT · Machine learning · Measurement · Neural network

1 Introduction

1.1 Matter of Study

It is observed that an individual employee follows the goals, targets and measurements cascaded to him/her from organisation level strategy and scorecard. These are enforced directives for an individual and may not be taken voluntarily. Organisations should encourage employees to take goals, targets and measurements which are self-motivated. This will be important for both individual and organisation as the desired goals and targets will be met with inner drive and simultaneously ensure personal satisfaction.

1.2 Important Topic for Further Study

It is observed that during the deployment, implementation of the assigned goals and targets by individual employee, there is little personal commitment seen and mostly 'out of necessity to continue the job'. This study looks at how to increase personal involvement and commitment in achieving goals and targets assigned. An individual need to achieve the goals and targets for the organisation but at the same time these should help in achieving personal targets and satisfaction. This necessitates to look at the current scorecard approach.

1.3 Knowledge About This Topic

We already know about various approaches such as Balanced Scorecard, Hoshin Kanri, Management point—Check point, etc. which are in use by many organisations for deploying their strategy to the lower levels. These approaches are used for long time and well-proven. But these approaches are mostly one-sided and looked from organisations point of view. While 'employee' is considered somewhat under

'Human Resource' and 'Learning and Growth' perspectives but the inclusion is limited. Employee personal growth to higher levels is not taken into account.

1.4 This Study Shows New Ways of Understanding

This study talks about moving from 'Financial, Customer, Internal Process and Learning & Growth (L & G)' perspectives to 'Intelligence, Physical, Emotional and most important is Spiritual (IPES)' perspectives. It is also looking at the personal involvement to help self and organisation to achieve goals and targets. We are suggesting herewith new way of using scorecard approach with new perspectives.

1.5 Literature Review

Dr. Robert Kaplan and Dr. David Norton, the creators of 'Balanced Scorecard' came out with the concept stating that most of the organisations were measuring their performance on the basis of top-driven numbers, i.e. volumes, numbers, sales, etc. which were helping organisations to make profits and sometime losses too. This type of tracking of performance was basically focused on the shareholders and investors. Such organisations were ignoring the vital parts of the organisations and most important stakeholders such as employees, society at large and customers. The measurement of performance was lope-sided and only focused on numbers. Such performance tracking results in corner cutting of processes, employee well-being and learning and development, less focus on society and other stakeholders. Such organisations could survive till the markets were driven by their products and services. But when the market changed or customer wants and desires changed, these organisations suffered. These organisations had no answers to the market change and why they are making losses.

Balanced Scorecard (Fig. 1) brought the fresh perspective to the organisations and showed them the way to manage performance in all areas of business. Over the period, the four perspectives—'Financial, Customer, Internal Process and Learning & Growth (L & G) of organisation' became popular and organisations learnt the new way of performance measurement and all-round improvement. Balanced Scorecard after successful implementation brought cultural change in most of the organisations.

It must be noted that one of the important role of the Balanced Scorecard (BSC) is to translate organisational strategy into more simple and implementable initiatives and action plans which has base of clear and accurate understanding of the organisations culture, mission, vision, strategy, goals and objectives, measurable action plans. Please refer Fig. 2 for this.

While organisations started using BSC for strategy cascade and performance measurement, this remained on the organisational levels. Many organisations found

Traditional BSC model

Balanced Scorecard provides a framework to translate strategy into operational terms.

- Strategy is 'described' using four 'perspectives
- Cause and effect is a key element
- Measures are developed to monitor performance

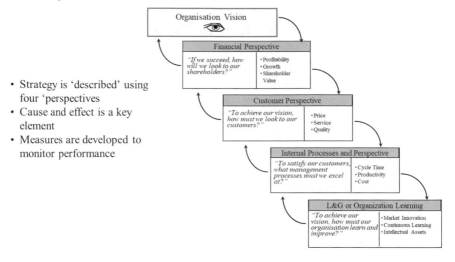

Fig. 1 Concept of traditional Balanced Scorecard

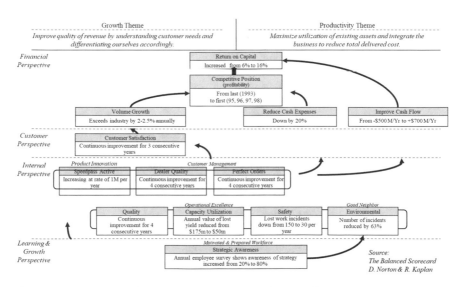

Fig. 2 Typical example of Balanced Scorecard and its 4 perspectives

way to measure individual employee performance measurement and management with the help of BSC cascade. Few organisations tweaked the BSC and used the relevant components of BSC for individual goal setting and direct alignment with

Strategic Theme: Operating Efficiency		Objectives	Measurement	Target	Initiative
Financial	Profitability, Fewer Planes, More Customers	• Profitability • More Customers • Fewer planes	• Market Value • Seat Revenue • Plane Lease Cost	• 30% CAGR • 20% CAGR • 5% CAGR	
Customer	Flight Is on Time, Lowest Prices	• Flight is on -time • Lowest prices	• FAA On Time Arrival Rating • Customer Ranking (Market Survey)	• #1 • #1	• Quality management • Customer loyalty program
Internal	Fast Ground Turnaround	• Fast ground turnaround	• On Ground Time • On-Time Departure	• 30 Minutes • 90%	• Cycle time optimization program
Learning	Ground Crew Alignment	• Ground crew alignment	• % Ground crew trained • % Ground crew stockholders	• yr. 1 -70% yr. 3 90% yr. 5 100%	• ESOP • Ground crew training

Source: Dr Kaplan and Dr Norton

Fig. 3 Balanced Scorecard example with objectives, measures, targets and initiatives

The Balanced Scorecard is part of a continuum of logic and action that translates a mission into desired outcomes

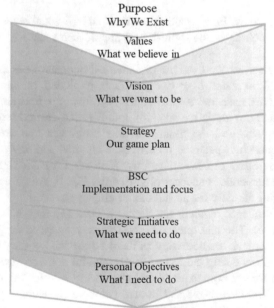

Purpose
Why We Exist

Values
What we believe in

Vision
What we want to be

Strategy
Our game plan

BSC
Implementation and focus

Strategic Initiatives
What we need to do

Personal Objectives
What I need to do

Strategic Outcomes

Satisfied Shareholders   Effective Processes
Delighted Customers   Motivated & Prepared Workforce

Fig. 4 Balanced Scorecard cascaded to individual level in the organisation

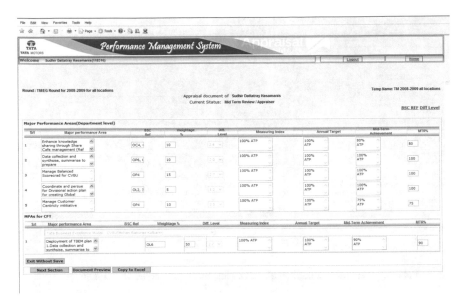

Fig. 5 Performance management system

organisational strategy. Few organisations derived individual level targets and goals from the BSC, looking at the right fit and appropriate measures for an individual level.

While Fig. 2 shows the example of Balanced Scorecard based on its four perspectives, Fig. 3 explains how one theme is exploded in all perspectives, its strategic objectives, measures to see that the objectives are tracked, targets taken for each measure and any initiative taken to complete the desired target.

Figure 4, explains how the organisational strategy gets converted into BSC and then it is cascaded into personal objectives. This methodology or framework has been adopted by many organisations. Tata Motors's CVBU (Commercial Vehicle Business Unit) has done it successfully. See the Fig. 5, which explains the direct connectivity with the BSc for an individual employee.

This whole sequence explains the Strategy, BSC and then how it gets cascaded to individual level. The real question arises when managers, despite of their hierarchical positions, lack the necessary skills to do the job (as they have grown over the period in the leadership ladder) more meaningful and less stressful. While they are responsible for getting results, they have no say in selecting their team. As L. Myland 1995, The supervisor and human resource management has stated that, 'The absence of power to make decisions in the human resource arena almost inevitably undermines the supervisor's responsibility and renders meaningless any attempt to praise, reward or get tough when things go wrong'.

2 Method

2.1 The New Framework—IPES for Measuring Individual Performance

2.1.1 Definition of Terms

Intelligence Quotient refers to the measured level of cognitive abilities such as problem solving and logical thinking that might be a factor in individual performance of employees.

Physical Quotient is a function of self-awareness at the most basic level. It's a function of how well one is attuned to his/her physical well-being, how one treats his/her mind, body and spirit.

Emotional Quotient refers to the measured emotional intelligence of an individual in understanding own emotion as well as understanding the emotions of other people.

Spiritual Quotient (Critical existential thinking, personal meaning production, transcendental awareness and conscious state expansion) refers to the measured spiritual abilities of a person such as being creative, insightful and believing in spiritual manifestation or religious or value-based beliefs as a guide in a way of life.

2.1.2 Intelligence Quotient (IQ)

Alfred Binet developed a method of measuring ability to learn and to detect children would face difficulties in the school system. He believed that intelligence increases through childhood. It was also believed that the rise in intelligence across childhood was not due to developments in sensory acuity or precision, nor was it a direct result of special education or training. The concept of Binet on IQ presents as IQ = Mental age/Chronological age × 100. Chronological age refers to how old is the child and Mental age refers to how old would the average child be who performed at this child's level of performance. Mental age is person's cognitive abilities relative to what others can do of different ages. Mental age is what one is capable of doing, scaled to what the average people of different ages can do. It means that, if one is capable of doing what an average 25-year old can do, then mental age is 25. An IQ score of 100 means average intelligence, that anything over 100 means advanced intelligence and that anything under 100 means delayed intelligence.

2.1.3 Physical Quotient (PQ)

Physical Quotient is inseparably linked to one's ability to perform to the best of his/ her potential. There isn't a single test of leadership and resolve that would not improve or made easier to navigate by having actively managed your work–life balance, health and well-being, and having an acute awareness of the PQ of those around you. In my own experience, it has been very instructive to observe the relationship between a leader's PQ and their resilience to events that stress them (tough times at work, personal issues, medical problems, the list is endless).

PQ is like any other resource in a biological system; it can be developed (saved up) and it can be eroded (spent). Developing PQ is analogous to training for a marathon. The more miles you grind out in training, the easier the race will be. So, it follows that the more PQ you can develop, the more you have available to spend when challenges face you. How able you are to react to those challenges is in no small part a function of how you treat your body over the long term.

2.1.4 Emotional Quotient (EQ)

Emotional Quotation (EQ) is more important than one's intelligence (IQ) in attaining success in their lives and careers. As individuals our success and the success of the profession today depend on our ability to read other people's signals and react appropriately to them.

Therefore, each one of us must develop the mature emotional intelligence skills required to better understand, empathise and negotiate with other people—particularly as the economy has become more global. Otherwise, success will elude us in our lives and careers.

'Your EQ is the level of your ability to understand other people, what motivates them and how to work cooperatively with them,' says Howard Gardner, the influential Harvard theorist. Five major categories of emotional intelligence skills are recognised by researchers in this area.

3 Understanding the Five Categories of Emotional Intelligence (EQ)

1. **Self-awareness**. The ability to recognise an emotion as it 'happens' is the key to your EQ. Developing self-awareness requires tuning into your true feelings. If you evaluate your emotions, you can manage them. The major elements of self-awareness are:

- Emotional awareness. Your ability to recognise your own emotions and their effects.
- Self-confidence. Sureness about your self-worth and capabilities.

2. **Self-regulation**. You often have little control over when you experience emotions. You can, however, have some say in how long an emotion will last by using a number of techniques to alleviate negative emotions such as anger, anxiety or depression. A few of these techniques include recasting a situation in a more positive light, taking a long walk and meditation or prayer. Self-regulation involves

- Self-control. Managing disruptive impulses.
- Trustworthiness. Maintaining standards of honesty and integrity.
- Conscientiousness. Taking responsibility for your own performance.
- Adaptability. Handling change; with flexibility.
- Innovation. Being open to new ideas.

3. **Motivation**. To motivate yourself for any achievement requires clear goals and a positive attitude. Although you may have a predisposition to either a positive or a negative attitude, you can with effort and practice learn to think more positively. If you catch negative thoughts as they occur, you can reframe them in more positive terms—which will help you achieve your goals. Motivation is made up of:

- Achievement drive. Your constant striving to improve or to meet a standard of excellence.
- Commitment. Aligning with the goals of the group or organisation.
- Initiative. Readying yourself to act on opportunities.
- Optimism. Pursuing goals persistently despite obstacles and setbacks.

4. **Empathy**. The ability to recognise how people feel is important to success in your life and career. The more skilful you are at discerning the feelings behind others' signals the better you can control the signals you send them. An empathetic person excels at:

- Service orientation. Anticipating, recognising and meeting clients' needs.
- Developing others. Sensing what others need to progress and bolstering their abilities.
- Leveraging diversity. Cultivating opportunities through diverse people.
- Political awareness. Reading a group's emotional currents and power relationships.
- Understanding others. Discerning the feelings behind the needs and wants of others.

5. **Social skills**. The development of good interpersonal skills is tantamount to success in your life and career. In today's always-connected world, everyone has immediate access to technical knowledge. Thus, 'people skills' are even more important now because you must possess a high EQ to better understand, empathise and negotiate with others in a global economy. Among the most useful skills are:

- Influence. Wielding effective persuasion tactics.

- Communication. Sending clear messages.
- Leadership. Inspiring and guiding groups and people.
- Change catalyst. Initiating or managing change.
- Conflict management. Understanding, negotiating and resolving disagreements.
- Building bonds. Nurturing instrumental relationships.
- Collaboration and cooperation. Working with others toward shared goals.
- Team capabilities. Creating group synergy in pursuing collective goals.

How well you do in your life and career is determined by both. IQ alone is not enough; EQ also matters. In fact, psychologists generally agree that among the ingredients for success, IQ counts for roughly 10% (at best 25%); the rest depends on everything else—including EQ.

4 Spiritual Quotient

Spiritual Quotient that measures the ability of a person to express, manifest and represent spiritual resources, values and properties to improve everyday performance Azizi and Zamaniyan (2013). In short, it is more on intuitive abilities and self-awareness thus; it will answer 'What person is or what I am' Selman et al. (2005), and Adversity Quotient® that measures the abilities of a person to respond positively in any adversities or difficulties experience in life and it also represents how well the individual deal and overcome the difficulties and the capacity to survive and conquer the challenges encountered along the way Huijuan (2009).

The term spiritual intelligence and spiritual quotient is mostly attributed to Danah Zohar and Ian Marshall based on their pioneering book, SQ: Connecting with Our Spiritual Intelligence. They describe SQ as 'our most fundamental intelligence. It is what we use to develop our capacity for meaning, vision and value. It allows us to dream and to strive. It underlies the things we believe in and the role our beliefs and values play in the actions that we take. Spiritual Intelligence explores how accessing our SQ helps us to live up to our potential for better, more satisfying lives.'

4.1 Three Aspects of Spirituality

Responsibility—You might have asked yourself what is your purpose in life or who are you responsible for. Thinking about these makes us realise that we should have a vision of how we are to spend our lives. After all, we are not here forever and we should make some form of contribution to the next generation.

Humility—When you think about it, we are just a speck in the universe. We are just one of the 7 billion people on Earth; just one among those 108 billion people who have ever lived. So, what makes us think that our existence is more important than others?

Happiness—The world has progressed and it has offered us convenience. But can we truly say that we are a lot happier now. What exactly are the things that make us happy? We all want to be happy. But how, exactly, do you go about it?

4.2 Developing Your SQ

When it comes to spirituality, it is quite common to divide people into two categories; believers and non-believers but this is a false categorisation because everyone believes in something. As an atheist, I personally believe that you do not need to believe in God, or read your Bible or Quran to lead a spiritually charged life. I believe that humans are built-in with spirituality within our core. We are spiritual beings, after all. Owen Water, in his book The Shift: The Revolution in Human Consciousness explains that there are six basic stages of human development followed by six spiritual stages. Once the basic stages are completed, you move into the spiritual tier and your progress becomes much more focused because, at that point, you have gained a sense of the reason for human existence.

Parallel to the growing interest in EI, there is an emerging interest in combining the constructs of spirituality and intelligence into spiritual intelligence (SI) (Emmons 2000a, b; Halama and Strizenec 2004; Noble 2000; Vaughan 2002; Zohar and Marshall 2000). There are many definitions and measures of spirituality (Elmer et al. 2003; Lopez and Snyder 2003; Stanard et al. 2000). And despite the overlap between religion and spirituality, there is general agreement on their distinction: religion is focused on the sacred within institutional organisations, and spirituality refers to experiential elements of meaning and transcendence (Worthington 2001). Elkins et al. (1988) identify several important dimensions of spirituality that include a sense of meaning and mission in life, a sense of sacredness of life, balanced appreciation of material values, and a vision for the betterment of the world. Friedman and MacDonald (2002) review many definitions of spirituality and identify several common themes. Based on these themes, spirituality can be defined as: (a) focus on ultimate meaning, (b) awareness and development of multiple levels of consciousness, (c) experience of the preciousness and sacredness of life, and (d) transcendence of self into a greater whole.

Spiritual quotient is going beyond your cognitive and emotional skills. It is acknowledging your mortality and thinking of what you could offer humanity. It is living in humility; bearing in mind that you are just a tiny compared to the vastness of the universe. A robust understanding of SQ motivates people to balance their work commitments, time with family and inner growth.

5   Discussion

5.1   IPES Balanced Scorecard

IPES Balanced Scorecard is a journey to explore inner self and unearth values, expectations, dreams and aspirations waiting for oneself. If one takes this journey as an individual seeker with stated objectives, it evolves a certain roadmap for one's dreams and aspirations converted into manageable and measurable milestones.

This IPES Balanced Scorecard (Fig. 10) will be very effective way for employees to achieve integrity and alignment between work and life. This can be used as an organisational and cultural change methodology as it can combine organisational goals too. As these goals are taken by the person on his own it will prove effective and that person or employee will strive to excel in the targets. There will be great commitment to self-responsibility.

By working on this concept is like getting an insight into oneself. This scorecard will allow people to manage their emotional and spiritual quotients and widen their horizons. It is one's moral duty and responsibility to develop oneself, for one's own good, for his family and colleagues, organisation and country.

This IPES Scorecard will be personal level coaching, an inner spiritual and emotional learning methodology. It will be good tool to do with IQ balance, emotional and spiritual quotients (intelligence), i.e. balance between the left-side brain and right-side brain. It is coaching yourself. Such implementation makes oneself richer and more fulfilling life. This integrated approach will pave the way for continuous improvement and development by using self-talent, self-learning and self-discipline.

It will be more beneficial for a person if PDCA cycle is coupled with this scorecard, to help in holding the benefits and further improvements. There will be no slide-back.

The factors involved in improving IPES quotients are given in Fig. 6.

Figure 7 shows the interdependence and hierarchy of IPES.

IPES actually creates a 'Life' quotient and it can be depicted as shown in Fig. 8.

Figure 9 shows the IPES quotients drives personal performance which will be detailed out in example of BSC (Fig. 10).

Intelligence	Physical	Emotional	Spiritual
Continuous, systematic and disciplines study and education. Application of learning. Cultivating self-awareness. Learning by teaching and doing.	Wise nutrition. Proper rest. Consistent exercises. Stress management. Relaxation.	Self-awareness. Personal motivation. Self-regulation. Social skills. Empathy.	Integrity, i.e., character building. Meaning, Purpose of life. Listen to inner voice. Self-enlightenment.

Fig. 6 Factors to improve IPES quotients

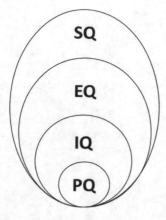

Fig. 7 IPES hierarchy shows the interdependence and hierarchy of IPES

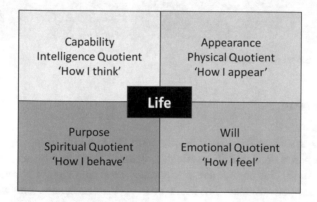

Fig. 8 IPES grid

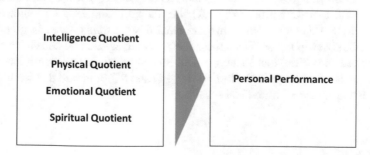

Fig. 9 IPES quotients drives personal performance

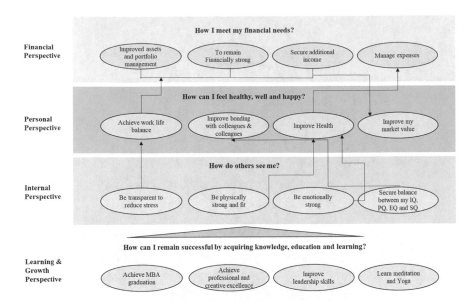

Fig. 10 IPES Balanced Scorecard

6 Conclusion

Based on the theory of Dr. Robert Kaplan and Dr. David Norton, the development of scorecard based on the IPES quotients will surely help in all-round development of a person. In regular assessment or appraisal process a person is reluctant to get assessed on the performance. But in the case of IPES scorecard, one will be happy to get assessed and improve as the goals are selected by that person and targets are also taken by that person.

While EQ deals with human side of life and how effectively and efficiently work in the environments we experience. IQ and EQ help us in present conditions but SQ helps in total transformation. When IQ, i.e. thoughts and EQ, i.e. emotions are governed by SQ, i.e. spirit, thoughts and emotions are transformed in quality and given a whole new purpose. Once these factors are analysed and correlated to each other, it becomes clear that mental, social and physical factors contribute to the holistic health of a person in the form of IQ, EQ and SQ. It is said that health of the society is dependent of individual health.

References

1. Villagonzalo, R. (2016, March). Intelligence quotient, emotional quotient, spiritual quotient, and adversity quotient® and the academic performance of students.

2. Vyas,P. Slideshare.
3. Dr Kaplan, & Dr Norton. The Balanced Scorecard: Translating Strategy into Action.
4. Amram, J. (Yosi). *Intelligence beyond IQ: The contribution of emotional and spiritual intelligences to effective business leadership*. Institute of Transpersonal Psychology.
5. www.lifeoptimiser.com.
6. The 8th Habit. www.ymresourcer.com.
7. https://www.peaklearning.com/documents/PEAK_GRI_Villagonzalo.pdf.
8. Scotland, K. Strategy deployment—https://availagility.co.uk/2016/02/05/what-is-strategy-deployment/.
9. Paul, P. Strategy deployment. https://peterpaul.com/capabilities/process-methods/strategy-deployment.
10. Goals and Goal setting. http://www.referenceforbusiness.com/management/Ex-Gov/Goals-and-Goal-Setting.html.
11. The role of personal purpose and personal goals in symbiotic visions. https://www.ncbi.nlm.nih.gov/pmc/articles/PMC4396129/.
12. Akers, M., & Porter, G. *What is Emotional Intelligence (EQ)*.
13. Goodfriend. W. *Methods of measuring intelligence: Interpreting IQ scores & score range*.
14. Umiya Career Development Centre. Multi Intelligence or Quotient, (IQ, EQ, CQ, PQ, AQ, SQ).
15. Rejolo, D. *Philosophy as a way of life: What is spiritual quotient?*.
16. Your Personal Balanced Scorecard, by http://open.lib.umn.edu.
17. Drucker, P. (1954). *The Practice of management*. New York: HarperCollins.
18. Manochio, M. (2008, September 30). http://www.dailyrecord.com/apps/pbcs.dll/article?AID=/20080930/COMMUNITIES12/809300311.
19. Wallace, C. *The personal balanced scorecard: A tool for your own end-of-year review*.
20. Elena Salazar, L. Creating your PBSC. (Personal Balanced Score Card).

Part III
Artificial Intelligence and Data Analysis

Consumer Engagement Pattern Analysis Leading to Improved Churn Analytics: An Approach for Telecom Industry

Amit Bharat

Abstract Telecom industry is ever evolving not only in terms of technology and services but also with new players entering the market, and providing consumers with attractive options. The churn in telecom has always been a concern for service providers. Churn may relate to prepaid or postpaid services. With postpaid services, the customer may voluntarily churn by raising a cancelation request, hence enabling the service provider to take retention action. However, in case of prepaid services, the customer may stop using the services abruptly without prior intimation to the service provider. In such cases, there is no way to retain, as the customer has already churned. The only way to handle such problem is by predictive modeling where the target variable is defined as customers who have churned in past, to train the model. However, the problem with such model is that they use "decline in usage" as independent variables while modeling. These variables are generally a weak indicator of churn. The main goal of this research is to provide a new mechanism to develop churn model by taking into account the shifts in activity patterns of service usages. In particular, we measured the customer activity by measuring the average length of inactive days and frequency of inactivity. By introducing such measurements of inactivity in defining the churn variable, we could: (1) Quantify churn risk for customers. (2) Attempt to retain the customers while they are still active.

Keywords Churn · Predictive modeling · Telecommunications

1 Introduction

Customer value management is important for better customer experience, prolonged relationships, assured revenues, and its growth. Churn is the last stage of the customer lifecycle but it may happen at any stage. Churn prevention is the most

A. Bharat (✉)
Tata Consultancy Services, Bangalore, India
e-mail: amit.bharat@tcs.com

© Springer Nature Singapore Pte Ltd. 2019
V. E. Balas et al. (eds.), *Data Management, Analytics and Innovation*,
Advances in Intelligent Systems and Computing 839,
https://doi.org/10.1007/978-981-13-1274-8_16

important stage of customer lifecycle not only because acquisition costs are much higher than retention costs but also because it leads to loss of perpetual revenue.

A customer may churn for numerous reasons, for example, service quality, signal quality, cost, network coverage, lack of add-on benefits, etc. A churning customer may end his service abruptly without showing any inherent pattern changes in his usage or he may slowly show signs of inactivity. The former is difficult to detect and model but latter can definitely provide sufficient indications of churn. However, decline in usage are not always related to churn. Because of such problem, the strong indicators of churn generally come out to be nonintuitive, for example, "age on network".

In this study, we try to achieve two complementary goals for effective churn management with a rather aggressive approach. First, we aim to analyze how the changes in the way we define churn targets, and also have an impact on target customers for retention. Second, we use the most suitable definition to best model the churning customers. The modeling is done with the available usage history and customer profiles. Prior to model building, the predefined customer segments are analyzed for utilizing the same set of models with different segment combinations. The customer segments were predefined on the basis of age on network and ARPU (Average Revenue Per User).

After having the appropriate target definition, data preparation, model building, and testing, the final model was deployed with marketing team to fetch them a list of customer targets for retention. Top three segments were picked for testing and campaigning. The model was then run every week for 8 weeks and risky customers were targeted with aggressive offers. Finally, the pre-post analysis was conducted which led to significant deduction in churn rate. In this paper, we have highlighted our approach with the support of various indicative figures, number, and results, which can help practitioners to leverage it for implementation perspective. Subsequent sections highlight literature reviews that encompass key research work in this area, end-to-end analytical methodology for practitioners, and results with analysis considering various outcomes and scenarios.

2   Literature Review

Literature reviews show that many researchers have worked in this area, and many frequently used techniques are cited along with various industry applications. For example, Abbas Al-Refaie in his paper about cluster analysis of customer churn in telecom industry [1], investigated the effect of switching cost, customer satisfaction, trust, communications, value-added services, customer expectations, complaint management systems, brand image, and price perception on customer churn. While such analysis is useful to understand the segments of customers at high risk, it neither provides actionable information nor is predictive in nature. For predictive modeling, researchers have tried basic as well as advanced algorithms such as logistic regression, decision trees, support vector machines, and neural network, and hybrid approaches such as combining clustering with artificial neural network [2]. Umayaparvathi et al. have attempted to solve the problem with a rather new and advanced machine learning techniques such as automated feature selection using deep learning to avoid time-consuming practice of feature engineering. By exploring further, Muhammad Azeem et al. have predicted prepaid churn using fuzzy classifier which provided reasonably good results for noisy data [3]. Researcher Hossam Faris et al. have tried revolutionary technique of genetic programming-based framework to build enhanced decision trees. Their results surpassed state-of-the-art classification methods [4]. Umayaparvathi et al. used deep learning for automated feature selection and predictive modeling [5]. Witold Gruszczyński1 et al. identified churning customer by building a social network by looking at customer's interaction among themselves [6].

While researchers have worked on a wide range of machine learning and heuristic-based techniques to predict churn, there has been limited discussion on defining the churn. The definition of churned customer even varies across geographies as well as network operators. Some define it as continuous inactivity of 15 days, some with 30 days, and some even with 60 days. This work attempts to take an entire different way of defining the churned customers by looking at the activity patterns which will be discussed in the next section.

3 Problem Description

The problem attempted to solve here consists of two different components: (1) Predict churn in advance. Meaning that the customer who are going to churn have to be identified earlier so that their retention can be attempted (2) Contact customer while he is in the system. Meaning that even if predictive model identifies set of churn risk customers, there should be sufficient time to reach out to them with attractive offers. Later, it is a more important aspect of the problem because the value of predictive model cannot be realized if it cannot be effectively implemented. In the next section, the methodology to solve such problem is described.

4 Proposed Methodology

The methodology followed consists of the following steps: (1) Data preparation, (2) Segment segregation, (3) Inactivity analysis, (4) Target customer definition, and (5) Model building.

4.1 Data

As described in Fig. 1, the data window considered for the exercise was of 6 months with 2 months for observation window and 1 month for prediction

	Month 1	Month 2	Month 3	Month 4	Month 5
Development Sample	Observation Window		Prediction Window	Validation Window	
Out of Time Sample		Observation Window		Prediction Window	Validation Window

Fig. 1 Observation, prediction, and validation windows for model development and for out-of-time validation

window. Another month was considered for validating the targeted customers. Similar data was created for out-of-time testing by considering an offset of 1 month for all the windows.

Data considered for model building is contained, and are as follows:

1. Call Data Records (CDR) of SMS, voice, and data usage.
2. Customer Profiles: Plan taken, date of activation, etc.
3. Secondary Information: Customer segments, recharges/top-up information.

4.2 Segments

Customers were classified into tight different segments based on ARPU and age on network as described in Fig. 2. For ease of this exercise, they were combined into three categories based on the similarity in usage and churn rates.

1. High: Ultra High, Very High.
2. Medium: High, Medium High, and Medium.
3. Low: Medium Low, Low, and Very Low.

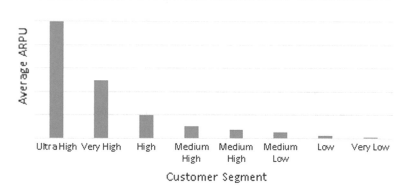

Fig. 2 Indicative graph of average ARPU for various customer segments

4.3 Inactivity–Activity Analysis

For analysis of customer activities spread across a month, two different variables were created based on the CDR data. For each day in a given month, for each type of service usage, SMS, voice, and data, a flag was created. Flag equals 1 when for a given day, usage of one or more of the services, i.e., either voice, SMS, or data is observed. Flag equals 0 when no activity is observed, i.e., the user did not use the phone for any activity. From these daily flags, two variables were created which indicates the customer's usage pattern in a given month:

1. Average length on inactivity in a month.
2. Count of inactivity events observed in a month.

For example, as shown in Table 1, the total inactive events observed are two. One that starts at day 2 and ends at day 4, and the other that starts at day 8 and ends at day 9. The average length of inactivity, in this case, is 2.5 days. Which is calculated as total inactive day, i.e., 5 divided by total inactivity events count, i.e., 2.

4.4 Target Variable Definition

Based on the average length of inactivity, a target variable was defined which was iterated for stability check for the purpose of validation.

Figure 3 indicates that of the customer who showed inactivity patter in a given month, only a proportion followed the similar or worse pattern in next month (represented by blue and redline, respectively). This study is useful in defining the churn target for modeling.

Some amount of inactivity is observed in almost 50% of the population. However, after a certain threshold, i.e., 7 as observed in this graph (indicative number), the population is consistent and 90% people maintain the same pattern.

Based on the various stability thresholds as observed in the graph above, the target variable was defined in the following way: A customer is considered at risk for churn if inactivity observed for him is greater than a chosen threshold. Such customers are expected to continue this pattern or go even more inactive in the next month with 90% probability.

4.5 Model Building

With target variable defined, the usual steps for model building and validation were conducted in the following order:

Table 1 Example of activity and inactivity events

	Day 1	Day 2	Day 3	Day 4	Day 5	Day 6	Day 7	Day 8	Day 9	Day 10
Activity flag	1	0	0	0	1	1	1	0	0	1
Activity status	Active	Inactive	Inactive	Inactive	Active	Active	Active	Inactive	Inactive	Active

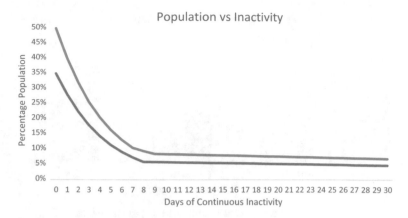

Fig. 3 Population versus inactivity

1. Variable creation: Various variables were created for the model building such as days since last activity, average revenue, usages in last 1, 2, 3 months, and last 7, 15 days, etc.
2. Splitting data as per the segment group as defined in sections above
3. Exploratory Data Analysis (EDA): EDA was conducted for finding association of target variable with variables created in step 1. It helped in choosing input variables for different models
4. Variable Selection: Information Value, Principal Component Analysis (PCA), and sufficiency analysis to pick the right variables
5. Sampling: Splitting the sampled data into train and test with 70% in test and 30% in train
6. Model building: Iterating over various models and input variables in various settings
7. Choosing the right model based on accuracy, specificity, sensitivity, Area Under Curve (AUC), gain chart, stability in train, test, out of time, etc.

5 Results and Discussion

The model developed using logistic regression gave a gain of 70% in the top decile for top segment. The model statistics were stable for training, validation, and out-of-time sample. The important variables came out to be: Inactivity days count in last 30 days, Inactivity days count in last 7 days, Average revenue in last 30 days, Average daily voice outgoing minutes in last 30 days, etc.

The model promised savings worth 0.1–0.5% of the total segment revenue by assuming a pessimistic response rate of 10% from targeted churners. To test such promising results, retention campaigns were planned with in collaboration with marketing team.

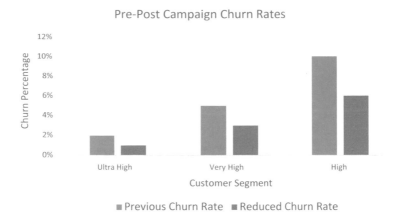

Fig. 4 Pre- versus post-churn rates for top segments

The variable creation methodology and model equations were shared with marketing team. The marketing team then ran weekly retention campaign for 8 weeks for three segments. After 8 weeks, the data was analyzed which included revenue, activity, and recharges/top up. It was then concluded that churn rate reduced by half after such campaigns in top segments.

In Fig. 4, the blue bar represents the churn rate prior to model deployment and orange bar represents churn rate after the model deployment. The churn rate reduced significantly in top segments, i.e., up to 50%.

To maintain the reduced churn rate, the model and hence, the campaigns have to be run frequently. The frequency may be decided by business or by keeping track of the churn rate every month. The type of offer given to customer will also have an impact on the retention rate. To match relevant offers to right set of customer is a problem in its own.

6 Conclusion

In this paper, we observed that using a nontraditional definition of target variable can make a lot of difference in retention strategies for a business. The campaign results were observed to be quite successful for top segments which contribute the most towards the business revenue. The approach observed in this paper can be used for other industries as well. The model selected for deployment was logistic, however, combining this approach with machine learning techniques might give even better results.

References

1. Al-Refaie, A. (2017). Cluster analysis of customer churn in telecom industry. In *World Academy of Science, Engineering and Technology International Journal of Social, Behavioral, Educational, Economic, Business and Industrial Engineering* (Vol: 11, No: 5).
2. Babu, S., & Ananthanarayanan, N. R. (2014). A review on customer Churn prediction in telecommunication using data mining techniques. *International Journal of Scientific Engineering and Research (IJSER)*. ISSN (Online): 2347–3878, Impact Factor (2014): 3.05
3. Azeem, M., Usman, M., & Fong, A. C. M. (2017). A churn prediction model for prepaid customers in telecom using fuzzy classifiers. *Telecommunication Systems, 66*(4), 603–614
4. Faris, H., Al-Shboul, B., & Ghatasheh, N. (2014). A genetic programming based framework for churn prediction in telecommunication industry. In D. Hwang, J. J. Jung, & N. T. Nguyen (Eds.), *Lecture Notes in Computer Science: Vol. 8733Computational collective intelligence. Technologies and applications.* ICCCI 2014. Cham: Springer
5. Umayaparvathi, V., & Iyakutti, K. (2017) Automated feature selection and Churn prediction using deep learning models. *International Research Journal of Engineering and Technology (IRJET), 04*(03), Mar-2017
6. Gruszczynski, W, & Arabas, P. (2016). Application of social network inferred data to Churn modeling in telecoms. *Journal of Telecommunications and Information Technology, 2*(2016), 77–86.

Writer-Independent Offline Signature Verification System

Malini Jayaraman and Surendra Babu Gadwala

Abstract Banking transactions are of multiple types and checks are commonly used method for business-to-business transactions. Checks are physical document of payment transfer authenticated by the signature of the account holder. To verify a check, banks manually compare the signature with signature template of the account holder. Signatures tend to have a variation based on the mood, health, etc., of the person; no two genuine signature of same person are identical. With advancement in technology, the forging of signatures have become more sophisticated. Due to these factors, the manual verification of a signature is very challenging. Also with increase in the volume of the transactions requiring verification, it has become a herculean task to manually check and process each signature leading to the need of an automated system to identify forged signatures with speed and accuracy. In this paper, we are proposing a classification methodology to automatically detect forged signatures from the genuine signatures with low FAR (False Acceptance Rate).

Keywords Offline signature verification · Random forest · Skilled forgery detection

1 Introduction

Security has always been a cause for concern and with advancement in technology, the easiness of breaking through a system has increased giving fraudsters access to very sensitive data. While different verification methods such as knowledge-based (passwords), token-based (such as ID, magnetic strips) are being used in industry for Identity and access management, and biometric-based signature verification

M. Jayaraman (✉) · S. B. Gadwala
Tata Consultancy Services, Bangalore, India
e-mail: malini.jayaraman@tcs.com

S. B. Gadwala
e-mail: surendra.5@tcs.com

© Springer Nature Singapore Pte Ltd. 2019   213
V. E. Balas et al. (eds.), *Data Management, Analytics and Innovation*,
Advances in Intelligent Systems and Computing 839,
https://doi.org/10.1007/978-981-13-1274-8_17

method remains as one of the widely accepted and commonly used. Signature has been used for a very long time to authenticate documents credibility. The ease at which it can be collected as well as its noninvasive nature has made signatures the most popularly used authentication metric in the banking industry. Checks, which are one of the common and widely used financial transaction methods, use signatures as the main verification metric. Though digital transactions have increased, checks still remain as the most commonly used method for business-to-business transactions. And as per a recent study, checks are the most targeted payment method by fraudsters with 71% of the participating companies experiencing actual/attempted fraud. To authenticate a check, traditionally a bank manually checks the signature on the check with the existing customer signature base.

With the help of high-level softwares and hardwares, the checks and signatures can now be printed and forged without much difficulty, leading to an increase in skill of the forgery. The forged signatures are also very near-to-original, making it highly difficult to tell apart from genuine signatures. Also unlike the other verification metrics, genuine signatures of a person tend to vary based on different aspects such as the environment, mood, and health of the individual. Two genuine signatures of the same person will rarely be the same. Due to both inbuilt variation in genuine signatures and high skill of fraudsters to forge signature, manual verification of signatures has become difficult, time-consuming, and inefficient. By forging signatures on checks, fraudsters have many times managed to bypass this manual checking process and swindle huge sum of money.

One of the ways to counter this is to create an automatic system that can read the signatures and identify forgeries based on its variation from the base of original signatures. There has been active research in this area. Based on the method of data capturing, signature verification system can be of two types: Offline and Online. Offline systems capture signatures on a paper and contain only the static information such as shape and structural information. Online systems capture signatures on an electronic tablet using pen and contain dynamic details such as the path and time information, pressure, speed, etc. Offline system does not need any additional hardware for data capture, and online signatures need additional hardware and software. Offline signatures are more difficult to analyze as it does not have dynamic information like the online signatures. But due to the relative simplicity in implementing it, the offline method is still the most preferred method. In this paper, we propose an automatic verification system for offline signatures.

2 Literature Review

There have been different approaches used to verify signatures. In all the methods, the hypothesis and the function remain the same. The hypothesis is that a given signature input X_Q of individual I does not come from the individual I. A similarity function S is used to measure the similarity between the input signature (X_Q) and the reference signature (X_t) of individual I based on the features extracted from both the signatures.

The hypothesis is either accepted or rejected by comparing this distance function with a threshold (t). The variation in the different signature matching approaches lies in the features extracted from the signatures and the similarity function.

Bansal et al. [1] proposed a contour matching algorithm for signature verification. Using a sample of eight genuine signatures, they read the interpersonal variation in genuine signature and the geometrical properties of the signature. A new signature was then verified on the basis of critical points position using a triangle matching algorithm. Ferrer et al. [2] extracted geometric features from the signatures and tested multiple classifiers such as such as hidden Markov models, Support Vector Machines, and Euclidean distance classifier. It was found that HMM worked better than SVM and Euclidean distance classifier. Drouhard et al. [3] used the directional Probability Density Function (PDF) of the signatures as the global feature and implemented a classification system using neural network. Baltzakis et al. [4] proposed a two-stage neural network classifier. Global, grid, and texture features were extracted from the signatures and fed into the first-level classifiers, which combined a neural network and Euclidean distance model. The results from the first-level classifier were fed into a radial base function neural network. Bajaj et al. [5] proposed a system based on global features, namely the horizontal, vertical projections, and the upper and lower envelopes of the signatures. A feed-forward neural network was used to classify the signature based on each of the input features. The output of the three classifiers was then combined using an ADALINE feed-forward neural network, which was used to classify these signatures. Ramachandra et al. [6] proposed a graph-based algorithm for signature verification. They developed a Signature Verification using Graph Matching and Cross-Validation Principle (SVGMC) algorithm which checks for a signature's similarity with genuine signature using a bipartite graph and calculating the minimum cost factor for complete matching is obtained. Then based on this, the Euclidean distance is measured. Kalera et al. [7] used statistical distance measures such as the Bayes classifier and the k-nearest neighbor (k-NN) classifier on both global and local features extracted from a signature. Jana et al. [8] used Euclidean distance between the texture and topological features as a measure of the similarity between a given signature and the template. Hanmandlu et al. [9] proposed a fuzzy model using the Takagi–Sugeno (TS) model with structural parameters taken into account for capturing the local variation in genuine signatures. Özgündüz et al. [10] extracted global, directional, and grid features from the signatures and used Support Vector Machine (SVM) to classify them into genuine and forged signatures.

The classification models mentioned above has been built as writer-dependent models where the distance function and threshold is writer dependent. Hence, one model is built for each individual. When a new signature comes, the features extracted are fed into the corresponding model for identifying the output class.

The other type of model is a writer-independent model. A single global model is trained and used for all the individuals in this approach. For this, the distance function and threshold are common across all individuals. Any incoming signatures need to be preprocessed and fed into this single model to identify the class of the signature. This will reduce the complexity involved in model implementation.

Santos et al. [11] proposed a neural network-based writer-independent approach. The features used by document experts were extracted from the signatures. The distance function was used to identify the variation of these features from the features of genuine signatures. These were then fed to a neural network to classify into genuine/forged. Oliveira et al. [12] proposed a writer-independent model where the partial decisions yielded by different SVM classifiers were combined, and then the ROC produced by different classifiers were combined using maximum likelihood analysis. The graphometric features such as slant and distribution of pixels are calculated for each signature. The feature of any incoming signature is then compared with the features of the set of genuine signatures. The distance function is used as a difference between the two features. Though these approaches have explored writer-independent models, the accuracy of these experimental results has not been very high. The neural network-based approach has a FAR of 15% for a skilled forgery. The SVM classifier model has a TRR of 5 and 2% FAR.

3   Problem Statement

The objective of this research is to build a writer-independent automatic signature verification system with low true rejection rate and False Acceptance Rate. The aim of the model is to reduce the manual verification and the error in the check verification. The proposed model would also be updated in real time to improve its accuracy.

4   Data Collection

The data for this study has been collected from 82 participants. A signature collection sheet having four signature boxes of size 3" * 1.2" is prepared. 82 participants were asked to sign in these four boxes. These signatures were then forged four times by a skilled artist. A sample of the collection sheet is shown below in Fig. 1.

a.Sample of genuine signature   b. Sample of forgedsignature sheet
collection sheet

Fig. 1 Sample of data collection

Fig. 2 Sample of final
digitalized signatures

a. Sample of genuine   b. Sample of forged signature
signature

Hence, a total of eight signatures per person (four genuine and four forged) was collected per person. All these sheets are scanned using a digital scanner and saved as PDF Files. The individual signatures are then cut using snipping tool and saved as PNG files (Fig. 2).

5 Definitions

5.1 Signature Forgery Types

Signature forgery is of majorly three types: Random forgery, unskilled forgery, and skilled forgery.

Random forgery is when the forger has no prior information about a person and randomly attempts to sign the person's signature on a document.

Unskilled forgery is when the forger has information about a person's signature but does not forge with precision.

Skilled forgery is when a person knows and can forge a person's signature with good precision. An example of these three types is shown in Fig. 3.

5.2 Model

The objective of a signature verification model is to classify a signature of a person as genuine or forged. It is a basic classification problem. According to Jain et al. [13], it can be formulated as Given a signature S_Q and claimed identity I, determine whether (S_Q, I) belongs to class W_1 or W_2 where W_1 indicates that the claim is true (genuine signature of person I) and W_2 indicates that the claim is false (forged

a. Original signature   b. Simple Forgery   c. Unskilled Forgery   d. Skilled Forgery

Fig. 3 Different types of forgery

signature of person I). Typically, the features extracted from signature $S_Q (X_Q)$ is compared with the features extracted from the base/reference signatures of I (X_t) from the database. Thus,

Function to classify signatures into genuine and forged.

$$(S_Q, I) \in \begin{cases} w_1 & \text{if } S(X_Q, X_t) \geq t, \\ w_2 & \text{otherwise} \end{cases} \tag{1}$$

where S is the function that determines the similarity between the features X_Q and X_t and t is the threshold for categorizing it into genuine signature.

5.3 Model Performance

The performance of the automatic signature verification system can be evaluated using different error metrics, namely, Type I and Type II errors. Type I or False Acceptance error is when a forged signature is classified by the system as a genuine signature. Type II or False Rejection error is when a genuine signature is classified by the system as a forged signature. The False Acceptance Rate (FAR) and the False Rejection Rate (FRR) can be controlled by varying the threshold, t. But there is as a trade-off: As FAR decreases, the FRR increases—i.e., when the cut-off is increased to prevent forged signatures being passed as genuine signatures, and more genuine signatures will be classified as forged signature. It has been mathematically represented by Jain et al. [13] as:

If X_t represents the features extracted from the base/reference signatures of I and X_Q is the features extracted from the input signature that needs to be validated, then the null hypothesis is:

H_o: Input X_Q does not come from the individual I.

Thus, when H_o is accepted, it means that the system classifies the input signature as forged and when H_o is rejected, the system classifies the input signature as genuine.

Type I error is when H_o is rejected even when it is true. (i.e.,) The signature X_Q is accepted as genuine signature of I even though it is forged signature, indicating that the threshold was low for this particular case. Type II error is when H_o is accepted even when it is false. (i.e.,) The signature X_Q is classified as fraudulent signature of I, even though it is a genuine signature, indicating that the threshold was high for this particular case. Hence, there is a trade-off between Type I and Type II errors.

Since the impact of a forged signature being classified into a genuine signature is much greater than a genuine signature being classified as forged, it is preferred that the FAR is optimized as low as possible.

6 Methodology

Signature verification systems consist of mainly three parts: Preprocessing, feature extraction, and model. Each of the steps is explained below.

6.1 Preprocessing

Due to scanning of the signature, there would be some noise introduced into the system. Also, there would variations of ink color and smoothness in the signatures. Hence, the signatures need to be preprocessed and standardized before giving it as an input to the model. The following steps have been used for converting all the images into standard format:

- **Binarization**: The signature color can vary based upon the pen used. It is necessary to convert all the signatures into monotone for easier processing. In binarization, the image is first converted into gray scale and then into binary scale. The final image contains only two types of color—black andddd white.
- **Smoothing**: Due to the shakiness of the individual's hand or the instability of the surface, there might be some disturbances in the signature contour. Smoothening technique tries to remove these disturbances and capture only the important patterns in the data. In this step, unconnected small dots/noises are also removed.
- **Thinning**: Due to pen variation, the thickness of the signature may vary. Using the thinning technique, a single pixel thick skeleton of the signature is obtained.
- **Cropping**: Out of the entire signing area, only a part of it might be occupied by a signature. It is required to remove the blank spaces so that the comparison between two different signatures is made only on the signatures and not on white spaces (Fig. 4).

6.2 Feature Extraction

Once the signatures have been preprocessed and standardized, the unique features of the signatures can be extracted. The features are of different types, namely:

1. *Global features*: The features of the signature as a whole is captured. This includes:

    i. Height, width, and area of the signature.

| Original | Binary | Smoothing & Thinning | Cropped |

Fig. 4 Sample of final digitalized signatures

Customer	File	Width	Height	Area	UpLi	LoLi	VCe	HCe	NSI	PSI	VSI	HSI	HtW	AtC	TtA	BtH	LtH	UtH	correla	contrast	dissim	homc	ASM	Energy
C101	G01	189	69	####	31	41	38	81	666	798	844	773	0.37	0.41	0.077	0.5942	0.594	0.5652	0.762	1874.757	7.352	0.97	0.851	0.92
C101	G02	217	79	####	25	42	40	102	619	757	817	723	0.36	0.409	0.057	0.5316	0.532	0.6962	0.758	1417.563	5.559	0.98	0.888	0.94
C101	G03	191	63	####	12	28	25	78	734	812	876	836	0.33	0.379	0.087	0.4444	0.444	0.8254	0.721	2303.308	9.033	0.96	0.839	0.92
C101	F01	157	70	####	20	37	35	69	692	795	827	778	0.45	0.474	0.088	0.5286	0.529	0.7286	0.737	2238.956	8.78	0.97	0.836	0.91
C102	G01	192	88	####	24	43	37	97	1095	1159	1221	1294	0.46	0.482	0.088	0.4886	0.489	0.7386	0.732	2336.691	9.163	0.96	0.831	0.91

Fig. 5 Sample of the extracted features

ii. Different height, width, and area ratios.
iii. Horizontal and vertical psition of projections limits (upper and lower).
iv. Horizontal and vertical center of projections.

2. *Slant features*: Score for Positive/Negative slants and Horizontal/Vertical slants.
3. *Texture Features*: Correlation, Contrast, Homogeneity, Angular Second Moment (ASM), and energy are found using Gray-Level Co-occurrence Matrix (GLCM).

A total of 23 features were extracted for each image. The snapshot of created dataset is shown in Fig. 5.

6.3 Classification Model

Once these features are extracted, the data transform from image to numeric. To measure the distance between input signature and genuine signature, we find the absolute difference of the different standardized features of the two signatures. As the genuine signature of same individual varies over time, standardization of the features helps in accounting for this variation. For this, the mean and standard deviation of each the feature is calculated individually from genuine dataset. This gives us the standardized value of all the features for each individual. Then for each genuine and forged feature, the following equation is applied.

Function to calculate the distance

$$x_{std} = \frac{(X - mean)}{std\ deviation} \tag{2}$$

This function identifies the distance of the given signature from the standardized signature. Ideally, the variation of the genuine signature from standardized base is lesser than the variation of a forged signature. This information is captured using the above formula. Once the distance has been calculated, we have trained a classification model to identify genuine and forged. For this, the data has been split into train and test. For training the model, 70% of the customer data, i.e., 57 customers' data (4 genuine and 4 forged) are used. The remaining 25 customers' data is used for testing the model. The 456 data points from train dataset is further split into 70–30 for modeling and verification. A random forest model is run on this

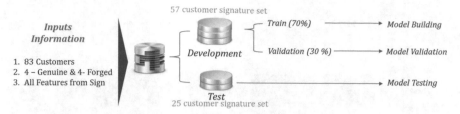

Fig. 6 Pictorial flow of classification model steps

70% (i.e., 320 data point dataset) and validated on the remaining 30% (136 data points). The model was then tested on the 25 customers' data (i.e., 200 data points). A model flowchart of the model is shown in Fig. 6.

7 Results

The fine-tuned random forest model was able to classify the validation sample set with 98.05% accuracy. The False Acceptance Rate (FAR) in this model as 2%. When the classification model was tested on the 25 new customers, n accuracy of 89.5% was obtained with a FAR of 4%. The results of the snapshots are shown in Fig. 7. As FAR poses a serious threat, we further tried to reduce the FAR of the model by introducing a cut-off for manual verification. Any signature having a probability of being a fraudulent signature in the cut-off range can be separated and manually checked. In this model, a cut-off for 0.45–0.65 when imposed for manual verification reduced the FAR to 0%. The signatures coming into this manual cut-off needs to be sent to an expert who will classify the signatures. These cases will again be fed into the model as a feedback to improve the model efficiency.

8 Model Implementation

Since the model is a writer-independent model, the single model can be hosted in a server and used for all incoming signatures. The only requirement is that a sample of genuine signatures needs to be available in the database.

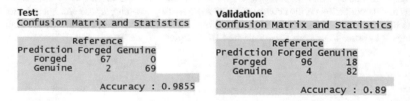

Fig. 7 Results of the snapshot showing accuracy on test and validation dataset

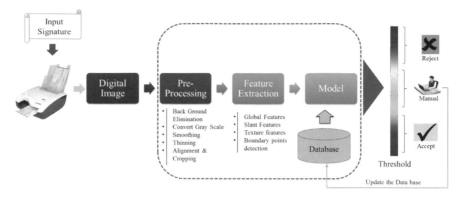

Fig. 8 Proposed architecture for implementation of the model

Once a signature is uploaded, basic preprocessing is performed and the features of the input signatures are extracted. From the database, the mean and standard deviation of genuine signatures of the corresponding individual is calculated. The distance of these features is then calculated using the mentioned distance function. Once the distance function is calculated for all the features, it is fed into the trained classification model which then provides a score. Based on this score, the signature is then labeled as genuine, forged, or sent for manual cut-off. Figure 8 shows the skeleton of the implementation plan.

9 Conclusion

In this paper, we have attempted to build an automatic system to detect fraudulent signatures. We have proposed a writer-independent classification model to detect forged signature. Our classification model has an accuracy of 89% with a FAR of 0 when a cut-off of 0.45–0.65 was introduced for manual verification. This research, though studied on offline signatures, can also be extended for online signatures. This system when implemented will help to reduce the manual effort in authenticating signatures in the industry such as banks and legal systems.

Acknowledgements It is our great pleasure to express our sincere thanks to all our colleagues of TATA Consultancy Services, Bangalore who has volunteered and helped us in our research in terms of data collection on the consent of using it purely for research purpose and not to be misuse these data points in any manner. We are also immensely grateful to our Infrastructure team for arranging logistics to convert these data into HQ images.

References

1. Bansal, A., Garg, D., & Gupta, A. (2008). A pattern matching classifier for offline signature verification. In *2008 ICETET'08 First International Conference on Emerging Trends in Engineering and Technology*. IEEE.
2. Ferrer, M. A., Alonso, J. B., & Travieso, C. M. (2005). Offline geometric parameters for automatic signature verification using fixed-point arithmetic. *IEEE Transactions on Pattern Analysis and Machine Intelligence, 27*(6), 993–997.
3. Drouhard, J.-P., Sabourin, R., & Godbout, M. (1996). A neural network approach to off-line signature verification using directional PDF. *Pattern Recognition, 29*(3), 415–424.
4. Baltzakis, H., & Papamarkos, N. (2001). A new signature verification technique based on a two-stage neural network classifier. *Engineering Applications of Artificial Intelligence, 14*(1), 95–103.
5. Bajaj, R., & Chaudhury, S. (1997). Signature verification using multiple neural classifiers. *Pattern Recognition, 30*(1), 1–7.
6. Ramachandra, A. C., Ravi, J., & Venugopal, K. R. (2009). Signature verification using graph matching and cross-validation principle. *International Journal of Recent Trends in Engineering, 1*(1).
7. Kalera, M. K., Srihari, S., & Aihua, X. (2004). Offline signature verification and identification using distance statistics. *International Journal of Pattern Recognition and Artificial Intelligence, 18*(07), 1339–1360.
8. Jana, R., Saha, R., & Datta, D. (2014). Offline signature verification using euclidian distance. *International Journal of Computer Science and Information Technologies, 5*(1), 707–710.
9. Hanmandlu, M., Yusof, M. H. M., & Madasu, V. K. (2005). Off-line signature verification and forgery detection using fuzzy modeling. *Pattern Recognition, 38*(3), 341–356.
10. Özgündüz, E., Şentürk, T., & Karslıgil, M. F. (2005). Off-line signature verification and recognition by support vector machine. In *2005 13th European Signal Processing Conference*. IEEE.
11. Santos, C. et al. (2004). An off-line signature verification method based on the questioned document expert's approach and a neural network classifier. In Ninth International Workshop on Frontiers in Handwriting Recognition, 2004. IWFHR-9 2004. IEEE.
12. Oliveira, L. S., Justino, E., & Sabourin, R. (2007). Off-line signature verification using writer-independent approach. In *International Joint Conference on Neural Networks, 2007*. IJCNN 2007. IEEE.
13. Jain, A. K., Ross, A., & Prabhakar, S. (2004). An introduction to biometric recognition. *IEEE Transactions on Circuits and Systems for Video Technology, 14*(1), 4–20.

Maximizing Pickup Efficiency and Utilization in Online Grocery: Two-Phase Heuristic Approach

Dharmender Yadav and Avneet Saxena

Abstract Online grocery shopping is getting popular in recent time due to convenient online shopping applications, doorstep delivery, and competitive discounts. Specially, this mode of shopping is getting more popularity in urban areas due to resulting saving consumer valuable time. Over the next decade, the demand for online grocery is expected to grow further at least fivefold. Major advantages of using online grocery are that consumer need not stand in long queue to buy grocery items, easy SKU selection and review process, book online their grocery items at discounted price and delivery at convenient time window. Due to this growing consumer demand for online grocery, many big retailers are attracted to invest and improve their services. However, with many aforesaid advantages, there are some challenges for retailers especially additional cost burden on portal management, picking of orders, delivery of orders, and managing seamless services. For instance, ordering online requires easy portal for ordering with the flexibility to cater different consumers requirements, competitive discounted, price, consumer service requirement, delivering product to consumer on given time window, additional workforce for picking SKUs, additional fleet for distribution, and return product mechanism. It has identified that efficient order picking is one of the major challenges where any improvement can provide a significant cost benefit to the company in form of resource minimization. In this paper, our focus is on efficient picking of SKUs. This will help the retailers in completing the pickup of items in minimum time, improving the overall efficiency in terms of picking the items in minimum time by the trolley, and also minimizing the number of buckets leading to reduced number of fleets. This solution methodology can be leveraged across retailer and has huge potential for future research.

Keywords Efficient picking · Online grocery · Closest cluster addition Clarke and Wright

D. Yadav (✉) · A. Saxena
TCS, Think Campus, Electronic City, Phase-2, Bangalore, India
e-mail: dharmender.yadav@tcs.com

A. Saxena
e-mail: avneet.saxena@tcs.com

© Springer Nature Singapore Pte Ltd. 2019
V. E. Balas et al. (eds.), *Data Management, Analytics and Innovation*,
Advances in Intelligent Systems and Computing 839,
https://doi.org/10.1007/978-981-13-1274-8_18

225

1 Introduction

With the emergence of the e-retailing or online grocery shopping, most of the key retail companies are focused on delivering differential value and signal a shift in the nature of retail competition through grabbing online consumers market. The process of online grocery ordering starts when customer places online order based on his grocery requirements. The order consists of basic details of each SKU, for example, quantity, weight/volume, and splitability in different buckets. It also contains where exactly the SKU can be found out. The position of SKU is defined in terms of aisle/ bay/shelf. Aisle is the row number as shown in Fig. 1 and aisle in the example is from A1 to A7. Within each aisle, there are subsections called bay. For example, in Fig. 1 and under aisle A1, there are four bays (B1, B2, B3, B4). Under each bay, there are height-wise multiple subsections called shelf. In Fig. 1, there are shelf S1 and S2. Each SKU is tagged with A/B/S so that it is easier to locate the SKU in warehouse.

Hence, once the customer's order is received, all the basic properties of the items belonging to that order (Splitability, temperature zone, A/B/S, item weight/volume, order quantity) are identified. If item is splittable, then it can be assigned to multiple buckets. First of all filtering of items is done based on their weight and volume. If the item weight/volume exceeds or equal to bucket's weight/volume, then such items are filtered out. Also if the item is non-splittable and such item's total volume/ weight (item order quantity multiplied by item weight/volume) exceeds buckets weight/volume, then such items are also filtered out. Further, items are divided into different temperature zones namely Ambient, Chilled, Freezer, and Security. All items which can be stored at normal temperature belong to Ambient and items that need to be kept in the temperature range of 4–8 °C belong to chilled category. Items which are kept at 0 or below 0 °C belong to Freezer category. Some items which are high value are usually kept at a specific place with more security. These items belong to security zone category. The next task is to assign items to buckets. However, allocation of items to buckets should be done in such a way that total pick

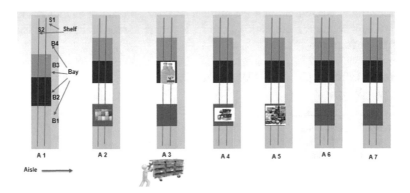

Fig. 1 SKU positioning in warehouse

time for each bucket is minimum and utilization of buckets is maximum and one bucket should never contain the items of another customer's order and if the item is not splittable, then item should be allocated only in one bucket. Once items are assigned to buckets, then buckets are allocated to trolley. Usually, a trolley can accommodate maximum of six buckets. The allocation to trolley has to be done in such a way that total trolley pick time is minimum. After the bucket assignment to trolley, the actual picking of items starts. RFID tag in buckets directs the next nearest item to be picked while moving the trolley. Once all the items are picked, the trolley is sent to fleet area where all the buckets will get loaded into vehicle for delivery to final customer.

2 Literature Review

Order picking optimization is vast area. So there are several strategy and levels at which order picking optimization can be done. It is further categorized into three parts. The first part is storage Policy [1]. Here, the focus is on finding the optimal location of all the items inside warehouse so that overall there is better picking efficiency. Next part is order consolidation policy. Here the key focus is on how to consolidate the customer's orders such that they are picked in minimum time to be precise, which orders to combine in a trolley so that overall trolley pick time is minimum [2, 3] and the last part is Routing policy. Routing policy is concerned with ordering a pick list in a sequence that will minimize the travel distance of an order picker.

Our focus in this paper is on order consolidation policy under order picking optimization (Fig. 2). There are three conventional approaches for solving order picking optimization problem. Priority rule-based algorithms consist of two phases. In the first stage, each customer is assigned a priority and in the second stage, customer orders are sequentially allocated to batches in the decreasing order of priority. The results obtained using next fit rule are usually used as a baseline solution [4]. The problem with priority-based approach is that being sequential in nature, it can miss some promising better solutions. Seed algorithms construct

Fig. 2 Classification of order picking optimization problem

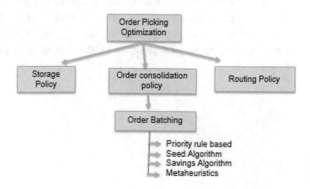

batches sequentially. A batch here is constructed in two stages. In the first stage, a seed order is selected using the seed selection rule. In the second stage, other orders get added to the particular batch according to the order selection rule until there is no order left which can be added to that particular batch without exceeding the picking device capacity [5]. The order selection rules are usually distance-based from the seed [6, 7]. Savings algorithms are usually based on Clarke and Wright savings algorithm for the vehicle routing problem [5, 8]. Customer orders which give the largest savings are added first to batch without violating the capacity constraint. Recalculation of savings is done every time as soon a customer order gets added to a batch and savings are recalculated.

3 Problem Description

In online grocery, out of many other challenges, order fulfillment in minimum time and with minimum operational cost is most important and challenging problem areas.

Order fulfillment in minimum picking time is needed to enable maximum number of order processing in limited time and limited available resources. So the problem is to identify the strategy of picking items in buckets in minimum time. But the problem is not limited to pick up orders in minimum time, because while doing this, we also need to ensure that the buckets are utilized up to maximum level considering the bucket weight and volume capacity so that lesser number of buckets are utilized to pick up all items of an order. Lesser number of buckets in picking an order will help in reducing the number of trollies to carry the bucket during the order picking process. Further, it will also help in reducing the vehicle trips because lesser number of buckets are being used and vehicle has limited capacity to carry the number of buckets. Some frequently used terminology is explained below.

Online Grocery: A system where customer can book grocery items online and get it delivered at doorstep. Order: A list of items ordered by a customer. Picking Efficiency: Time taken to pick up items of an order in warehouse. Trolley: A trolley is used to carry multiple buckets so that items located at different positions can be picked up easily within a warehouse. Bucket Utilization: Percentage of total bucket utilized by allocation items to bucket. Bucket utilization can be in terms of total bucket utilized volume and total bucket utilization weight. Individual pick score: Approximate time to pick an item in warehouse which is based on A/B/S location of item.

4 Our Analytical Approach

Our analytical solution consists of two-phase heuristic approach to improve overall item picking efficiency without compromising with standardizing order picking process. The first phase explains orders items to buckets allocation process and the second phase cites methodology to deal with bucket to trolley allocation. Following is a solution phase 1 where item to bucket or trolley allocation approach is highlighted.

4.1 Items to Bucket Allocation (Phase-1)

Items to bucket assignments problem have two contradictory objectives.

1. Maximize the picking efficiency.
2. Maximize utilization of buckets.

Picking efficiency will be higher if items which are closer are assigned in the same bucket. So it may happen that a solution having best picking efficiency can utilize more number of buckets. Similarly, it may also happen that in the lieu of best utilizing the buckets, the items which are apart are picked in same bucket leading to lower picking efficiency. It is an NP-hard optimization problem with complexity of approximately O (2^n). So it is very difficult to solve this problem using optimization modeling approach.

We have come up with a novel heuristic approach called closest cluster addition to solve this problem and also reduced the complexity to a great extent by developing a new multistage heuristic approach. We have divided this problem into two stages. Stage one is called bucket allocation and the second stage is called bucket minimization. In stage one, the objective is to assign the closest items to buckets in such a way that nearest items are placed in the same bucket if the capacity of buckets allows. For this assignment, individual pick score of each item is calculated which is based on the A/B/S location and assign one bucket to each item. After that, items which are placed closer inside the warehouse will have least difference between their individual pick score. Item to item individual pick score difference matrix is calculated and sorted in descending order and then one by one item-item pair is picked and assigned in buckets. This process continues until there is there is no item bucket left having only one item. There can be only one item in one bucket only if other items cannot be accommodated in such bucket due to capacity constraint. At the end of the first stage, closest items are assigned in buckets.

In the second stage, which is bucket minimization step, the buckets are minimized by completely assigning all items from one bucket to another bucket. For this, first of all, those item-item pairs are identified which are closest and are placed in different buckets. Such item-item pairs are identified and sorted based on minimum pick time difference. One by one item-item pair is picked up and their corresponding buckets are identified and then tried one bucket is to merge into another bucket if capacity permits.

At the end of this stage, what we get is the minimum possible number of buckets containing items closer to each other.

4.1.1 Items to Bucket Allocation: Closest Cluster Addition Heuristic

Let us assume that we have an order $O1$ which is having n different items ($i1, i2....$ in) under temperature zone as Ambient. Also, let us assume that we have already filtered out those items which cannot fit inside the buckets. Let $q_i, A_i, B_i, S_i, w_i,$ and v_i represent the item's order quantity, aisle location, bay location, shelf location, weight, and volume, respectively. Let BW and BV represent the maximum weight and volume capacity of bucket, respectively.

Step 1: Bucket Picking Efficiency Maximization

Step 1. Calculate total volume and weight of each item.

Let TV_i and TW_i denote total volume and the total weight of an item, respectively. Then it can be calculated as

$$TV_i = q_i v_i \quad \forall i \in O1$$
$$TW_i = q_i w_i \quad \forall i \in O1$$

Step 2. With the given input data, the first step is to calculate the individual pick score (IPS) for each item. It is actually the approximate relevant time taken to reach from starting location (Aisle 0) to a particular item in the warehouse.

Let one aisle be equivalent to x bay and y shelf values. So we need to do the standardization of these three values since items exact location is described by these three parameters. (For example, 1 Aisle is supposed to be equivalent to 10 bay value)

$$IPS_i = A_i + (B_i/x) + (S_i/y) \quad \forall i \in O1$$

(Since we do not have the actual distance/time taken, we are calculating approximated time taken by using this formula as approximation).

Step 3. Calculate item to item pick score matrix (without duplication). Let us denote it by PS_{ij}, where i represents ith item and j represents jth item.

It is the approximate relevant time taken that a trolley will take to reach from item i to item j. Here, we are taking the absolute value of the difference to remove negative value. It is calculated by the following formula:

$$PS_{ij} = Absolute(IPS_i - IPS_j) \quad \forall i, j \in O1$$

Step 4. Sort the PS_{ij} matrix in descending order.

Step 5. Assign to each item a bucket and calculate the remaining weight and volume for each bucket after assignment. So in this case if there are n items, then n ($b1, b2...bn$) corresponding buckets will be assigned to $i1, i2...$ in items, respectively. Let the remaining weight volume after assignment for each bucket be RBW_{bi} and RBV_{bi}, respectively. Then,

$$RBW_{bi} = BW - TW_i \quad \forall\, bi\, bi$$

and ch bucket is RBW (*bi*) by these three parameters. (Aisle is supposed to be equivalent to 10 of individual pick score of

$$RBV_{bi} = BV - TV_i \quad \forall\, bi$$

Step 6. Now iterate over all item-item pair one by one over the matrix PS_{ij} obtained in Step 4.

There can be three cases.

Case (1). Item *I* and item *j* both belong to different buckets and both buckets are having exactly one item. Case (2). Item *I* and item *j* both belong to different buckets and only one of the bucket is having exactly one item. Case (3). Item *I* and item *j* both belong to the same bucket. Case (4). Item *I* and item *j* both belong to different buckets and both buckets already have more than one item.

Now the following steps will be executed. Get the Item *I* and item *j* from matrix PS_{ij}

Identify the case by counting how many items are there in each bucket.

If case = case 1

Get the bucket details and remaining weight/volume of item *i*. RBW_{bi} and RBV_{bi}

Now allocate anyone item to another item's bucket and recalculate the remaining weight volume of bucket after allocation. If $RBW_{bi} - w_j$ or $RBV_{bi} - w_j < 0$

Then remove that item-item pair from PS_{ij} Move to next item-item pair

Else Assign item *j* to bucket *bi* and update remaining weight and volume of the bucket.

$$RBW_{bi} = RBW_{bi} - w_j \text{ and } RBV_{bi} = RBV_{bi} - w_j$$

Delete item-item pair from the PS_{ij}. Delete bucket *bj*

If case = case 2

Get the bucket details and remaining weight/volume of item *i*. RBW_{bi} and RBV_{bi}

Identify the bucket having exactly one item. (Say *bi*)

If $RBW_{bi} - w_j$ or $RBV_{bi} - w_j < 0$

Then remove that item-item pair from PS_{ij}

Move to next item-item pair

Else

Assign item *j* to bucket bi and update remaining weight and volume of the bucket.

$$RBW_{bi} = RBW_{bi} - w_j \text{ and } RBV_{bi} = RBV_{bi} - w_j$$

Delete item-item pair from the PS_{ij}. Delete bucket *bj*

If case = case 3, delete item-item pair from the PS_{ij}

If case = case 4, Move to next item-item pair

At the end of this step, we will be left with item-item pair referring only to case 4 and reduced number of bucket having items which are closest to each other.

This is the end of **Bucket Picking Efficiency Maximization step**.

After this, next stage begins.

4.1.2   Stage 2: Bucket Minimization Stage

Now we have reduced number of buckets and

PS_{ij} Matrix having item-item pairs referring only to case 4.

Now iterate over all remaining item-item pair one by one over the matrix PS_{ij}

Start from the first item-item pair and for each item-item pair, do the following

Get the bucket details and remaining weight/volume of item i and item j.

Let item i is in bucket bi and item j in bucket bj

If $RBW_{bi} + RBW_{bj} > BW$ or $RBV_{bi} + RBV_{bj} > BV$

Then remove that item-item pair from PS_{ij}. Move to next item-item pair

Else

Assign all items of bucket bj to bucket bi and update remaining weight/volume of the bucket bi.

$$RBW_{bi} = BW - RBW_{bi} - RBW_{bj} \text{ and } RBV_{bi} = BV - RBV_{bi} - RBV_{bj}$$

Delete item-item pair from the PS_{ij} and delete bucket bj

Repeat this until the last row (item-item pair) in the PS_{ij} matrix.

At the end of this step, we will get the minimum possible number of buckets to be utilized.

Also, calculate the pick score of each bucket. Let PSB_k represent pick score of bucket k. It is basically the approximate time a bucket will take to collect all items assigned to it. To calculate the pick score, first all the items assigned to a bucket are sorted in order of increasing individual pick score. Then the first item in that order is given 0 as pick score assuming the time will start after we pick the first item in bucket. Then in the sorted list, first item to next item pick score is calculated by subtracting second item individual pick score from first item individual pick score. This difference represents the approximate time to reach from the first item to the second item. Similarly, the second item to the third item pick score is calculated and so on till there is a last item in the bucket.

Suppose there are n items in bucket k then, $bk = \{i1, i2, \ldots in\}$

Total pick score is the summation of all such pick score calculated above.

$$PSB_k = PSD_{i1,i2} + PSD_{i2,i3} + \cdots + PSD_{in-1,in}$$

where $PSD_{i,i+1}$ is the time taken for going from item I to next nearest item $i + 1$ assigned in bucket and PSB_k is the total pick score of kth bucket, and PSD_{ij} is calculated using the formula below.

$$PSD_{i,i+1} = IPS_{i+1} - IPS_i$$

Items to Bucket Allocation: Mathematical Formulation

For the same problem mentioned in Sect. 4.1, we have also proposed a mathematical optimization model with the objective to minimize total number of bucket count. This model will be used to compare the results of closest cluster heuristic.

Mathematical model:

Indices:

i denotes set of items, where $i = i_1, i_2 \ldots i_n$, j denotes alias of index i, k denotes set of buckets

Parameters

W_i, V_i, S_i, C_i, are the weight, volume, splitability, and count of ith item
BV is the volume of the bucket and BW is the weight of the bucket

Decision Variables

X_{ik} is the number of quantities of ith item to be allocated to kth bucket (integer)

Y_k Bucket used (Binary), Z Bucket count (>0)

Objective function

ZM where M is very large number

Constraints

(1) $Z< = \sum_k Y_k$
(2) $\sum_k X_{ik} = C_i \quad \forall i$
(3) $\sum_i X_{ik} W_i = Y_k BW \quad \forall k$
(4) $\sum_i X_{ik} V_i = Y_k BV \quad \forall k$

We used this mathematical model to know the minimum possible number of buckets that will be utilized for an order and it served as a benchmark for our proposed heuristic closest cluster addition for item to bucket allocation.

Other possible approaches (Next Fit and Optimization Model)

In order to compare the performance of our results, we analyzed our results based on two more approaches. One of them is formulating this problem as mix integer optimization problem to maximize utilization of buckets. The number of buckets used by this approach will be the benchmark for optimal number of buckets utilization but not the total pick time of buckets. Another approach is the next fit approach where all items are sorted by their aisle, bay, and shelf location and then one by one all the items are taken and assigned in the bucket. If an item is not able to assign to a bucket due to bucket capacity, then next items are assigned in the basket and so on. The approach works on the principle of nearest neighbor. In next phase, bucket to trolley allocation methodology is explained in detail.

4.2 Bucket to Trolley Allocation: Modified Clarke and Wright Algorithm

Once the items are assigned to Bucket, the next step is to assign these buckets to trollies such that trolley total pick up time is minimum. The actual Clarke and Wright algorithm is based on distance/time savings when two nodes are connected. It is basically used for creating routes so that total distance is minimum. The savings are calculated as the difference between individually covering the nodes minus total distance covered/time taken when both nodes get combined.

$$S_{ij} = D_{0i} + D_{i0} + D_{0j} + D_{j0} - D_{0i} - D_{ij} - D_{j0} \ldots \tag{1}$$

$$S_{ij} = D_{i0} + D_{0j} - D_{ij} \ldots \tag{2}$$

(0 is the origin/destination and i and j are nodes to be connected.)

Equation (2) is the actual savings equation of Clarke and Wright [9]. Once the savings are calculated, savings matrix is sorted based on savings in decreasing order. After that one by one not pair is taken and is added to the route and this process continues till there are no capacity violations. We have modified this algorithm to make it suitable to solve our problem. Here, we need to join two buckets instead of nodes and each bucket will contain at least one item and need to have a different formula for calculating the savings while combining two buckets.

Let $B_1, B_2 \ldots B_k$ be the set of buckets and for each bucket, let l_i represent the set of items inside each bucket. So $B_{1l1}, B_{2l2} \ldots B_{nln}$ represents buckets along with items inside each bucket and we already have calculated the total pick time of each bucket in item to bucket allocation step represented by PSB_k. Detail steps are:

1. Calculate for each possible bucket pair the total pick score by combining both buckets all items and sorting them in increasing order of individual pick score of items inside both the buckets using the logic it was calculated for PSB_k
2. Let the final pick score calculated be $PSPOB_k$(Pick score for pair of bucket)
3. Now to calculate the savings in getting two buckets in the same trolley, we need to have individual pick score (individual time taken PSB_k) of each bucket and combined pick score (combined time taken $PSPOB_k$) and savings can be calculated as

$$\text{CWS}_{kk'} = \text{PSB}_k + \text{PSB}_{k'} - \text{PSPOB}_k$$

4. Sort the $\text{CWS}_{kk'}$ matrix obtained in Step 3 in decreasing order of savings and apply the following logic.
5. Let x = bucket carrying capacity of trolley (bucket count).
6. Let N row = number of rows in $\text{CWS}_{kk'}$ matrix.

5 Results and Practitioner Implications

Following sample input data set (Table 1) is considered for this analysis and compared to optimization model, next fit, and our improved heuristic.

Parameter: Bucket max volume = 35000 cc and bucket max weight = 13,000 (g).

Item to Bucket allocation: Key Outputs Tables 2, 3, 4, 5, and 6.

The number of buckets used by optimization model and closest cluster approach is same as 3 but next fit has used 5 buckets. Also, the total pick score is least (4.35) in closest cluster addition approach (Stage-1) and after completion of stage 2, the number of buckets is reduced to 3 but pick score has increased to 11.5. This is likely to happen because if buckets are reduced, then more items are placed in a bucket and the total time to pick all items will increase. The key feature of this heuristic is that it tries to minimize the number of buckets at minimum pick score. Optimization model pick score is 15.75 (higher than closest cluster addition) because the objective of the optimization model was to reduce the number of buckets only. Similar to this result, we run these three approaches on total 15 different orders and the result shows buckets used by optimization, next fit, and closest clusters are 80, 103, and 88, respectively. Now after completing the item to bucket allocation with 15 orders mentioned above, we passed the outcome of results obtained in Table 7 to bucket to trolley allocation using the modified Clarke and Wright approach mentioned in this paper. Bucket carrying capacity of the trolley was taken to be 6 for this analysis. The result obtained the from bucket to trolley allocation is mentioned in Table 4.

All in all, by using our approach of closest cluster addition, we are not only maximizing the picking efficiency and bucket utilization but also its impact is carried forward to the next stages also by using lesser number of trolley and vehicles thereby reducing the total transportation cost. This order SKU to bucket till trolley allocation methodology can be leveraged by across all retail managers and practitioners and provide ample opportunity for researcher for future research.

Table 1 Sample input

Order No	Item Id	Order quantity	Bay location	Aisle location	Shelf location	PICKING ZONE	Item volume (CC)	Item weight (gm)	Individual pick score	Total weight	Total volume
11710101	I1	1	7	13	2	Ambient	35,000	1960	13.07	1960	35,000
11710101	I2	1	9	7	3	Ambient	581.14	200	7.09	200	581.14
11710101	I3	1	23	7	2	Ambient	311.11	137	7.23	137	311.11
11710101	I4	1	32	8	1	Ambient	1000	7590	8.32	7590	1000
11710101	I5	2	41	9	3	Ambient	1370.02	1000	9.41	2000	2740
11710101	I6	1	47	9	3	Ambient	7060.03	40	9.47	40	7060
11710101	I7	1	24	6	5	Ambient	1022.24	520	6.24	520	1022.2
11710101	I8	1	41	8	1	Ambient	7672.25	2549	8.41	2549	7672.3
11710101	I9	2	6	2	6	Ambient	171.18	98	2.06	196	342.36
11710101	I10	4	28	3	4	Ambient	2391	1370	3.28	5480	9564
11710101	I11	1	24	5	2	Ambient	2004.91	500	5.24	500	2004.9
11710101	I12	1	24	5	2	Ambient	839.16	500	5.24	500	839.16
11710101	I13	1	24	5	1	Ambient	2573.64	500	5.24	500	2573.6
11710101	I14	1	56	5	3	Ambient	725.52	202	5.56	202	725.52
11710101	I15	1	7	1	2	Ambient	11,607.7	185	1.07	185	11,608

Table 2 Output of heuristic—next fit

Buckets utilized	Weight utilization	Volume utilization	Picking score	List of items inside bucket								
B1	15.1	100	0	I1								
B2	76.7	33.4	2.38	I2	I3	I4	I5	I6				
B3	23.6	24.8	2.17	I7	I8							
B4	56.8	45.9	3.5	I9	I9	I10	I10	I11	I12	I13	I14	
B5	10	30	0	I15								

Table 3 Output of optimization model

Buckets utilized	Weight utilization	Volume utilization	Picking score	List of items inside bucket									
B1	15.1	100	0	I1									
B2	90.7	75.1	8.4	I9	I10	I3	I4	I1	I6				
B3	67.8	62.1	7.35	I9	I10	I11	I12	I13	I14	I7	I2	I8	I5

Table 4 Closest cluster addition—stage-1

Buckets utilized	Weight utilization	Volume utilization	Picking score	List of items inside bucket							
B1	15.1	100	0	I1							
B2	15.7	28	0.06	I5	I6						
B3	78	24.8	0.09	I9	I9	I10	I10	I15			
B4	45.1	61.5	2.21	I8	I4						
B5	19.7	23	1.99	I11	I12	I13	I14	I2	I3	I7	

6 Conclusions

In this paper, we proposed a unique and practically implementable approach to improve order picking efficiency in two phases, viz. 1. Item to bucket allocation and 2. Bucket to trolley allocation. We have broken down this problem (which is originally NP hard in nature) into two phases and outcome of the first phase is used as an input to the second phase. We have also taken care of most of the operational and business constraints in consideration, for example, items splitability into multiple buckets (wherever applicable), etc. Our approach was aligned to existing standard picking process mechanism of retail warehouses and looks at the closest items to minimize total distance traveled. In this way, our suggested approach (i.e., execute operation in parallel mode) performs better than traditional sequential algorithm like next fit algorithm. This has clearly demonstrated in results and analysis section where we have shown comparative results. This approach helps in reducing the complexity of this problem from $O(2^n)$ to $O(n!)$. Suggested solution

Table 5 Closest cluster addition—stage-2 (final result)

Buckets utilized	Weight utilization	Volume utilization	Picking score	List of items inside bucket									
B1	15.1	100	0	I1									
B2	64.7	83.4	7.3	I8	I4	I5	I6						
B3	93.8	53.9	4.2	I12	I11	I13	I14	I7	I2	I3	I9	I15	I10	I10	I9

Table 6 Comparison of all the approaches

Approach	KPIs	
	Bucket used	Total pick score
Optimization model	3	15.75
Heuristic-next fit	5	8.05
Heuristic (closest cluster addition)-stage 1	5	4.35
Heuristic (closest cluster addition)-stage 2	3	11.5

Table 7 Resource utilization

Resource utilized	Optimization	Next fit	Closest cluster
Trolley	14	17	15
Vehicle	2	3	2

methodology opens up a new avenue for the researchers and practitioners working in this direction and enables them to enrich this domain with robust solution implementation.

References

1. Becerril-Arreola, R., Leng, M., & Parlar, M. (2013). Online retailers' promotional pricing, free-shipping threshold, and inventory decisions: A simulation-based analysis. *European Journal of Operational Research, 230*, 272–283.
2. Wascher, G. (2004). Order picking: A survey of planning problems and methods. In *Supply chain management and reverse logistics* (pp. 324–370). Berlin: Springer.
3. Berman, B., & Evans, J. R. (2010). *Retail management—A strategic approach* (11th edn). Pearson Prentice Hall (11th edn).
4. Croxton, K. L. (2002). The order fulfillment process. *The International Journal of Logistics Management, 14*(1), 19–32.
5. Henn, S., Koch, S., & W¨ascher, G. (2012). Order batching in order picking warehouses: A survey of solution approaches. In *Warehousing in the global supply chain—Advanced models, tools and applications for storage systems* (pp. 105–137). London: Springer-Verlag.
6. Ho, Y.-C., & Tseng, Y.-Y. (2006). A study on order-batching methods of order-picking in a distribution centre with two cross-aisles. *International Journal of Production Research, 44*(17), 3391–3417.
7. De Koster, R., Le-Duc, T., & Roodbergen, K. J. (2007). Design and control of warehouse order picking: A literature review. *European Journal of Operational Research, 182*(2), 481–501.
8. Dung, H., Nguyen, L., de Leeuw, S., & Dullaert, W. E. H. (2016). Consumer behaviour and order fulfilment in online retailing: A systematic review. *International Journal of Management Reviews, 00*, 1–22. https://doi.org/10.1111/ijmr.12129.
9. Clarke, G., & Wright, J. W. (1964). Scheduling of vehicles from a central depot to a number of delivery points. *Operations Research, 12*, 568–581.

Review of Various Techniques Used for Automatic Detection of Malignancy in Pap Smear Test

Priya Chaudhari and Sharad Gore

Abstract Cervical cancer is the second-largest toll-taking disease in women all over the world. This paper takes the review of how it is diagnosed and what methods have been developed in computer science for analysing the Pap smear images. We suggest a new method for diagnosing the presence of cervical cancer. This will help classifying the benign and malign cases automatically and only the doubtful cases can be observed by the pathologist, which reduces the burden over him.

Keywords SVM · Pap smear · Cervical cancer

1 Introduction

Cancer is a word that threatens any person. Though medical science is advanced a lot, fear of death prevails in the thoughts of a person suffering from 'cancer'. It is said that cancer, if detected early, is curable. As it advances, it might spread and becomes difficult and painful to cure. WHO (World Health Organisation) has reported that Cervical cancer is the second most common cancer in developing and less-developed countries [1].

If we try to find the statistics in India regarding cervical cancer, it is found that here too in India, cervical cancer is most frequent among woman between 15 and 44 years [2, 3]. Loss of family member to an individual family cannot be reimbursed. Also the amount that individual family needs to pay during the treatment of cervical cancer is a lot.

P. Chaudhari (✉)
Sinhgad Institute of Business Administration and Research, Pune, India
e-mail: priya.abhijit@gmail.com

S. Gore
Department of Statistics, SPPU, Pune, India
e-mail: sharaddgore@gmail.com

© Springer Nature Singapore Pte Ltd. 2019
V. E. Balas et al. (eds.), *Data Management, Analytics and Innovation*,
Advances in Intelligent Systems and Computing 839,
https://doi.org/10.1007/978-981-13-1274-8_19

Around 80% of cervical cancer can be cured if diagnosed in earlier stage. Regular screening programme is available in developed countries. This has reduced female deaths by 50–80% which otherwise might have caused due to cervical cancer [1]. It takes around thirty years to develop invasive cancer from the earlier stage of infection. If the infection is detected earlier then it is easy to treat the patient. With small scale procedures, the infections can be removed and the patient's life be saved. Thus this screening test has a significant role in reducing the death rate.

1.1 Cervical Cancer

Cervical cancer is the cancer at Cervix. The signs and symptoms of cervical cancer are: 1. Abnormal vaginal bleeding 2. Vaginal discomfort 3. Malodorous discharge (foul smelling) 4. Dysuria (Painful and difficult urination).

1.2 Pathophysiology of Cervical Cancer

It is said that HPV (Human Papilloma Virus) infection must be present for cervical cancer to occur. There are more than 100 types of HPVs. From which 12 are considered to be of 'high risk' or oncogenic (causing development of tumour) in nature. There are very few chances that the infection turns into cancer. Also it may take as much as 30 years for this conversion.

There are other factors that are involved in the process of carcinogenesis (initiation of cancer formation) [4–6]. These factors may reduce the number of years of cancer formation. These factors are like poor immunity, environmental factors like smoking, vitamin deficiencies, no routine cytology screening, early age of first intercourse and higher number of sexual partners, oral contraceptives.

1.3 Screening and Biopsy

Screening is the process in which tissues at the cervix are scraped and checked under microscope. If any abnormalities are found in the sample cells gathered, then the amount of abnormality is checked. If required biopsy is done, this includes cutting a small part of the tissue and then examined to find out the stage of invasion.

1.4 Pap Test

Pap smear test is the test which is used for screening purpose in cervical cancer.

In 1928 Dr. Georgios Papanicolaou proposed for the first time that the presence of cancerous cells give warning about cervical cancer. Smear is the material on the outer layer of cervix. This material consists of the superficial layer cells. In this test, cells are exfoliated from the transformation zone of the cervix. These cells are examined microscopically for detection of cancerous or precancerous lesions [7–9]. It is found that more than 50% of deaths that might have had caused by cervical cancer are reduced due to regular screening using Pap test.

1.5 Execution of Pap Test

There are two methods of collecting the sample for Pap smear test [8, 10]. First older method is to directly transfer the cervical cells to the microscope slide. The sample slide may contain blood cells and other debris. This might lead misinterpretation. The second, new method is liquid-based cytology. The sample is transferred to liquid preservative and then the slide is produced. There are three types of cells that can be present in the smear slide. Ectocervix cells, Endocervix cells and Endometrial cells. Presence of these cells is identified. These cells are observed to give result, i.e. whether the cells are normal or no.

The two methods mentioned above are (1) ThinPrep and (2) Surepath [8]. These were the byproduct while developing the automated methods. These were developed to minimize the cell overlap to increase the accuracy of the automated methods. Below we describe these methods in brief.

ThinPrep:

This method was developed by Cytyc Corporation, Boxborough, MA. This was given approval by Food and Drug Administration (FDA) in 1996. The sample is obtained with a 'broom', or a plastic spatula, or an endocervial brush. The sample is preserved using methanol-based preservative. The sample is then evenly distributed in a circle of 20 mm diameter. It is also approved by FDA for direct testing for human papilloma virus (HPV). Figure 1 shows the diagrammatic representation of ThinPrep method.

Fig. 1 ThinPrep
diagrammatic representation

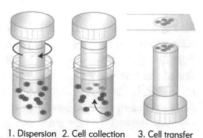

1. Dispersion 2. Cell collection   3. Cell transfer

Fig. 2 ThinPrep
diagrammatic representation

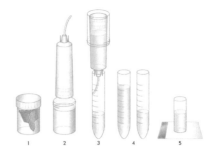

SurePath:

SurePath was developed by TriPath Imaging. The preservative used here is ethanol-based. This method gives better results to identify the lesions that the ThinPrep method. Figure 2 shows the diagrammatic representation of SurePath method.

The steps are follows:

1. Sample is vortexed
2. Syringing the sample through a small orifice
3. Sample poured to centrifuge tube
4. Sedimentation
5. Tubes transferred to PREPStain instrument.

These cells are diagnosed on the following variables, size, shape and N:C ratio. If these are not found normal then further variables are studied so that the severity can be understood. If the abnormality is found then other factors like infection, chronic irritation, and increase in Progesterone hormone—if any hormone therapy is given, artifacts (quality of the slide) are also checked. Lastly the possibility of cancer is checked. The results are reported in terms of variables. For dysplasia (enlargement) and pleomorphism (changes in size and shape), grades are mentioned, while for nucleoli the number found is mentioned.

1.6   Bethesda System

The system was named as Bethesda after the location name Bethesda in Maryland where a conference was organized, in which the system was established. This system says that the specimen should be adequate. Initially in 1988 Bethesda system mentioned the adequacy as 'satisfactory', 'less than optimal' and 'unsatisfactory'. Later on, in 2001, the second category was omitted, as it was thought to create confusion. Adequate sample also mentioned the presence of Squamous component in a particular percentage. It was said that this component should be 'well-preserved and well-visualized' and it should cover more than 10% of the slide surface.

Bethesda system is the system used for giving the result. Depending on the abnormalities seen in the cells, one of thc following result is given [9–11].

Depending on the severity the results are given as one of the type—Atypical Squamous cells (ASC-US and ASC-H), Low-grade Squamous intraepithelial lesion (LGSIL or LSIL) termed as CIN 1, High-grade Squamous intraepithelial lesion (HGSIL or HSIL) termed as CIN 2 or CIN 3, Squamous cells carcinoma, Atypical Glandular cells (AGC or AGC-NOS), Atypical Glandular cells—suspicious for AIS (Adenocarcinoma in situ) or cancer (AGS-neoplastic), Endocervical Adenocarcinoma in situ (AIS). Here CIN stands for Cervical intraepithelial neoplasia. It is the abnormality in the cells of ectocervix and endo cervix. Sometimes the cells from endometrium are also found in the sample.

Below we show few images for cervical cells (Figs. 3, 4, 5, 6 and 7):

Both the methods of gathering sample have similar sensitivity and specificity for moderate dysplasia or worse lesions. Colposcopy is suggested when abnormal cells are found in the Pap test. Diagnosis of dysplasia via colposcopic biopsies is then executed. Cervical cancer can be prevented by treatment of these cervical cancer predecessors.

Limitations of the Pap test: The sensitivity of a Pap smear test for cervical dysplasia is said to be in the range of 30–87%. The average is approximately 58%. It depends on the observer. False negative results for Pap smears test are up to 30%.

Fig. 3 Normal Superficial and intermediate Squamous cells

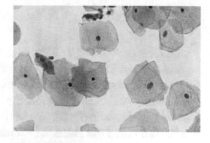

Fig. 4 HSIL—High-grade Squamous intraepithelial lesion

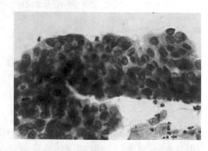

Fig. 5 SQC—Invasive Squamous Cells Carcinoma

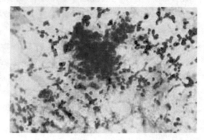

Fig. 6 AIS—
Adenocarcinoma in situ

Fig. 7 Small Cell Carcinoma

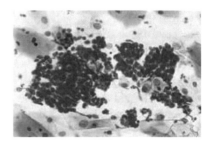

1.7 Requirement of Automation in This Field

As we have seen through all the data, conditions and seriousness of cervical cancer, screening programmes can help to reduce the deaths due to cervical cancer. If large number of screening is done then it would be overload on the pathologists. Hence we recommend that if first screening is done through automation then, the load on pathologists can be reduced. We have also seen that interobserver and intraobserver results also differ. An automated screening system will be able to reduce this error. This system would be complementary to the manual method currently in use.

2 Understanding SVM and Its Applications

2.1 SVM

SVM was introduced by Boser, Guyon and Vapnik. It was first popular in Neural Information processing system (NIPS) community. It is now important part of machine learning research, optimization, statistics, neural networks, functional analysis, etc. [12].

SVM i.e. Support vector machines are supervised learning models. These are associated with learning algorithms that analyse data and recognize patterns. SVMs are used for classification and regression analysis. In simple language SVMs are used for classification of data. These algorithms can solve the computational and statistical problems. The computational problems are like clustering, classifying,

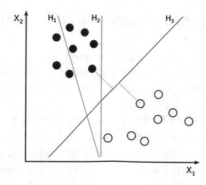

Fig. 8 Depicting Linear separation of class with SVM

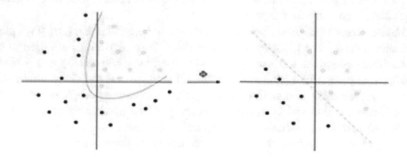

Fig. 9 Depicting nonlinear separation of class with SVM

ranking, cleaning, etc. the statistical problems include representation of complex pattern, excluding spurious, i.e. over fitting problems. SVM uses various parameters of interest and relevance for classification. It can be trained to classify certain data using already classified data, called the training set. It can be then used to classify the unclassified data, called the testing set. This makes SVM a non-probabilistic binary linear classifier. Kernel is a function used in the training algorithm [13–15].

SVM algorithm finds a hyperplane (a line or a plane) to classify the elements. There can be various number of hyperplane in a given data those classify. SVM finds this hyperplane. This hyperplane maximizes the distance between the closest elements. Hyperplane can be a straight line, a circle, etc. Optimisation of the distance is applied to the closest point to the classifier (hyperplane). These points are called as Support Vectors. No other points should be present beyond this boundary towards the hyperplane. These Support Vectors are useful for training. Patterns that are nonlinearly separable are converted to Feature space using some function, so that they can be easily separated. SVM can be easily used as classifier. It can be used as binary as well as multi-class classifier [16, 17] (Figs. 8 and 9).

2.2   Use of SVM for Pap Smear Test in Literature Review

Zhang and Liu have presented a new method for screening features. This is SVM-based method. In this paper the magnitude of the weights of a linear discriminant classifier is used as an indicator of feature relevance. They gave extension to the linear scheme [18]. SVM boundary is used to improve performance. The method mentioned in paper tries to globally characterize the discriminative information, which is embedded in the SVM decision boundary. They consider it to be more efficient as it generates global feature relevance measures. The proposed method was applied to multispectral Pap smear image classification for cervical cancer detection. Results of various experiments were compared. These results show that the improvements in pixel-level classification accuracy are significant using the new feature screening method. It was shown by the authors that the effectiveness of image feature screening/selection in cancerous cell detection on a novel image modality (multispectral image).

Maykel Orzco et al. used nuclei segmentation and classification [19]. Two phase approach is used. First phase is for segmentation. Two algorithms are used in this phase. These algorithms are morphological algorithm (Watershed) and a hierarchical merging algorithm (Waterfall). Waterfall uses spectral and shape information along with their class information. Second phase is classification phase. In the first phase Nucleus region and cytoplasm areas are obtained by classifying the regions. These two areas, i.e. nucleus and cytoplasm areas are obtained on the basis of spectral and shape features, and merging the adjacent regions belonging to the same class.

Individual regions are obtained using SVM classifier. The pathologists are requested to examine the same images used for classification using algorithms. The segmentation and classification results given by experts are compared to the results given by proposed methods. This comparison demonstrates the efficacy of the proposed method. First step of detecting abnormalities or precancerous cells is to extract nucleus. To detect abnormalities nucleus regions and other regions including cytoplasm and background are used. It is difficult to segment nuclei due to inherent colours and shape variability. In this paper author has proposed hybrid two-step approach to cell segmentation.

First phase includes creating a nested hierarchy of partitions. A hierarchical segmentation is produced which uses the spectral and shape information and class information.

Most meaningful hierarchical level is selected by using a segmentation quality criterion. Hierarchy of nested partition is constructed using Waterfall model.

In second phase nucleus and cytoplasm areas are classified using multiple spectral and shape features. Neighbouring regions belonging to same class are merged. Further the spectral and shape features are identified. Each region is obtained using the SVM classifier.

Boundaries surrounded by other boundaries are obtained. To form the hierarchy of boundaries, Watershed algorithm is used to obtain the starting boundary.

Waterfall algorithm is used to successfully removing the outer boundaries. Thus, by removing the boundaries surrounded by outer boundaries the hierarchy of boundaries if formed.

The method proposed in this paper cuts most of the wrongly segmented cells. It avoids the over or under segmentation. Thus, it takes the partitioning closer to the pathologists segmentation.

Plissiti and Nikou studied that the cervical cells can be classified on the basis of nucleus features [20]. They say that the nuclei areas can be extracted automatically from Pap smear images. They represented the features in low-dimensional spaces. Nonlinear dimensionality reduction techniques are used for the same. The classification was performed in a supervised manner, using support vector machines (SVM) with several kernel functions. The results depict high performance in the classification, done using the features extracted only from the nuclei areas. The features considered for this study are area, brightness, shorter diameter, longest diameter, elongation, roundness, perimeter, maxima and minima. Maxima and minima is the number of pixels with the maximum/minimum intensity value.

Plissitti and Nikou also developed a fully automatic robust and accurate method to identify the cell nuclei [21]. This method can then be used as basis for the classification of cervical cells. Every pixel in the cell image is considered as either a nuclei pixel or a cytoplasm pixel. The method is suitable for images with isolated cells and with high degree cell overlapping.

Yung-Fu Chen and et al. developed a PC-based cellular image analysis system for segmenting nuclear and cytoplasmic contours and for computing morphometric and textual features [22]. The features obtained are used to train the support vector machine (SVM). SVM now classifies four different types of cells. It also differentiates the dysplastic cells from the normal cells. A software programme incorporated functions for image reviewing. Files were given standard names. This naming helped the standardization of workflow of cell analyses. The classification was verified by two experiments. The cross-validation results of the first experiment showed that average accuracies of 97.16 and 98.83%, respectively, for differentiating four different types of cells (one dysplastic and three normal). The discrimination between dysplastic and normal cells is achieved using various salient features. The number of features used was eight for four-cluster and seven for two-cluster classifiers. This was done with SVM recursive feature addition. In the second experiment included 70% (837) of the cell images. These were used for training and 30% (361) for testing. This achieved an accuracy of 96.12 and 98.61% for four-cluster and two-cluster classifiers, respectively. According to the author, the proposed system provides a feasible and effective tool in evaluating cytologic specimens.

New radiating gradient vector flow (RGVF) algorithm is proposed by Kuan Li et al. [23]. Extraction of nucleus and cytoplasm in a single cervical smear cell is done with the help of this algorithm. Firstly the cervical image is extracted. The pre-processing is done with the help of k-means algorithm. This helps to locate the nucleus and cytoplasm roughly. Locating and extracting the nucleus is done by the SVM. The new RGVF algorithm is used with a new edge map computational stack based refinement algorithm which neglects the false negative rates and locates the obscure boundaries effectively in nucleus and cytoplasm.

Xue et al. and Tulpule et al. [24, 25] classified the uterine cervical cancer with mosaic patterns. This pattern is one of the visual symptoms that are used as an abnormality in the cervical tissues. The mosaic regions are divided into two regions. These are fine and coarse. The fine region is composed of a network of thin calibres and the coarse region is composed of a network of large calibres. A supervised learning approach is used to segment the mosaic vascular regions. A binary map is formed by classifying each segment as mosaic tile. The Support Vector Machine is used to classify these segments. SVM is used along with a Gaussian kernel function. This validates the parameters.

Kumar et al. [26] demonstrate an effective mass screening to avoid the quantity of false positive rates. Some artifacts have similar size and shape. This leads to misclassification. Misclassification can be avoided by using SVM to obtain a new pattern strategy. SVM removes the artifacts from the epithelial cells. The image is acquitted by the E-smear software. The pre-processing is carried out by the median filter, the nuclear segmentation is performed by the edge based Laplacian of Gaussian. The classification is used with the SVM which is located at the middle of both support hyper planes.

Wang et al. [27] demonstrate an automated computer based system for the diagnosis of cervical intraepithelial Neoplasia (CIN) which is the cause of cervical cancer. The support vector machine (SVM) is used as the classifier for the normal and abnormal CIN. The classification is identified by the hierarchy of four binary classifications in different texture forms as perpendicular lines into normal, CIN1, CIN2 and CIN3. For each classifier the optimal set of features is used by systematic testing.

Kashyap et al. have proposed an automatic method to detect and classify cervical cancer. Authors have used geometric and texture features of Pap smear images to classify. Segmentation is applied to nucleus and cytoplasm. GCLM texture is defined. Combination of PCA and SVM used for classification, which, yields accuracy of 95% [28].

Orozco-Monteagudo et al. [29] have proposed a two phase method to classify pap smear cells. The first phase yields a nested hierarchical partition based on spectral and shape. In the second level, best hierarchical level based on unsupervised criterion is selected. This step also refines the segmentation obtained in first phase. This refining is done by a SVM classifier. After refining the adjacent regions, belonging to same class are merged. The results are compared with the results given by pathologists.

2.3 Discussion

After studying the cytology and also studying various papers in which it is mentioned that SVM is used for detecting cervical cancer, the following list of features is found:

1. Nucleus—colour, shape, contour
2. Cytoplasm—colour, nucleus to cytoplasm ratio
3. Nucleus texture
4. Any benign infections
5. History of the patient
6. Cervigrams.

Also it is found that almost every paper uses only one or two features at the most. This can mislead to the diagnosis. After reading these papers and understanding the situation of a patient, it is concluded that all of the features should be considered while giving the diagnosis. Thus we have decided to use all of the above given features in our study.

3 Other Methods Used Apart from SVM for Pap Test

Vijayshree and Ramesh Rao used image morphometric software along with its some of the plugins. This is used to create macro for analysing large number of cells at a time [30]. Plugins used are BEEPS, Kuwahara filter and Mexican Hat filter. These are used to create macros to analyse normal, reactive and neoplastic Pap smears. This study found strong correlation among the results obtained with macros and that of manual.

In the method nuclear measurements were taken. Nucleus plays an important role in finding the dysplacia. This was done on normal and abnormal Pap smears. The abnormal smear cells were classified as follows: (1) Low-grade Squamous intraepithelial lesion (LSIL), (2) High-grade Squamous Intraepithelial lesion (HSIL), and (3) Squamous cell carcinoma, (4) Reactive smears and (5) Smears with atypical Squamous cells of uncertain significance (ASCUS). Reactive smears show changes due to various organisms like Herpes, Trichomonas, Chlamydia, Gardnerella and Candida.

The authors have applied morphometric analysis. This was done using Imagej. Three different plugins BEEPS (Bi-exponential edge preserving smoother), Kuwahara filter and Mexican hat filter were used in Imagej. BEEPS filter is used to remove unnecessary details. This does not affect the edges. Kuwahara filter is used to reduce the noise along with preserving the edge. Lastly, Mexican Hat Filter is used to separate signal from the noise by applying Laplacian of Gaussian filter. Along with this inbuilt median filter of Imagej was also used. Macro was developed for automated analysis.

Algorithm has the steps as given below:

1. Split channels
2. Choose red or green channel
3. Apply 'Median', BEEPS, Kuwahara filters
4. Apply Mexican Hat filter
5. Thresholding and Binary conversion. Filling the holes and watershedding
6. Analyse with overlay
7. Result.

Along with the macro, a tool called 'Cell Magic Wand Tool' was also used. Through this tool when a nucleus or cell is selected by clicking, delineation of its perimeter is obtained. This method is to be applied on each cell separately. The values are obtained by the above method. The values are also obtained by the macro. Both these values are obtained on the same nuclei. These two values are then correlated. The correlation coefficient was determined. After studying the correlation the authors came to conclusion that the two methods can be used interchangeably.

Macro and tool measured area, perimeter, diameter, feret, width and height of nucleus. A chart was prepared depicting mean and standard deviation of above parameters for various types of cells and lesions. The cells and lesion types are—superficial cells, Intermediate cells, parabasal cells, reactive smear, ASCUS, LSIL, HSIL, SCC. The presence of asinonucleosis is prominently seen in lesion cells. It has high standard deviation for all the parameters due to large variety of cells. Thus this method can be used to differentiate between normal cells and lesion cells.

Before classifying the nucleus using SVM, Plissiti and Nikou classified the nucleus data using PCA, K-PCA (Polynomial, Gaussian), Isomap, LLE, Laplacian eigenmaps [31]. In this study too they considered the same nucleus features as area, brightness, shorter diameter, longest diameter, elongation, roundness, perimeter, maxima, and minima. Maxima and minima is the number of pixels with the maximum/minimum intensity value. They got better classification using SVM.

Song et al. carried out the experiment on the basis of cervigrams. Cervigram are uterine cervix images. They said that in case of limited resources available to study the Pap smear images for detecting the cervical cancer, then a method based on study of cervigrams can be used. They have developed and explained the same, a data-driven computer algorithm in [32] that works on cervigrams. When high-grade cervical lesions are differentiated from low-grade cervical lesions and normal tissue, 74% sensitivity and 90% specificity was obtained respectively. They compared the result with experiments carried on same datasets for Pap smear tests and HPV tests. For Pap test they found the sensitivity of 37% and specificity of 96%. For HPV test they found sensitivity of 57% and specificity of 93%. A comprehensive algorithmic framework was developed. This was based on Multimodal Entity Co reference for combining various tests to perform disease classification and diagnosis. Multiple tests were integrated. Information gain and gradient-based approaches were adopted for learning the relative weights of different tests. In the evaluation, authors presented an overall algorithm that integrates cervical images, Pap, HPV and patient age. This integration gives 83.21% sensitivity and 94.79% specificity. These statistics show significant improvement over using any single source of information alone.

In [33] feature analysis and classification detection is done with the help of segmentation. After segmentation Affinity Propagation algorithm is used to cluster the cells of same type. This research is mainly based on the cervical epithelium cells. Features are considered in the combination of size of nucleus, Nucleus to Cytoplasm ratio, Circularity, Compactness, centroid position and nucleus boundary. The method combines help from pathologists. Authors say that the method is fast and accurate as the pathologists are involved in it.

There are various methods used for study of cervical cells in Pap smear [33–38] colour models are used, 8-bit gray levels are used. This method uses the pre-processing of colour images.

The other method uses colour model HSI. This is used for image segmentation. Segmentation of cell images is done using following methods in various papers:

1. Water immersion algorithm
2. An active contour model or snake and improved snake model
3. Unsupervised nucleus segmentation method based on dual active contour
4. Methods based on Hough transform, generalized Hough transform, compact Hough transform
5. Fuzzy logic engine
6. Genetic algorithm. Multifractal algorithm
7. Seed-based region growing algorithm
8. Automatic colour image segmentation
9. Moving k-means clustering
10. Modified seed-based region growing algorithm
11. Watershed transformation.

Classification of cervical cells is done with following methods

1. Dual wavelength—For isolating cytoplasm and cell nuclei
2. Data analysis—consists of manual classification of cell classes, feature extraction, cell classification.

Pap smear classification is done with following methods

1. Technique based on the nearest neighbour classification rule
2. Tabu search
3. Feature subset selection
4. Ant colony optimisation is also used in feature subset selection problem
5. Using hyperchromasia and texture for analysis of malignancy in Pap smear [39]
6. Nuclei shape analysis and structure based segmentation [40].

This work is based on the analysis of abnormal cervical cells based on nucleus/cytoplasmic (N/C) ratio. N/C ratio is one of the most important morphological features. This feature is the first which is checked by pathologists. It is used to distinguish between normal and abnormal cervical cells. Authors use MATLAB to analyse the Pap smear images. Pap smear test is a simple harmless test that is used in diagnosis of cervical cancer. This will be great help to pathologist in identifying dysplatic cells and help in prevention of cervical cancer.

4   Neural Network and Pap Smear Test

In the paper [41, 42] authored by Xin et al. cervigrams are studied. These are then classified using various decision tree models and SVM, MLP—MultiLayer Perceptron (i.e. feed forward Neural Network), Logistic Regression and k-Nearest Neighbours. When we go through the results obtained for the results of SVM and Neural Network it can be seen that SVM gives better results.

In the papers [43, 44] authored by Athinarayanan and Srinath, nucleus features are used to classify the normal and abnormal cervical cells. The proposed method has three stages, nucleus detection, feature extraction and classification. Authors used SVM for classification. Along with that the same data was also classified with the help of neural network. They used feed forward neural network and radial basis function neural network. The results obtained with SVM were better than that of the neural network. SVM gave better accuracy, sensitivity and specificity than neural network.

Samuel et al. in paper [45] uses Genetic Algorithms with SRC algorithm for classifying the cervical cells. They have obtained twenty (20) numerical features for the cervical cell. They also compared the results obtained with Artificial Neural Network, Sequential Minimal Optimisation, Adaboost, Naive Bayes Classifier. With Genetic algorithm and SRC algorithm 93% of accuracy is obtained.

There is paper authored by Osareh et al. [46] that studies the intraretinal fatty exudates. Intraretinal fatty exudates are the visible sign of diabetic retinopathy. It confirms the presence of co-existent retinal oedema. The patient may lose vision if these symptoms are present. If detected earlier ophthalmologists can treat the patient so that it does not spread. These images are studied and classified with the help of Support Vector and Neural Network. Two different learning methods, Back Propagation (BP) and Scaled Conjugate Gradient (SCG), are used with neural network. Whereas with SVM, different values for soft margin, C, are used. After comparing the results it is found that SVM gives robust output, better classification, with good trade-off between false positive and false negative cases.

4.1   Discussion

It is found that many-a-times neural network and Support Vector Machines (SVM) give similar classification results, but when we consider the parameters as sensitivity and specificity, a good balance is maintained by SVM. When the classification problem from medical field is considered, only accuracy is not sufficient. The classifier should also maintain balance between sensitivity and specificity. From various papers which studied the comparison between the working of Neural Network and SVM, it is found that SVM works (classifies) better. Also SVM can achieve a trade-off between false positive and false negative cases. This is done by using soft margin. SVMs always converge to the same solution for a given data set regardless of initial conditions; also they remove the chance of over fitting. Thus we say that Support Vector Machines (SVM) is better choice than the Neural Network.

5   Automated Methods Review

Various devices were manufactured since 1950s for automatic image scanning of Pap smear. These are Cytoanalyser, CYBEST, TI-CAS, Quantimet, BIOPEPR, CERVIFIP, DIASCANNER, FAZYTAN and LEYTAS, AutoCyte, Cytyc, Neopath, and Neuromedical Systems. BioPEPR, FAZYTAN, Cerviscan, LEYTAS, and the Diascanner [47–57]. Unfortunately, tests with the Cytoanalyzer showed that the special purpose fixed logic pattern recognition produced too many false alarms on the cell level. In preclinical experiments it did not perform well and the project was stopped [58].

Cytology automation experiments conducted in US did not give satisfactory results. Hence it was said that this process cannot be automated. All experiments were too expensive. After US, the experiments on cytology automation were also done in Europe and Japan. In Britain a one-parameter (nuclear size) automatic screener was developed in the late sixties.

In Japan prototypes were used in large field trials. These were used in the Japanese screening programme. These results were promising. Still it could not produce an automated product for cytology screening.

The Technical Challenges in Automation:

1. Specimen preparation
2. Scanning
3. Segmenting cells and nuclei
4. Artifacts rejection
5. Feature Extraction.

6   Discussion

After reading and understanding various papers as mentioned above we come to the conclusion, till date various methods have been devised to classify the Pap smear test images into benign and malign. These papers have used either one or two parameters. What can be done is a model can be devised which uses all the parameters that are involved in the study of Pap smear test which will classify the images into benign and malign classes. We say that the dependability and reliability will increase when more and more parameters are added.

After the study of Pap smear cells, following are the parameters that can be involved in developing the model. Model can be based on SVM or NN or tree structure, etc. It can be studied which method is better suitable to develop the model.

1. Nucleus—colour, shape (circularity), contour, size, texture, compactness, centroid position, nucleus boundary, number of nucleoli
2. Cytoplasm—colour
3. Nucleus to cytoplasm ratio (N/C ratio)
4. Any benign infections
5. History of the patient
6. Cervigrams.

References

1. WHO. (2015). Human papillomavirus (HPV) and cervical cancer [Online]. Available at: http://www.who.int/mediacentre/factsheets/fs380/en/#.
2. Kaarthigeyan, K. (2012). Cervical cancer in India and HPV vaccination. *Indian Journal of Medical and Paediatric Oncology, 33*(1), 7–12 Jan–Mar 2012. Available at: http://www.ncbi.nlm.nih.gov/pmc/articles/PMC3385284/.
3. ICO Information Centre on HPV and Cancer (HPV Information Centre) 2014. (2014). *Human Papillomavirus and Related Diseases Report*. Available at: http://www.hpvcentre.net/statistics/reports/IND.pdf.
4. Medscape Journal. (2015). *Cervical Cancer* by Cecelia H Boardman, MD. Available at: http://emedicine.medscape.com/article/253513-overview#aw2aab6b2b2.
5. National Cancer Institute. (2015). *Cervical cancer treatment*. Available at: http://www.cancer.gov/types/cervical/patient/cervical-treatment-pdq.
6. Cancer research UK. *Cervical cancer stages*. Available at: http://www.cancerresearchuk.org/about-cancer/type/cervical-cancer/treatment/cervical-cancer-stages.
7. Medscape Journal. (2015). *Pap Smear* by Nicole W Karjane, MD. Available at: http://emedicine.medscape.com/article/1947979-overview.
8. Cibas, E. A. Cervical and Vaginal Cytology. Cervical and Vaginal Cytology.pdf.
9. National Cervical Screening Program (Australia). An abnormal Pap smear result—What it means for you. ISBN: 0 642 82959 442 82959 4.
10. PAML Cytology reference manual. (2012). Available at: www.paml.com.
11. Cervical abnormalities: CIN3 and CGIN. (2015). Available at: http://www.healthtalk.org/peoples-experiences/cancer/cervical-abnormalities-cin3-and-cgin/what-cin.
12. SVM. Nello Cristianini BIOwulf Technologies. Available from: http://www.support-vector.net/tutorial.html.
13. SVM Tutorial. Available from: http://svm-tutorial.com/svm-tutorial.
14. Burges, C. J. C. (1998). A tutorial on support vector machines for pattern recognition. *Data Mining and Knowledge Discovery, 2,* 121–167.
15. SVM Tutorial. Available at: http://svms.org/tutorials.
16. VC Dimension. Available at: http://www.svms.org/vc-dimension/vc-dimension.pdf.
17. Karatzoglou, A., Meyer, D., & Hornik, K. (2006). Support vector machines in R. *Journal of Statistical Software, 15*(9) Apr 2006. Available at: http://jstatsoft.org.

18. Zhang, J., & Liu, Y. (2004). *Cervical cancer detection using SVM based feature screening*. Available from: https://www.ri.cmu.edu/pub_files/pub4/zhang2004/zhang_jiayong_2004_4.pdf.
19. Orozco-Monteagudo, M., Mihai, C., Sahli, H., & Taboada-Crispi, A. (2012). Combined hierarchical watershed segmentation and SVM classification for pap smear cell nucleus extraction. *Computacion y Sistemas, 16*(2), 133–145.
20. Plissiti, M., Nikou, C. (2012). *On the importance of nucleus features in the classification of cervical cells in Pap smear images*. Available from: https://cs.uoi.gr/~ cnikou/Publications/C047_IWPRHA_Tsukuba-2012.pdf.
21. Plissiti, M., Nikou, C., & Charchanti, A. (2011). Automated detection of cell nuclei in pap smear images using morphological reconstruction and clustering. *IEEE Transactions on Information Technology in Biomedicine, 15*(2), 233–241.
22. Chen, Y., Huang, P., Lin, K., Lin, H., Wang, L., Cheng, C., et al. (2014). Semi-automatic segmentation and classification of pap smear cells. *IEEE Journal of Biomedical and Health Informatics, 18*(1), 94–108.
23. Li, K., Lu, Z., Liu, W., & Yin, J. (2012). Cytoplasm and nucleus segmentation in cervical smear images using radiating GVF snake. Pattern recognition 2012, pp. 1255–1264.
24. Xue, Z., Long, L. R., Antani, S., & Thoma, G. R. (2010). Automatic extraction of mosaic patterns in uterine cervix images. In *IEEE Symposium on Computer-Based Medical Systems*, (pp. 273–278). http://doi.org/10.1109/CBMS.2010.6042655.
25. Tulpule, B., Yang, S. Y. S., Srinivasan, Y., Mitra, S., & Nutter, B. (2005). Segmentation and classification of cervix lesions by pattern and texture analysis. In *The 14th IEEE International Conference on Fuzzy Systems, 2005* (pp. 173–176). FUZZ '05.
26. Kumar, R. R., Kumar, V. A., Kumar, P. N. S., Sudhamony, S., & Ravindrakumar, R. (2011). Detection and removal of artifacts in cervical Cytology images using Support Vector Machine. *IEEE International Symposium on IT in Medicine and Education, 1*, 717–721. https://doi.org/10.1109/ITiME.2011.6130760.
27. Wang, Y., Crookes, D., Eldin, O. S., Wang, S., Hamilton, P., & Diamond, J. (2009). Assisted diagnosis of cervical intraepithelial neoplasia (CIN). *IEEE Journal of Selected Topics in Signal Processing, 3*(1), 112–121.
28. Kashyap, D., Somani, A. et al. (2016). Cervical cancer detection and classification using independent levels sets and multi SVMs. In *39th International conference Telecommunications and Signal Processing (TSP)*.
29. Orozco-Monteagudo, M., Sahli, H., Mihai, C., & Taboada-Crispi, A. (2011). A hybrid approach for pap-smear cell nucleus extraction. In *Mexican Conference on Pattern Recognition 2011* (pp. 174–183).
30. Vijayashree, R., & Ramesh Rao, K. (2015). A semi-automated morphometric assessment of nuclei in pap smears using Imagej. *Journal of Evolution of Medical and Dental Sciences 2015, 4*(31), 5363–5370 April 16. https://doi.org/10.14260/jemds/2015/784.
31. Plissiti, M., & Nikou, C. (2012). Cervical cell classification based exclusively on nucleus features. Available from: https://link.springer.com/chapter/10.1007/978-3-642-31298-4_57.
32. Song, D., Kim, E., Huang, X., Patruno, J., Muñoz-Avila, H., Heflin, J., Rodney Long, L., & Antani, S. (2015). Multimodal Entity Coreference for Cervical Dysplasia Diagnosis. *IEEE Transactions on Medical Imaging, 34*(1), 229–245 Jan 2015.
33. Zhao, M., Chen, L., Bian, L., Zhang, J., Yao, C., & Zhang, J. (2015). Feature quantification and abnormal detection on cervical squamous epithelial cells. *Computational and Mathematical Methods in Medicine, 2015*, 9. Article ID 941680. Hindawi Publishing Corporation. Available at: http://dx.doi.org/10.1155/941680.
34. Duanggate, C., Uyyanonvara, B., & Koanantakul. T. (2008). A review of image analysis and pattern classification techniques for automatic pap smear screening process. In *The 2008 international conference on embedded systems and intelligent technology* (pp. 212–217), Feb 2008.
35. Lu, Z., Carneiro, G., Bradley, A., Ushizima, D., Nosrati, M., Bianchi, A., et al. (2015). Evaluation of three algorithms for the segmentation of overlapping cervical cells. *IEEE Transactions on Image Processing, 24*(4), 1261–1272.

36. Plissiti, M., & Nikou, C. (2012). Ovelapping cell nuclei segmentation using a spatially adaptive active physical model. *IEEE Transactions on Image Processing, 21*(11), 4568–4580.
37. Beucher, S. (1992). *The Watershed Transformation Applied to Image segmentation.* Available from: https://www.researchgate.net/publication/2407235_The_Watershed_Transformation_Applied_To_Image_Segmentation.
38. Nor Ashidi Mat Isa. (2005). Automated edge detection technique for pap smear images using moving K-means clustering and modified seed based region growing algorithm. *International Journal of the Computer, the Internet and Management, 3*(3), 45–59 Sept–Dec 2005.
39. Mahanta, L., & Bora, K. (2013). Hyperchromasia and texture as effective features for analysis of malignancy in Pap smear images. *International Journal of Signal Processing, Image Processing and Pattern Recognition, 6*(4), 451–466.
40. Mahanta, L., Nath, D., & Nath, C. (2012). Cervix cancer diagnosis from pap smear images using structure based segmentation and shape analysis. *Journal of Emerging Trends in Computing and Information Sciences, 3*(2), 245–249.
41. Mahanta, L., & Bora, K. (2012). Analysis of malignant cervical cells based on N/C ratio using pap smear images. *International Journal of Advanced Research in Computer Science and Software Engineering, 2*(11), 341–346.
42. Xu, T., Xin, C., Rodney Long, L., Antani, S., Xue, Z., Kim, E., & Huang, X. (2015). A new image data set and benchmark for cervical dysplasia classification evaluation. In *Machine Learning in Medical Imaging, 2015* (pp. 26–35). Springer International Publishing, LNCS 9352.
43. Athinarayanan, S., & Srinath, M. V. (2016). Robust and efficient diagnosis of cervical cancer in pap smear images using textures features with RBF and kernel SVM classification. *APRN Journal of Engineering and Applied Sciences, 11*(7), 4504–4515.
44. Athinarayanan, S., & Srinath, M. V. (2016). Classification of cervical cancer cells in pap smear screening test. *ICTACT Journal on Image and Video Processing, 06*(04), 1234–1238.
45. Samuel, S., Mathew, A., & Sreekumar, S. (2014). Comparative study between sparse representation classification and classical classifiers on cervical cancer cell images. *International Journal of Advanced Research in Computer and Communication Engineering, 3*(8) Aug 2014.
46. Osareh, A., Mirmehdi, M., Thomas, B., & Markham, R. (2002). Comparitive exudate classification using support vector machines and neural networks. In *5th International Conference* (pp. 413–420). Tokyo, Japan Sept 2002.
47. Cibas, E. S., & Ducatman, B. S. *Cytology Diagnostic Principles and Clinical Correlates* (4th ed.) 2014.
48. Nordin, B. (1989). *The development of an automatic prescreener for the early detection of cervical cancer: algorithms and implementation* (Ph.D. thesis). Uppsala, Sweden: Uppsala University.
49. Tolles, W. E., & Bostrom, R.C. (1956). Automatic screening of cytological smears for cancer: The instrumentation. *Annals of the New York Academy of Sciences, 63*(6), 1211–1218. [PubMed].
50. Spencer, C. C., & Bostrom, R. C. (1962). Performance of the cytoanalyzer in recent clinical trials. *Journal of the National Cancer Institute, 29*, 267–276. [PubMed].
51. Spriggs, A. I., Diamond, R. A., & Meyer, E. W. (1968). Automated screening for cervical smears. *The Lancet, 1*(7538), 359–360. [PubMed].
52. Watanabe, S. (1974). An automated apparatus for cancer prescreening: Cybest. *CGIP, 3*(4), 350–358.
53. Tanaka, N., Ueno, T., Ikeda, H. et al. (1987). CYBEST Model 4. Automated cytologic screening system for uterine cancer utilizing image analysis processing. *Analytical and Quantitative Cytology and Histology, 9*(5), 449–454. [PubMed].
54. Zahniser, D. J., Oud, P. S., Raaijmakers, M. C., Vooys, G. P., & van de Walle, R.T. (1979). BioPEPR: A system for the automatic prescreening of cervical smears. *Journal of Histochemistry and Cytochemistry, 27*(1), 635–641. [PubMed].

55. Erhardt, R., Reinhardt, E. R., Schlipf, W., & Bloss, W. H. (1980). FAZYTAN: A system for fast automated cell segmentation, cell image analysis and feature analysis and feature extraction based on TV-image pickup and parallel processing. *Analytical and Quantitative Cytology, 2*(1), 25–40. [PubMed].
56. Tucker, J. H., Husain, O. A. (1981). Trials with the Cerviscan experimental prescreening device on polylysine-prepared slides. *Analytical and Quantitative Cytology, 3*(2), 117–120. [PubMed].
57. Ploem, J. S. V. N., & van Driel-Kulker, A. M. J. (1987). Leytas—A cytology screening system using the new modular image analysis computer (miac) from leitz. In G. Burger, J. S. Ploem, & K. Goerttler (Eds.), *Clinical cytometry and histometry*. London, UK: Academic Press.
58. Bengtsson, E., & Malm, P. (2014). *Screening for cervical cancer using automated analysis of pap-smears, Pubmed*. Available at: https://www.ncbi.nlm.nih.gov/pmc/articles/PMC3977449/.

Queue Classification for Fraud Types: Banking Domain

Archana Trikha and Sandip M. Khant

Abstract This white paper describes the concept of Queue Management for different types of Fraud and how it can be leveraged using Analytics for retail banking [1–3]. Not managing queues properly poses challenges in terms of operational efficiency as well as there is cost involved in training fraud reviewers for particular fraud types. Data science is vastly used in banking and finance industry for extracting insightful information which can be used for strategic decisions in fraud management [4–7]. This document explains a methodology to implement this concept using Machine Learning technique of Adaptive Boosting which takes care of multinomial classification problem. Analytics can help in classifying the fraud alerts to most likely queue/ types and channelize the alert to relevant investigation team. This helps to optimize the business process by responding to Fraud incident in efficient and effective manner; reduce misclassification cost, providing better customer service by preventing frauds with shorter turnaround time and eventually higher stakeholder value [8–20].

Keywords Fraud risk · Application fraud · Account take over
Counterfeit fraud · Classification · Analytics

1 Background and Business Challenge

Fraud alerts classification and queue management process has become big challenge for financial organizations. The challenges include misclassification cost, additional investigation, operational cost, late discovery of fraud cases and reputation damage. In order to understand better, we can take example of three prime Fraud Types in Banking Industry.

A. Trikha (✉)
Analytics and Insight, TCS, Mumbai, India
e-mail: archana.rakesh@gmail.com

S. M. Khant
Analytics and Insight, TCS, Pune, India
e-mail: smkhant@gmail.com

© Springer Nature Singapore Pte Ltd. 2019
V. E. Balas et al. (eds.), *Data Management, Analytics and Innovation*,
Advances in Intelligent Systems and Computing 839,
https://doi.org/10.1007/978-981-13-1274-8_20

(1) **Application Fraud (AF)**: Application Fraud occurs when fraudster used his own name but provided false information on application—example, false income, false address, incorrect employment details, false income. The use of a false document (such as a pay slip, bank statement or driving licence) when applying for an account, policy, service or insurance claim would also fall within this category.

(2) **Account Take Over (ATO)**: account could be taken over by fraudsters, including bank, credit card, email and other service providers. Online banking accounts are usually taken over as a result of phishing, spyware or malware scams. This is a form of Internet crime or computer crime.

(3) **Counterfeit Fraud (CFT)**: It is the crime of getting private information about somebody else's credit/debit card used in an otherwise normal transaction. The thief can procure a victim's card number using basic methods such as photocopying receipts or more advanced methods such as using a small electronic device (skimmer) to swipe and store hundreds of victims' card numbers. Common scenarios for skimming are restaurants or bars where the skimmer has possession of the victim's payment card out of their immediate view.

In the absence of any fraud classification mechanism, the alerts are classified under single operation analyst bucket with default Fraud type. For example-assume that all alerts are classified as ATO and then investigation is initiated. However, Fraud reviewers are trained to handle only specific type of Fraud, here, ATO. They have limited knowledge about how to handle cases of CFT and AF. This leads to higher time to resolve cases and increases operations cost.

This may lead to higher misclassification cost as well as additional delay to reclassify/channelize the case to proper queue for treatment. Time is crucial to resolve the case and to avoid further fraud losses. Again re-classification of the case to correct queue results in lack of quicker response to manage it, which increases time window for fraudsters to extract more money from deposit account leading to additional fraud losses.

2 Fraud Risk Management Process

Figure 1 describes the components of fraud management followed in banking industry.

(1) **User Activity/Transactions**: The first step describes the customer transactions and their activities across channels which could trigger the potential fraudulent incident in next steps. The user activities may include suspicious activities such as changing password/PIN, etc. This involves—Data Collection, Cleansing and Quality analytics.

(2) **Analytics Engine**: This component helps to score each User activity/ transactions. Scores are generated using significant predictors based on internal customer behavior information like historical patterns of fraud, Transaction history, application history, etc.

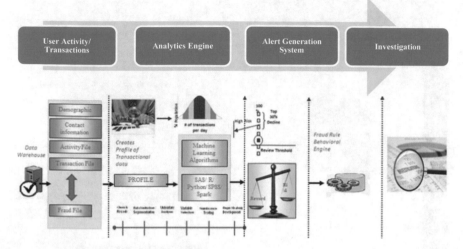

Fig. 1 Fraud risk management process

(3) **Alert Generation System**: High risky fraudulent transactions/cases based on the certain cut-off (threshold) score needs to be prioritized based on sequence and triggers the alert. After evaluation on daily Volume of Incidents, actual fraud, expected False Positive Ratio and Staff capacity, cut-off score is decided. Any incident generating SCORE > CUT-OFF score is passed forward to operations team. This takes into account-
  - Queue Prioritization and performance of strategies
  - Volume Monitoring
  - Operations Capacity Planning
  - Customer Impact Management

(4) **Investigation**: This is the final component of Fraud Management Process where set of alerted incident is processed through the queue for further investigation. Investigation involves reviewing the transactions of customer, checking scores and strategies which triggered it, talking and trying to find whether he/she has done this transaction and gathering artefacts.

3 Analytics Framework for Classification of Multinomial Categories of Fraud Types

Today's complex and evolving financial environment calls for robust frame work and strategies to handle risk management by segregating the fraud alerts in different fraud classes to avoid associated misclassification cost. Broadly following steps are involved in this framework-

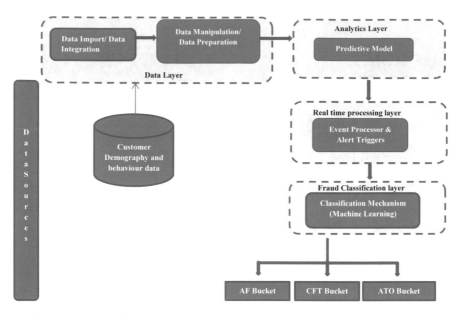

Fig. 2 Analytics modelling framework for fraud alert classification

- Fraud Trend Identification
- Data mining using Statistical/Machine learning Algorithms
- Multinomial classification of different fraud types

Fraud classification leverage analytics by consolidating data from a range of sources, and explore the patterns in the acquired data to come up with Rule sets. These rule sets will help to efficiently classifying the fraud alert and navigate to most appropriate channel for investigation, results in quick turnaround time, smoothen the fraud managing operations and eventually reduce the overall cost of misclassification.

Figure 2 represents the Analytics Modelling Framework for Fraud Alert classification.

In Data layer, data from multiple sources is imported into common database, integrated and standardized for further use. The base data to be imported includes Transaction data and out of pattern behavior, demographic, activity by customer, etc. Once data is imported, data is altered, processed, manipulated. The predictor variables are created at end of the Data preparation stage. Analytics layer involves the use of data mining rules or algorithms using predictive analytics techniques. Predictive analytics models can be built by integrating data mining techniques.

Some examples of activity based features are Number of Times log-on through different channels in recent days, Number of balance inquiry, etc.

Cut-off is decided after evaluating model performance for certain duration. This involves use of application of decision analytics or decision engineering techniques and makes the system robust and dynamic.

Fraud classification layer is the final stage where Statistical/Machine learning techniques are applied in order to classify alerts in different Fraud Category. Algorithm to be considered could be Decision trees, Rule-based classification, SVM or machine learning algorithm such as adaptive boosting or KNN (K Nearest Neighbour), etc. We will discuss further in next section.

The challenge is to come up with classification Rule set/Model which could most accurately classify the fraud types based on limited number of predictor variables such transactions, Non-Monetary activities, Customer demography, etc. Some of the most common predictors used for fraud alert classification are listed below.

The classification Rule set/Model must balance both Prediction Accuracy and Coverage (% fraud captured) for each Fraud Type. Business can set the target in terms of minimum Prediction Accuracy expected in percentage.

The performance of the Fraud classification is measured using performance metric representing both Prediction Accuracy versus % Fraud Captured.

	Actual AF	Actual CFT	Actual ATO	% Predicted fraud
Predicted AF	4700	551	5000	46
Predicted CFT	2212	3411	704	54
Predicted ATO	1534	684	31,756	93
% Actual fraud	56	73	85	

3.1 Type of Classification Techniques

Classification is the data mining technique that uses pre-classified examples to classify the required results. Some of the well-known classification techniques are

- Decision Trees
- Bayesian classification
- Rule-based classification
- SVM (Support Vector Machine)
- Adaptive boosting

After comparing different methods adaptive boosting gave best result to us. In our study and dataset, adaptive boosting method provided best results in terms of accuracy, recall and precision.

There is wide use of *adaptive boosting machine learning* technique for classification. Several rule sets or trees get generated. In first step, a single decision tree is built from the development data. This will contain some errors, say, error for 10 cases. When the second tree is constructed, more weight is given to error cases to get them right. Again, it also will make errors on some cases, and these cases are given more importance during construction of the third classifier. This continues for a finite number of trials, but stops when most recent classifier is either extremely accurate or inaccurate.

Each Rule consists of

- A rule number—this is quite arbitrary and serves only to identify the rule.
- Statistics (n, lift x) or (n/m, lift x) that summarize the performance of the rule. Similarly to a leaf, n is the number of training cases covered by the rule and m, if it appears, shows how many of them do not belong to the class predicted by the rule. The rule's accuracy is estimated by the Laplace ratio $(n-m + 1)/(n + 2)$. The lift x is the result of dividing the rule's estimated accuracy by the relative frequency of the predicted class in the training set.
- One or more conditions that must all be satisfied if the rule is to be applicable.
- A class predicted by the rule.
- A value between 0 and 1 that indicates the confidence with which this prediction is made.

Note: The boosting option described below employs an artificial weighting of the training cases; if it is used, the confidence may not reflect the true accuracy of the rule.

If the applicable rules predict different classes, there is an implicit conflict that could be resolved in several ways: for instance, we could believe the rule with the highest confidence, or we could attempt to aggregate the rules' predictions to reach a verdict. C5.0 adopts the latter strategy.

Each applicable rule votes for its predicted class with a voting weight equal to its confidence value, the votes are totalled up, and the class with the highest total vote is chosen as the final prediction. There is also a default class, which is used when none of the rules apply.

Performance of Individual Rule in rule sets can be measured in terms of coverage and accuracy. An example is provided below

Rule	Predicted fraud class	Actual AF class	Actual CFT class	Actual ATO class	Total	Actual AF %	Actual CFT %	Actual ATO %
AF_RULE_01	AF	14	0	0	14	100	0	0
AF_RULE_02	AF	9	0	0	9	100	0	0
AF_RULE_03	AF	23	0	2	25	92	0	8
AF_RULE_04	AF	34	4	0	38	89.47	10.53	0
AF_RULE_05	AF	699	7	164	870	80.34	0.8	18.85
AF_RULE_06	AF	372	86	48	506	73.52	17	9.49
CFT_RULE_01	CFT	0	45	2	47	0	95.74	4.26
CFT_RULE_02	CFT	0	6	0	6	0	100	0
CFT_RULE_03	CFT	0	18	4	22	0	81.82	18.18
CFT_RULE_04	CFT	627	917	441	1985	31.59	46.2	22.22
ATO_RULE_01	ATO	0	0	27	27	0	0	100
ATO_RULE_02	ATO	0	0	15	15	0	0	100
ATO_RULE_03	ATO	124	23	540	687	18.05	3.35	78.6
ATO_RULE_04	ATO	1623	536	8996	11,155	14.55	4.81	80.65

Important features of the classification Rule sets are listed as below.

- Rule sets are generally easier to understand than trees since each rule describes a specific context associated with a class.
- Furthermore, a rule set generated from a tree usually has fewer rules than the tree has leaves, another plus for comprehensibility. (In this example, the decision tree is so simple that the numbers of leaves and rules are the same.)
- Rule sets can be more accurate predictors than decision trees for very large datasets.
- The attribute usage for a decision tree and for a rule set can be a bit different. In the case of the tree, the attribute at the root is always used (provided its value is known) while an attribute further down the tree is used less frequently. For a rule set, an attribute is used to classify a case if it is referenced by a condition of at least one rule that applies to that case; the order in which attributes appear in a rule set is not relevant.

3.2 Advantage of Using C5.0

- The classification process generates fewer rules compare to other techniques so the proposed system has low memory usage. The large decision tree can be viewed as a set of rules which is easy to understand.
- Error rate is low so accuracy in result set is high and pruned tree is generated so the system generates fast results as compare with other technique.
- Feature selection technique assumes that the data contains many redundant features. So remove that features which provides no useful information in any context. Select relevant features which are useful in model construction.
- Cross-validation method gives more reliable estimate of predictive. Overfitting problem of the decision tree is solved by using reduced error pruning technique. With the proposed system achieve 1–3% of accuracy, reduced error rate and decision tree is construed within less time.
- C5 acknowledges the problem of noise and missing data at considerable extent.

4 Business Benefits of Fraud Classification

Fraud classification at the same stage of fraud detection enables swift detection of patterns in fraud classification. Here are a few other benefits of leveraging this framework.

- Fraud Classification enables high probability prediction for proactively identifying and countering fraud.
- It enables organizations to continuously and comprehensively track the latest fraud trends, customers and their behavior, and any likely attempt at fraud.

- Appending Fraud classification logic with Fraud score generation logic will avoid the additional cost of building separate new system of fraud classification from scratch.
- Based on sensitivity for each fraud class, separate queue, investigation team could be formed to improve turn round time and overall efficiency of fraud risk management.
- This framework frees investigators and data analysts from tedious work, while simplifying the process of visualizing and drawing insights from the data.

5 Conclusion

After comparing different methods for multinomial classification, we got better results using Adaptive boosting which were easy to understand by business due to option of providing rules, easy to implement due to limited features being used and provided more accurate results in terms of True Positives and True Negatives and reasonable fraud capture rate. We also tried using other methods like SVM which had better accuracy but was providing overfitting results across samples. Decision tree did not provide robust accurate results while Bayesian classification was not valid due to assumption of independent predictors.

References

1. Veeramachaneni, K., Arnaldo, I., et al. (2016). AI^2: Training a big data machine to defend. In *IEEE 2nd International Conference on Big Data Security on Cloud*.
2. Sankhwar, S., & Pandey, D. (2016). A safeguard against ATM fraud. In *IEEE 6th IACC*.
3. Coppolino, L., D'Antonio, S., Formicola, V., et al. (2015). Applying extensions of evidence theory to detect frauds in financial infrastructures.
4. Hooi, B., Shah, N., Beutel, A., et al. (2016). Birdnest: Bayesian inference for ratings-fraud detection. In *SIAM International Conference on Data Mining*.
5. PWC. (2016). *Global economic crime survey*. https://www.pwc.com/gx/en/services/advisory/forensics/economic-crime-survey.html.
6. Najafabadi, M. M., Villanustre, F., et al. (2015). Deep learning applications and challenges in big data analytics. *Journal of Big Data*.
7. Jiang, M., Beutel, A., Cui, P., Hooi, B., Yang, S., & Faloutsos, C. (2015). A general suspiciousness metric for dense blocks in multimodal data. In *ICDM 2015 IEEE International Conference*.
8. Sullivan, R. J. (2014). Controlling security risk and fraud in payment systems. *Economic Review—Federal Reserve Bank of Kansas City*; Kansas City (Third Quarter), 5–36.
9. Dal Pozzoloa, A., et al. (2014, August). Learned lessons in credit card fraud detection from a practitioner perspective. *Expert Systems with Applications, 41*(10).
10. Halvaiee, N. S., & Akbari, M. K. (2014). A novel model for credit card fraud detection using Artificial Immune Systems. *Applied Soft Computing*.
11. Gunnemann, S., Unnemann, N. G., & Faloutsos, C. (2014). Detecting anomalies in dynamic rating data: A robust probabilistic model for rating evolution. In *Proceedings of the 20th ACM SIGKDD (International Conference on Knowledge Discovery and Data Mining)*.

12. Jiang, M., Cui, P., Beutel, A., Faloutsos, C., & Yang, S. (2014). Inferring strange behavior from connectivity pattern in social networks. In *Advances in Knowledge Discovery and Data Mining* (pp. 126–138). Springer.
13. Pavía, J. M., Veres-Ferrer, E. J., & Foix-Escura, G. (2012). Credit card incidents and control systems. *International Journal of Information Management*.
14. Soltani, N., Akbari, M. K., & Javan, M. S. (2012). A new user-based model for credit card fraud detection based on artificial immune system. In *CSI International Symposium on Artificial Intelligence and Signal Processing (AISP)*.
15. Jyotindra, N. D., & Ashok, R. P. (2011). A data mining with hybrid approach based transaction risk score generation model (TRSGM) for fraud detection of online financial transaction. *International Journal of Computer Applications, 16*.
16. Krivko, M. (2010). A hybrid model for plastic card fraud detection systems. *Expert Systems with Applications, 37*.
17. Cheng, H., Tan, P.-N., Potter, C., & Klooster, S. A. (2009). Detection and characterization of anomalies in multivariate time series. *SDM* (pp. 413–424). SIAM.
18. Yue, D., Wu, X., Wang, Y., Li, Y., & Chu, C.-H. (2007). A review of data mining-based financial fraud detection research. In *International Conference on Wireless Communications Networking and Mobile Computing* (pp. 5514–5517), September 2007, IEEE.
19. Gao, J., & Tan, P.-N. (2006). Converting output scores from outlier detection algorithms into probability estimates. In *Proceedings of the Sixth International Conference on Data Mining*, US, IEEE.
20. Nikam, S. S. (2015). A comparative study of classification techniques in data mining algorithms. *Oriental Journal of Computer Science and Technology*. http://www.computerscijournal.org/?p=1592.

Portfolio Optimization Framework— Recommending Optimal Products and Services in a Dynamic Setup

Dibyendu Mukherjee

Abstract This paper is aimed at designing an optimal product portfolio recommendation framework for banking or financial institution customers factoring individual situations and member financial readiness aspects. The analytical framework is expected to factor in not only member behavior and related dimensions but also external dynamics like demographic changes, economic landscape, and information technology advancements in a dynamic setup. The framework could be used to empower product line leaders with the right strategy to provide personalized customer experience with right product mix. It could also be used to report portfolio level revenue changes over time and demonstrate customer migration across segments and product adoption changes. These insights are expected to help organizations achieve enhanced customer retention and engagement and quickly adapt to varying external and internal trends.

Keywords Portfolio optimisation · Benchmarking · Enhancement
Transformation · Next best action

1 Introduction

This framework is aimed at developing optimal product portfolio recommendations for banking or financial institution customers factoring individual situations and member financial readiness aspects. The analytical framework is expected to factor in not only member behavior and related dimensions but also external dynamics like demographic changes, economic landscape and information technology advancements in a dynamic setup.

This is expected to help in improving customer experience and marketing effectiveness and that could translate into enhanced customer relationship and

D. Mukherjee (✉)
Analytics and Insights, BFSI Practice Team, TCS, Kolkata, India
e-mail: Dibyendu.Mukherjee1@tcs.com; dibs291275@rediffmail.com

© Springer Nature Singapore Pte Ltd. 2019
V. E. Balas et al. (eds.), *Data Management, Analytics and Innovation*,
Advances in Intelligent Systems and Computing 839,
https://doi.org/10.1007/978-981-13-1274-8_21

loyalty. The framework would also provide the flexibility to fine tune strategy in response to varying external and internal trends.

2 Solution Applications

The analytical solution is expected to provide product recommendations for members accompanied by customized CRM activity. The solution would help in demonstrating the evolution of the portfolio mix at a segment level over time. In turn this would translate in showing portfolio level revenue changes over time and average member level specific value changes.

3 Framework Evolution

The evolution of the framework could be broadly categorized into Benchmarking—Enhancements—Transformation and finally the Optimized state. A dynamic time range stretching from 2016 to 2022 has been considered from an illustrative perspective.

Illustration—Framework Evolution

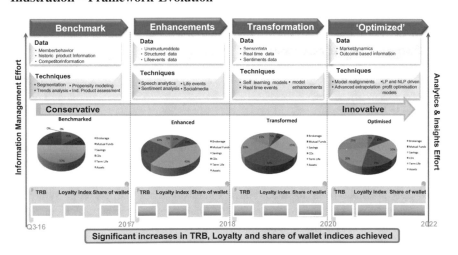

3.1 Benchmarking

Phase Objective The Benchmarking phase of the exercise is aimed at exploring what has been the historic product adoption trend across portfolio segments and identifying product combinations associated with high total relationship balance growth and low member attrition. A simultaneous exercise is also carried out to analyze product adoption trends across the industry and identify product combinations preferred by similar segments in competition portfolios. Historic insights and industry inputs are combined to identify benchmark product sequences and combinations. The phase then focuses on finalizing what is the most suitable product to offer next at a member level so that segment specific winning product sequences can be replicated across all members.

Specified data elements could be leveraged in the Benchmarking phase.

Demographics—Age, Geographic Details, Income, Family Size, Spouse information, Net worth, Education Details, Marital Status, Number of dependents.

Loyalty—Tenure with organisation, No. of product holdings with organisation, Preferred product holding information, e.g., preferred card holding, Share of wallet information, Net promoter score.

Risk Application risk score, Behavior score, Bureau score, Delinquency status and charge-off information.

Customer Needs—Financial Readiness score, Share of relationship, Satisfaction index, Investment objective, risk tolerance level, Service affinity indicators.

Customer Profitability—CLTV Score, Revenue, Acquisition cost, servicing cost, and other costs.

Phase Steps

3.1.1 Historical Trend Analysis is the Very First Step of the Benchmarking Phase and is Effected Through Three Specific Sub-steps

- Segmenting the portfolio using K-means clustering or hybrid clustering approaches.
- Tracking product adoption trends and evaluating them from Total Relationship Balance (TRB) and loyalty perspectives.
- Identifying winning product combination for every segment.

Detailed elaborations of the specified steps are provided in following sections.

The segmentation exercise could be carried out using either K-means or hybrid clustering approaches. The standard data treatment steps pertaining to default treatments, missing value replacements, and outlier treatments would be carried out on the extracted data. On the treated data dimension reduction would be next effected using principal component analysis or variable clustering methods. Using the refined set of dimensions the K-means clustering algorithm would be operational. The K-means approach is a non-hierarchical partitioning approach which is initiated with specification of a set of starting nodes followed by the assignment of observations to those nodes. In the process the centroids are recomputed and iterations follow till convergence. Post clustering development detailed validations and visualizations follow and eventually the segments are finalized. Hybrid clustering approach would also be explored as an alternative methodology for developing the segments.

All data elements specified in the section above would be evaluated as probable dimensions while finalizing the segmentation schema.

In the second step for each and every of the segments so identified over a historic finite (e.g., 3 years) time horizon product relationship opening date, closure date and balance change information are extracted. Moreover total member relationship closure dates in the specified horizon are also tracked. This information representation provides a comparative picture of TRB changes and attrition percentages across product combinations historically.

The third step essentially aims at evaluating the comparative visualization and identifying the winning product sequence for every segment from TRB and loyalty perspectives.

This winning product combination sets an aspirational portfolio goal for the representative member.

3.1.2 Analyzing Industry Wide Product Adoption Trends

A simultaneous exercise needs to be carried out to analyze current product holdings of demographic generation and lifestyle segments in competition portfolios. The probable source of such information could be competitor insight reports and product holding statistics. This information would help in providing realistic inputs about changing product preferences across comparable segments like say app based products for millennial segments and help shape aspirational portfolio goals.

Overall insights from steps 1 and 2 would be used to finalize benchmark product combinations.

Illustration—Benchmark Product Combination Finalization

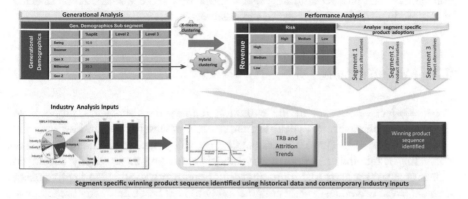

Benchmark product combination is finalized based on historic and industry wide assessment inputs.

It is anticipated that conservative product mixes adopted historically leaves a lot of improvement opportunity from Total Relationship Balance, Loyalty and Share of Wallet aspects.

Armed with insights obtained so far on historic and current product holdings of members and benchmark combinations the objective would now narrow down to identifying what is the most suitable product to offer next at a member level so that segment specific winning product sequences can be replicated across all members.

3.1.3 Next Best Offer Approach

This approach would focus on identifying the next key product priority at member level. This identification would be carried out by evaluating aligned product propensity scores and identifying the product with the highest score.

Similar to the segmentation development exercise data treatments comprising default, missing value replacement and outlier treatment would be carried out for the propensity model development exercise as well. On the treated data variable reduction would be carried out initially through multi-collinearity control using Variance Inflation factor and collinearity diagnostics. In a controlled collinearity state using stepwise regressions the set of final significant predictors would be identified.

Model fine tuning exercises would be next initiated ensuring stability, strength and accuracy benchmarks are duly attained. After finalization of product specific propensity scores an alignment exercise would be carried out to make the scores comparable. The product with the highest propensity score would then be selected as the product to be offered next to the customer.

3.2 Enhancements

Phase Objective As part of the evolution journey the framework proposes to enhance data elements, analytical techniques and processing capabilities over the period of 2016–18. These would entail revamping the segmentation schema, realigning or rebuilding the propensity models and using these freshly developed inputs to recast the next best offer approach. Framework would also involve using member event driven insights in conjunction with next best offer recommendations. These are expected to significantly enhance the customization of product solicitations. Finally the framework would also incorporate a functional relationship to incorporate the influence of changing external dynamics like economic changes, climate pattern variations, and technology advancements on member product preferences. These framework modifications and details are explained in the following sections.
Key incremental data elements that would be incorporated in this phase are as specified.

Data category	Dimensions
Unstructured data	Member call data
	e-mail message data
	Web page visits data
	Web page stay duration
	Social/feedback data
	Free text from customer surveys
Life events	Life stage score
	Job change
	Marriage/divorce
	Pregnancy
	Child birth
	Kid's education
	Major illness
	Bereavement in family
Transactional events	Outlier credit and debit trends
	Bonus payment
	Salary hike
	Salary stop

The enhanced data universe would now be composed of structured, unstructured, and events information.

Using this enriched data analytical framework modifications and incorporation of functional relationship would follow the specified sequence.

1. The segmentation schema would be rebuilt using hybrid clustering approach employing a two stage process. First starting with a K-means* followed by a single link methodology. This would help in achieving quick solution convergence without overlooking any underlying data distribution changes.
2. The propensity models** would be realigned or rebuilt using the enriched data universe.
3. The redeveloped propensity scores would feed as inputs to the revamped Next Best offer approach. Details on Next Best Offer have been explained in the prior section.
4. Triggers would be developed using life stage and transactional member specific event information.
5. These triggers would be overlaid as heuristic rules on propensity based recommendations.

The five staged changes are expected to enhance member targeting customization significantly.

A prime feature of the enhancement phase the functional relationship between external changes and internal product preferences would follow the specified methodology.

1. Average propensity scores guided by the models rebuilt using new data are available at a segment level.
2. Propensity score trends over time are computed.
3. These score trends are then correlated with external economic indicator trends like GDP, inflation or climatic dimensions like temperature change, precipitation trends, etc.
4. Varying correlation coefficients thus computed across product propensity series could help depict varying impacts on product take-ups.
5. These results could also be extrapolated ahead using time series forecasting approaches like Markov Chain to predict future product combinations at time points like 2022.
1. *—Details on K-means clustering methodology are explained in the benchmarking section.
2. **—Details on the propensity model development methodology and related aspects have been explained in the benchmarking section.

The Markov Chain methodology is used for describing systems that follow a chain of linked events where what happens next depends only on the current state of the system. In this framework preferred product combinations at a future state say 2022 could be predicted using information available currently say 2016. Moreover using this same methodology it is also possible to predict states of transition namely

2017–2018–2019, etc. This would enable to depict changing product preferences over time and link it up with distribution changes across portfolio segments.

Enhanced customization and incorporating dynamic influences in predicting member product preferences are expected to result in diversifying product mix for representative members for each and every segment. This in turn would translate into improved TRB, Loyalty, and Share of Wallet indices.

3.3   Transformation

Phase Objective The transformation phase could be viewed as a harmonization between data lake architecture, real-time information driven events, real-time processing capability and self-learning or machine learning models to be executed over the 2018–20 period.

Components

1. The data lake architecture would entail the housing of both structured, unstructured (covering call, e-mail text, web linked information) and real-time member interaction information under a single umbrella. Moreover it would greatly expand the capability to store a much broader window of information and thereby improve the insights from a historical perspective also.
2. Real-time member interaction driven insights could be used to develop triggers and could drive multi-channel contextual marketing efforts. This would imply

– Driving real time and interactive marketing guided by member digital experience
– Moving beyond next best offer to next best action bringing servicing and relationship management customizations at a member level into focus.

A key factor driving the transformation from next best offer to next best action would be propensity scores in conjunction with channel preference indexes. This would help in deciding not only the preferred product but also the suitable channel through which that product has to be solicited or member serviced.

These solicitations could be further customized using real-time events, e.g., environmental (sudden weather developments), geographical data (store location) and operational events (store product stock levels), could play an important in soliciting product offers through mobile-based apps.

3. Real-time processing aspects could be influenced by aligning with in-database analytics concepts that would essentially refer to the integration of data analytics into data warehousing functionality.
4. Self-learning or machine learning models—In the transformation phase random forest as a machine learning modeling technique could be adopted. Random forests are ensemble learning methods for classification, regression, and other tasks. They operate by constructing a multitude of decision trees at training time and outputting the mean prediction of the individual trees. These algorithms are

expected to have distinct advantages over traditional predictive modeling frameworks for example in tackling situations where a pre-defined hypothesis is lacking. Machine learning is less likely to overlook unexpected predictor variables or potential interactions.

Specified data elements could be used incrementally in the transformation phase.

Data category	Dimensions
Real-time events (financial)	**Abrupt increase in account credits/debits**
	Stark increase in ATM withdrawals
	Sharp drop in credit card purchases
Real-time channel usage	**Rapid increase in electronic transfers**
	High frequency web page visits
	Mobile app based purchases
Environmental events	**Stark increase in precipitation trends**
	Change in temperature trends
Sensor data	**Inputs from multiple sensors**
	Accelerometers
	Gyroscopes
	Magnetometers
	Pressure sensors

3.4 Optimized

Phase Objective The optimized state is the final culmination of the analytic framework evolution. This phase focuses on using outcome and monitoring data from prior operationalised analytic frameworks to fine tune next best action strategies. This would result in fine tuning of product offerings through member specific channels of choice. Moreover the phase also looks at profitability projections for member segments to get an idea of which all segments are above and below thresholds. Insights from profitability projections could help in segment specific product channel and pricing strategies.

The following information elements would be critical to support the phase and achieve balanced and optimal portfolio mix across member segments.

Information category	Dimensions
Propensity scorecard monitoring	Population stability index reports
	Characteristic stability index reports
	Post performance monitoring reports
	Scorecard alignment shift indexes

(continued)

(continued)

Information category	Dimensions
Segmentation monitoring reports	Member population distribution across segments
	Product adoption trends across segments
	Extent of deviation from winning product sequences
Profitability projections	Segment specific revenue, cost and loss trends
	Tracking segments above and below Contribution RWA * thresholds

The key objective in this phase would be to fine tune next best action strategies on the basis of insights generated from the specified sequence of actions.

1. Segmentation distribution trends would indicate the shift of member populations across segments. This would be supplemented with segment specific modal portfolio mix values.
2. Deviations of segment specific modal portfolio mixes from the projected optimum as per extrapolations from the enhancement phase will be marked out.
3. Propensity model pre-performance and post performance reports would be instrumental in carrying out health checks on the next best action framework.
4. Based on propensity model monitoring traffic lights realignment/recalibration or rebuild actions would be initiated.
5. Segment specific profitability projections would highlight segments above and below thresholds marking out cases for tailored promotion, price and channel strategies.

The Optimized state is expected to result in a balanced portfolio mix across segments translating into maximized TRB, Loyalty, and share of wallet indices.

Outcomes and Benefits of Framework

The Framework implementation is expected to generate a whole range of positive outcomes.

- Starting off with recommending Benchmark Product portfolio based on statistical models.
- Enhancement phase will provide improved product portfolio considering more attributes (from unstructured data).
- Greater understanding on the customer segmentation; and related customer insights.
- Increase in the Total Relationship Balance, Loyalty Index, Share of Wallet.
- Increase in profitability through better retention plans.
- Efficient cash flow management and increased profitability per unit of capital employed.
- Reduction in portfolio risk.
- Better return on marketing investments through the right marketing mix, pricing and promotion models, and loyalty programs.

4 Conclusion

Overall the proposed framework is expected to provide need based product recommendations and advise for customers and provide the flexibility to fine tune strategy in response to varying external and internal trends. This framework could be associated with business transformation initiatives and is expected to result in increased organizational competitiveness reduced risk of business disruption and greater returns on investment through proactive and forward looking product and service recommendations.

Movie Recommender System Based on Collaborative Filtering Using Apache Spark

Mohammed Fadhel Aljunid and D. H. Manjaiah

Abstract Recently, the building of recommender systems becomes a significant research area that attractive several scientists and researchers across the world. The recommender systems are used in a variety of areas including music, movies, books, news, search queries, and commercial products. Collaborative Filtering algorithm is one of the popular successful techniques of RS, which aims to find users closely similar to the active one in order to recommend items. Collaborative filtering (CF) with alternating least squares (ALS) algorithm is the most imperative techniques which are used for building a movie recommendation engine. The ALS algorithm is one of the models of matrix factorization related CF which is considered as the values in the item list of user matrix. As there is a need to perform analysis on the ALS algorithm by selecting different parameters which can eventually help in building efficient movie recommender engine. In this paper, we propose a movie recommender system based on ALS using Apache Spark. This research focuses on the selection of parameters of ALS algorithms that can affect the performance of a building robust RS. From the results, a conclusion is drawn according to the selection of parameters of ALS algorithms which can affect the performance of building of a movie recommender engine. The model evaluation is done using different metrics such as execution time, root mean squared error (RMSE) of rating prediction, and rank in which the best model was trained. Two best cases are chosen based on best parameters selection from experimental results which can lead to building good prediction rating for a movie recommender.

Keywords Recommender systems · Collaborative filtering · Alternating Least Squares · Apache Spark · Big data · MovieLens dataset

M. F. Aljunid (✉)
Mangalore University, Mangalore, Karnataka, India
e-mail: Ngm505@yahoo.com

D. H. Manjaiah (✉)
Department of Computer Science, Mangalore University, Mangalore
Karnataka, India
e-mail: drmdh2014@gmail.com

© Springer Nature Singapore Pte Ltd. 2019
V. E. Balas et al. (eds.), *Data Management, Analytics and Innovation*,
Advances in Intelligent Systems and Computing 839,
https://doi.org/10.1007/978-981-13-1274-8_22

283

1 Introduction

In recent times, big data is becoming one of the newest research interests in the areas of computer science and other related areas. With the possibility of a radical change in companies and organizations that use the information for improving the customer experience and transform their business models. Big data has several features which are volume, velocity, variety, value, and veracity. Big data is facing difficulties in managing using conventional tools, techniques, and procedures. Big data analytics is used for handling bulk quantities of data. It is used to mine and extract patterns, information, and knowledge from the data in an effective way. Big data analytics become an important trend for organizations and enterprises that are interesting in providing innovative ideas for enhancing and increasing their business performance and decision-making. RS are a group of techniques that allow filtering through large samples and information space in order to give suggestion to users when needed. Currently, RS are becoming highly popular and utilized in different areas such as movies, research articles, search queries, news, books, social tags, and music. Furthermore, there are other essential RS basically applicable for specialist, collaborators, funny story, restaurant and hotels, dresses, monetary services, life insurance, passion associates which give online dating services and several other social media such as Twitter, LinkedIn, and Facebook.

RS use a number of different technologies to filter out best suit results and provide to users to satisfy their information need. RS are classified into three broad groups which are content-based systems, collaborative filtering systems, and hybrid recommender system [1]. Content-based systems which try to test the behavior of the item which is labeled as recommended one. It works by learning the behavior of the new users based on their information need presented in objects whereby the user has rated. It is a keyword-specific RS where the keywords are used to illustrate the items. Thus, in a content-based RS, models work in such a way that they recommend users' comparable items that have been liked in the past or is browsing currently. For instance, if a MovieLen user has to browse several comedies movies, then, the RS will classify those movies into the database as getting the most ratings on the comedy varieties. Collaborative filtering system is based on similarity measures between user's information need and the items. The items recommended to a new user are those which were liked by other similar users in previous browsing history. Collaborative filtering algorithm uses an average rating of objects, recognizes similarities between the users on the basis of their ratings, and generates new recommendations based on inter-user comparisons. However, it faces many challenges and limitation such as data sparsity whose role is to the evaluation of large item set. Another limitation is hard to make prediction based on nearest neighbor algorithm, third is scalability in which number of users and number of items both increases, and the last one is cold start where poor relationship among like-minded people. To solve encounters, above mentioned, we moved to other approaches of collaborative filtering, and we landed up on model-based collaborative filtering [2]. Hybrid RS performs their tasks by

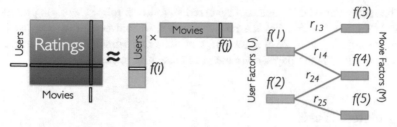

Fig. 1 Low rank factorization matrix [3]

considering the combining behavior of content-based and collaborative filtering techniques in such a way that it suits a particular item. Hybrid recommended system is regarded as the most frequently used RS system considered by many companies due to its ability to eliminate any weakness that might have arose when one RS is employed and in addition, its strength is the composite of more than two RS.

The main focus of this work is collaborative filtering system. It is well known that collaborative filtering could be described as a procedure whereby automatic prediction (i.e., filtering) about the interests of a user is made by gathering taste or preferences information from many users. The unexpressed assumption of the collaborative filtering approach can be best explained, viz., supposing a person A has similar opinion with person B on a particular issue, the assumption is that person A will be more likely to have the same opinion as person B on a different issue X did the opinion on X of a person chosen randomly [3]. Take for an instance the movie "RS" depicted in Fig. 1 which started with a matrix whose entries are movies rated by users. Both user (shown in green) and a particular movie (shown in blue) are represented each by column and rows respectively. Owing to the fact that not all users have rated all movies, all the entries in the matrix are unknown, which necessitate the need for collaborative filtering. There are ratings for only a subset of the movies for each user. With collaborative filtering, the idea is to approximate the rating matrix by factorizing it as the product of two matrices. That is the one that describes properties of each user (shown in green), and the other describing properties of each movie.

The minimization of the error for the users/movies pairs was chosen as the basis for the selection of the two matrices. The alternating least squares algorithm (ALS) which achieves this by randomly filling the user's matrix with values before optimizing the value of the movies was used for this purpose. The value of the user's matrix is optimized with the movie's matrix being kept constant (Fig. 1). Owing to a fixed set of user factors (i.e., values in the user's matrix), known ratings are employed to find the best values by optimizing the movie factors, written on top of the figure. The best user factor with the fixed movie factors is sleeted. This paper, reports for the first time, a movie recommendation system based on collaborative filtering using apache spark. The performance analysis and evaluation of proposed approach are performed on a MovieLens dataset. From the results obtained, it is concluded that the selection of parameters of ALS algorithms can affect the performance of recommender engine to be used.

The remainder of this paper is organized as follows: related work is provided in Sect. 2. Section 3 introduces the proposed movie recommender system using collaborative filtering with ALS algorithm while the experimental study is introduced in Sect. 4. Finally, the paper conclusion is presented in Sect. 5.

2 Related Work

So far, several researchers introduced and presented research in the area of building recommendation systems. Wei et al. [4] proposed a hybrid recommender model to address the cold start problem, which explores the item content features, learned from a deep learning neural network and applies them to the timeSVD++ CF model. A hybrid recommendation model is proposed which combines a time-aware model timeSVD++ with a deep learning architecture SDAE to address the cold start problem of collaborative filtering recommendation models. Kupisz and Unold [5] developed and compared item-based collaborative filtering algorithm using two cluster computing frameworks normally Hadoop's disk-based MapReduce paradigm and Spark's in-memory based RDD paradigm. In order to enhance the reliability, scalability, and to improve processing ability of large-scale data, Zeng et al. [6] proposed PLGM. In their work, two matrix factorization algorithms were considered, which are ALS and SGD. The parallel matrix factorization based on SGD was implemented on spark and was compared with ALS in MLib for its performance. The advantage and disadvantage of each model based on test results were analyzed. A variety of profile aggregation approaches were studied and the model which gives the best result was adopted. Models such as PLGM and LGM were studied in terms of efficiency and accuracy. Dianping, Lakshmi et al. [7] used item-based collaborative filtering techniques. In this method, they first inspect the user item rating matrix and they categorize the relationships among different items, and they utilize these relationships so as to figure out the recommendations for the user. A new concept namely movie swarm mining was proposed by Halder et al. [8] using format frequent item mining and two pruning rules. It addresses the problem of item recommendation and thus gives an idea about the user interests and famous movies trend. This technique can be very helpful for movie producers to manage their new movies. In addition to this, a new algorithm was proposed to recommend movies to a new user. A scalable method for building recommender systems based on similarity join has been proposed by Dev et al. [9]. MapReduce framework was used to design the system in order to work with big data applications. The unnecessary computation overhead such as redundant comparisons in the similarity computing phase can significantly be reduced by the system using a method called extended prefix filtering (REF). Chen et al. [10] used co-clustering with augmented matrices (CCAM) to design several methods including a heuristic scoring, traditional classifier, and machine learning to build a recommendation system and integrate content-based collaborative filtering for a hybrid recommendation system. Similarly, a collaborative filtering algorithm based on the ALS, as a powerful matrix decomposition algorithm, has been proposed by Wilkinson and Schreiber [11]. They found out that it can be awesome to extend to the distributed computing and solve the data sparse problem.

3 Proposed Movie Recommender System

This section provides the idea of the proposed system. The proposed system is a movie recommender system based on ALS using Apache Spark. The novelty of this work is based on the selection of parameters of ALS algorithms that can affect the performance of building of a movie recommender system.

3.1 Proposed System Block Diagram

In this work, we apply user's ratings from the datasets the popular website like IMDB, Rotten Tomatoes, MovieLen, and Time Movie Ratings. This dataset is available in many formats such as CSV file, text file, and databases. We can either stream the data live from the websites or download and store them on our local file system or HDFS. Spark streaming is used to stream real-time data from the various source like Twitter, the stock market, and geographical system and perform powerful analytics to businesses. It used for processing real-time streaming data. We use collaborative filtering (CF) to predict the ratings of users for particular movies based on their ratings for other movies. Then collaborate this with another user's rating for that particular movie. We train the ALS algorithm using MovieLen data and get the results from the machine learning model. We use spark SQL's data frame, dataset, and SQL service to store the data. The result of the machine learning model is stored in RDBMS so that the web application can display the recommendation to a particular use. The results of the movie recommendation system are stored in our local drive. We store the recommendation movies along with the ratings in a text file and CSV file formats. We prefer storing the result into an RDBMS system so as to access it directly from the web application and display recommendation and top movies as shown in Fig. 2.

3.2 Proposed System Steps

This subsection provides the steps of applying the ALS algorithm on MovieLens datasets for train and test the selection of best parameter when building a movie recommendation system.

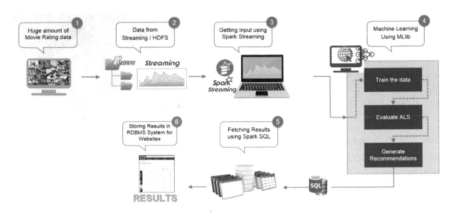

Fig. 2 Proposed movie recommendation system using CF with ALS

Movie Recommendation System using CF with ALS
_____

Input: MovieLens Dataset
Output: Top Recommended Movies.
 Procedures:
Procedure 1:Parsing and loading datasets
Procedure 2: Recognize the user as new or regular.
                    If new user goto **Procedure** 5
Procedure 3: Load training and test data into the table (userId, movieId, rating)
                    def parse_the_rating(line):
                        x = line.split()
                        return (int (x [0]), int (x [1]), float (x [2]))
                    training = sc.TrainingFile("__").map(parse the Rating).cache()
                    test = sc.Testfile("__").map(parse_the_Rating)
 Procedure 4: Train the recommender model.
                    New_model= ALS.train (rank, train, iteration)
Procedure 5:Create predictions on (user, movie) pairs from the test data
            Predict = New_model.predictAll (test.map(lambda x: (x[0], x[1]))
Procedure6: Adding new user ratings
Procedure 7: Display top N recommended movies.
Procedure 8: Save the New_model

4   Experimental Study

This section presents the experimental setup and results in discussion and analysis.

4.1   Apache Spark

Apache Spark [12] is a rapid and general-purpose cluster computing system. It introduces   high-level   application   programming   interfaces   (APIs)   using programming languages such as Java, Python, Scala, and R, and has an engine that supports general execution graphs. It also supports a good set of higher level tools involving Spark SQL for structured data processing, MLlib for machine learning, GraphX for graph processing, and Spark Streaming for real-time applications. It was built on top of Hadoop and MapReduce and extends the MapReduce Model to efficiently use more types of computations. Spark application runs as a separate set of process on the cluster. All of the distributed processes are coordinated by a SparkContext object in the drive program. SparkContext connects to one type of cluster manager (standalone/Yarn/Mesos) for resource allocation across clusters. Cluster manager provides executors, which are essentially JVM process to run the logic and store application data. Then the SparkContext object sends the application code (jar files/python scripts) to executors. Finally, the SparkContext executes tasks in each executor.

4.2 Data Preprocessing

The dataset which is used in this work is MovieLens dataset. This dataset contains 24 million ratings and 670,000 tag applications applied to 40,000 movies by 260,000 users. This dataset contains three files called ratings.csv, movies.csv and tags.csv. ratings.csv contains tree column (userId, movieId, rating). While movies. csv contains movieId, title, genres. The genres have the format: Genre1, Genre2, Genre3. The tags file (tags.csv) has the format: userId, movieId, tag, timestamp and finally, the links.csv file has the format: movieId, imdbId, tmdbId. We can split the data into three portions which are training, validation, and test data to parse their lines once they are loaded into RDDs. Parsing the movies and rating files yields two RDDs: For each row in the ratings dataset, we have created a vector of (userId, movieId, rating). During preprocessing, we have dropped the timestamp attribute because we do not need it for this recommender. Similarly, each row in the movies dataset, we have created a vector of (movieId, title). We have dropped the genres attribute because we do not use it for this recommender.

In order to determine the best ALS parameters for our experiments, we need to break up the ratings RDD dataset into three pieces as follows: a training set which we will use 60% of the data to train models, validation set, which used 20% of the data to choose the best model and test set, which used 20% of the data for our experiments to randomly split the dataset into the multiple groups.

4.3 Experimental Environment

The test has been done on a machine which contains the subsequent descriptions P. A machine with Ubuntu 14.04 LTS, 4 GB memory, and Intel® Core™ i5-2400 CPU @ 3.10 GHz × 4 processor as well as a hard disk of 500 GB. In this machine,

Apache Spark with version 2.1.1 is installed and is used to develop the proposed system. The dataset which is used in research work is MovieLens dataset [13]. In the proposed model, root mean squared error (RMSE) is used as a performance measure. RMSE works by measuring the difference between error rate a user gives to the system and the predicted error by the model. Equation (1) depicts how RMSE works on movie recommender system.

$$\text{RMES} = \sqrt{\sum_{i=0}^{n} \frac{(x_{ui} - y_{gi})^2}{n}} \tag{1}$$

whereby x_{ui} is the rating that user u gives to an item i in the experimental data, y_{gi} is a predicted rating that the movie that user u gives to an item and where n is the number of ratings in the test data.

4.4 Experimental Results Analysis and Discussion

Recommender system (RS) is becoming growingly popular. In this work, Apache Spark is used to demonstrate an efficient parallel implementation of a collaborative filtering method using ALS. ALS is used for dimensionality reduction purpose which helps in overcoming the limitations of collaborative filtering such as data sparsity and scalability. The challenges of data sparsity are appearing in numerous situations, specifically, another problem, when a new an item or user has just added to the system, it is difficult to find similar ones since there is no sufficient information, this problem is called cold start problem [14, 15]. When selecting the ALS algorithm as a part of building the proposed movie recommender system, there is basic parameter through them can determine the best rating of users for given movies. These parameters are Rank, Iterations, and Lambda.

The contribution of this paper is to study and determine the selection of parameters that affect the performance of ALS model in building a movie recommender system because from literature study, it is found that little research work focused on the study of the selection of ALS's parameters that can affect its performance in building a movie recommender engine using Apache Spark. The parameters, lambda, and iterations are used in order to control and adjust the predicting capability of matrix factorization which is depending on ALS technique which in turn affect the evaluation of movie RS. The iterations and lambda parameters are used as follows: Lambda which specifies the regularization parameter in ALS and iterations in which the proposed model should run the specified number of iterations. The ALS algorithm achieves its optimal solution between 5 and 20 iterations.

The parameters lambda and iteration in ALS model are used with different thresholds to realize the effects of matrix factorization performance on the performance of recommendation results and thus take the most appropriate parameters for the following test setups. Tables 1, 2 and 3 show the performance of movie recommendation engine based on ALS under different values of lambda and iteration. Table 1 illustrates the execution of time with the changes of lambda with iterations parameters of ALS model, while Table 2 the rank of best-trained model with the changes of lambda with iterations parameters of ALS algorithm, finally Table 3 indicates the RMSE with the changes of lambda with iterations parameters of ALS model. The results presented in Table 1 indicate that when lambda is set to **0.6** and iteration set is **10**, the time value is minimum which is **1.41323 s**, and rank value is **8** as shown in Table 2. Moreover, the RMSE register for this rating is **1.07424** as indicated in Table 3. On the other hand, as it is indicated in Table 1 when lambda is set to **0.2** and iteration is **15**, running time becomes **1.463743** and rank is **12** for this item as shown in Table 2. The RMSE value for this item is the minimum, which is **0.9167**, as presented in Table 3.

As mentioned above, the analysis for movie recommendation system is done using three quality metrics which are RMSE, time, and rank. Using these three metrics, two cases are achieved as shown in Table 4, case 1 with high time and low RMSE rate while the case 2 with low time and high RMSE rate. According to results in Table 4, the prediction for Top 25 movies is shown in Figs. 3 and 4.

Table 1 Time of matrix factorization using lambda and iteration parameters

Lambda	Iteration				
	5	10	15	20	25
0.1	1.489	1.454	1.473	1.469	1.512
0.2	1.485	1.437	**1.464**	1.438	1.514
0.3	1.472	1.481	1.4671	1.441	1.494
0.4	1.658	1.486	1.476	1.495	1.473
0.5	1.431	1.492	1.468	1.478	1.528
0.6	1.615	**1.413**	1.442	1.459	1.480
0.7	1.443	1.475	1.471	1.446	1.543
0.8	1.554	1.470	1.459	1.449	1.527
0.9	1.491	1.478	1.482	1.471	1.446

Table 2 Rank of matrix factorization using lambda and iteration parameters

Lambda	Iteration				
	5	10	15	20	25
0.1	4	12	4	4	4
0.2	8	12	**12**	12	12
0.3	4	8	8	8	8
0.4	4	8	8	8	8
0.5	8	8	8	8	8
0.6	8	**8**	8	8	12
0.7	12	12	4	4	12
0.8	12	12	4	4	4
0.9	12	4	4	4	4

Table 3 RMSE of matrix factorization using lambda and iteration parameters

Lambda	Iteration				
	5	10	15	20	25
0.1	0.947	0.942	0.940	0.938	0.938
0.2	0.919	0.917	**0.9167**	0.917	0.917
0.3	0.941	0.941	0.941	0.941	0.941
0.4	0.975	0.980	0.980	0.981	0.981
0.5	1.018	1.024	1.024	1.024	1.024
0.6	1.069	**1.074**	1.074	1.074	1.074
0.7	1.127	1.130	1.131	1.131	1.131
0.8	1.192	1.193	1.193	1.193	1.193
0.9	1.261	1.261	1.261	1.261	1.261

Table 4 Two cases for selecting parameters for ALS

Metrics	Case	
	Case 1	Case 2
Time	1.41323	**1.463743**
Rank	8	**12**
RMSE	1.07422	**0.9167**

```
My 25 highest rated movies as predicted (for movies with more than 75 reviews):
+-------+------+----------+-------+-----+-----------------+--------------------+
|movieId|userId|prediction|movieId|count|          average|               title|
+-------+------+----------+-------+-----+-----------------+--------------------+
| 159817|    0| 3.8603234| 159817|  193| 4.450777202072539| Planet Earth (2006)|
|    318|    0| 3.8078098|    318|84455|  4.43308862707951|Shawshank Redempt...|
| 160718|    0|  3.797955| 160718|   88| 4.232954545454546|        Piper (2016)|
| 134849|    0|  3.745416| 134849|  120| 4.041666666666667|   Duck Amuck (1953)|
|    858|    0| 3.7449322|    858|53547| 4.343623358918333|Godfather, The (1...|
|  26048|    0| 3.7382393|  26048|   76|3.9539473684210527|Human Condition I...|
|     50|    0| 3.7262065|     50|56348| 4.308635621494996|Usual Suspects, T...|
| 100044|    0| 3.722707| 100044|   97| 4.15979381443299| Human Planet (2011)|
| 142115|    0| 3.7207592| 142115|  150|             4.31|The Blue Planet (...|
|   2019|    0| 3.7023807|   2019|13394| 4.256196804539346|Seven Samurai (Sh...|
|  94466|    0| 3.7008467|  94466| 2477|4.2426322163907955| Black Mirror (2011)|
| 162376|    0| 3.6991944| 162376|  703| 4.312944523470839|     Stranger Things|
|  77658|    0| 3.6968207|  77658| 1740|4.1818965517241375|        Cosmos (1980)|
|    527|    0| 3.6902485|    527|63889| 4.2759629983252205|Schindler's List ...|
|    904|    0| 3.6900668|    904|20443| 4.238590226483393|   Rear Window (1954)|
|  26082|    0| 3.6886785|  26082|  542| 4.100553505535055|Harakiri (Seppuku...|
|   1178|    0| 3.6881678|   1178| 4064|  4.20435531496063|Paths of Glory (1...|
|    922|    0| 3.6870744|    922| 7606|  4.2058900867736|Sunset Blvd. (a.k...|
|   7926|    0| 3.6856618|   7926|  703|4.105263157894737|High and Low (Ten...|
|   1212|    0| 3.6842163|   1212| 7412| 4.214854290339989|Third Man, The (1...|
|   6669|    0| 3.6837091|   6669| 1340|4.1093283582089555|        Ikiru (1952)|
|   3030|    0| 3.6817052|   3030| 3988| 4.186434302908726|       Yojimbo (1961)|
| 127052|    0| 3.675985| 127052|  160|              4.1|Operation 'Y' & O...|
|   3435|    0| 3.672548|   3435| 5421| 4.203836930455635|Double Indemnity ...|
|   2920|    0| 3.6651812|   2920| 1115|4.1246636771300444|Children of Parad...|
+-------+------+----------+-------+-----+-----------------+--------------------+
                          only showing top 25 rows
```

Fig. 3 Prediction of top 25 movies for case 1

```
My 25 highest rated movies as predicted (for movies with more than 75 reviews):
+-------+------+----------+-------+-----+------------------+--------------------+
|movieId|userId|prediction|movieId|count|           average|               title|
+-------+------+----------+-------+-----+------------------+--------------------+
| 159817|     0| 4.6047864| 159817|  193| 4.450777202072539|  Planet Earth (2006)|
|    318|     0| 4.5063915|    318|84455|  4.43308862707951|Shawshank Redempt...|
| 160718|     0| 4.466501| 160718|   88| 4.232954545454546|         Piper (2016)|
| 134849|     0| 4.4416428| 134849|  120| 4.041666666666667|    Duck Amuck (1953)|
|    858|     0| 4.429425|    858|53547| 4.343623358918333|Godfather, The (1...|
|    527|     0| 4.413663|    527|63889|4.2759629983252205|Schindler's List ...|
| 100044|     0| 4.395683| 100044|   97| 4.15979381443299| Human Planet (2011)|
| 100553|     0| 4.3911767| 100553|  239|4.062761506276151|Frozen Planet (2011)|
|     50|     0| 4.387128|     50|56348| 4.308635621494996|Usual Suspects, T...|
|    912|     0| 4.383373|    912|29001| 4.220164821902693|    Casablanca (1942)|
|    904|     0| 4.382594|    904|20443| 4.238590226483393|   Rear Window (1954)|
|  77658|     0| 4.3616266|  77658| 1740|4.1818965517241375|         Cosmos (1980)|
|  77177|     0| 4.3589344|  77177|   91| 3.791208791208791|    Wild China (2008)|
|   1203|     0| 4.3545713|   1203|15938| 4.224777261889823| 12 Angry Men (1957)|
|   2019|     0| 4.3540235|   2019|13394| 4.256196804539346|Seven Samurai (Sh...|
| 108583|     0| 4.3512716| 108583|  921| 4.074918566775244|Fawlty Towers (19...|
| 127052|     0| 4.349377| 127052|  160|               4.1|Operation 'Y' & O...|
| 142115|     0| 4.3470507| 142115|  150|              4.31|The Blue Planet (...|
|  44555|     0| 4.343796|  44555| 8241|4.1996723698580265|Lives of Others, ...|
|   1207|     0| 4.3423305|   1207|16738| 4.162474608674872|To Kill a Mocking...|
|    908|     0| 4.339527|    908|18467| 4.208669518600748|North by Northwes...|
|  93040|     0| 4.3393526|  93040|  380| 4.044736842105263|Civil War, The (1...|
|  94466|     0| 4.3387938|  94466| 2477|4.2426322163907955|  Black Mirror (2011)|
|  82143|     0| 4.3382115|  82143|  295| 3.906779661016949|Alone in the Wild...|
| 162376|     0| 4.3322854| 162376|  703| 4.312944523470839|     Stranger Things|
+-------+------+----------+-------+-----+------------------+--------------------+
                        only showing top 25 rows
```

Fig. 4 Prediction of top 25 movies for case 2

In general, the lowest value of the RMSE is considered the best case for prediction in building recommendation system. Therefore, we will adopt the second case because the value of the RMSE is smaller compared to the value in the first case as well as adopt the second case as the best case because there is no significant difference in the amount of time execution between the two cases. Now, we can get the top recommended movies by using the second case. Finally, we concluded that from these results the best case is the second case which has the best value for RMSE, which can be useful for building recommendation engines for predicting the top 25 ranked movies.

5 Conclusion and Future Work

Movie recommender system plays a significant role in identifying a set of movies for users based on user interest. Although many move recommendation systems are available for users, these systems have the limitation of not recommending the movie efficiently to the existing users. This paper presented a movie recommender system based on collaborative filtering using Apache Spark. From the results, the selection of parameters of ALS algorithms can affect the performance of building of a movie recommender engine. System evaluation is done using various metrics such as execution time, RMSE of rating prediction, and rank in which the best model was trained. Two best cases are chosen based on best parameters selection from experimental results which can lead to building god prediction rating for a movie recommender engine. From these cases, the lowest value of the RMSE is considered the best case for prediction in building movie recommendation system. Therefore, the second case is recommended to be used since the value of the RMSE is smaller compared to the value in the first case as well as adopt the second case as the best case, because there is no significant difference in the amount of time execution between the two cases. Finally, we concluded that from these results that the best case is the second case which has the best value for RMSE, which can be useful for building recommendation engines for predicting the top 25 ranked movies. In the future work, we plan to develop and improve a new loss function because of the shortcomings of the recommender system algorithm based on ALS model based on the parameter of the best case which has the best value for RMSE using Apache Spark.

References

1. Verma, J. P., Patel, B., & Patel, A. (2015). Big data analysis: Recommendation system with Hadoop framework. In *2015 IEEE International Conference on Computational Intelligence & Communication Technology (CICT)*. IEEE.
2. Katarya, R., & Verma, O. P. (2016). A collaborative recommender system enhanced with particle swarm optimization technique. *Multimedia Tools and Applications, 75*(15), 9225–9239.
3. https://docs.databricks.com/_static/notebooks/cs100x-2015-introduction-to-big-data/module-5–machine-learning-lab.html.
4. Wei, J., et al. (2016). Collaborative filtering and deep learning based hybrid recommendation for cold start problem. In *2016 IEEE 14th International Conference on Dependable, Autonomic and Secure Computing, 14th International Conference on Pervasive Intelligence and Computing, 2nd International Conference on Big Data Intelligence and Computing and Cyber Science and Technology Congress (DASC/PiCom/DataCom/CyberSciTech)*. IEEE.
5. Kupisz, B., & Unold, O. (2015). Collaborative filtering recommendation algorithm based on Hadoop and Spark. In *2015 IEEE International Conference on Industrial Technology (ICIT)*. IEEE.
6. Zeng, X., et al. (2016). Parallelization of latent group model for group recommendation algorithm. In *IEEE International Conference on Data Science in Cyberspace (DSC)*. IEEE.
7. Ponnam, L. T., et al. (2016). Movie recommender system using item based collaborative filtering technique. In *International Conference on Emerging Trends in Engineering, Technology, and Science (ICETETS)*. IEEE.
8. Halder, S., Sarkar, A. M. J., & Lee, Y.-K. (2012). Movie recommendation system based on movie swarm. In *2012 Second International Conference on Cloud and Green Computing (CGC)*. IEEE.
9. Dev, A. V., & Mohan, A. (2016). Recommendation system for big data applications based on set similarity of user preferences. In *International Conference on Next Generation Intelligent Systems (ICNGIS)*. IEEE.
10. Chen, Y.-C., et al. (2016). User behavior analysis and commodity recommendation for point-earning apps. In *2016 Conference on Technologies and Applications of Artificial Intelligence (TAAI)*. IEEE.

11. Zhou, Y. H., Wilkinson, D., & Schreiber, R. (2008). Large scale parallel collaborative filtering for the Netflix prize. In *Proceedings of 4th International Conference on Algorithmic Aspects in Information and Management* (pp. 337–348). Shanghai: Springer.
12. https://spark.apache.org/docs/latest/. Accessed March 10, 2017.
13. https://grouplens.org/datasets/movielens/. Accessed May 15, 2017.
14. Delgado, J. A. (2000, February). *Agent-based information filtering and recommender systems on the internet* (Ph.D. thesis). Nagoya Institute of Technology.
15. Mooney, R. J., & Roy, L. (1999). Content-based book recommendation using learning for text categorization. In *Proceedings of the Workshop on Recommender Systems: Algorithms and Evaluation (SIGIR '99)*. Berkeley, CA, USA.

Design a Smart and Intelligent Routing Network Using Optimization Techniques

Joyce Yoseph Lemos, Abhijit R. Joshi, Manish M. D'souza and Archi D. D'souza

Abstract The Routing Problem (RP) is designed to find the minimal set of routes in order to deliver the services to a set of customers. There are many techniques available to solve RP problems such as Ant Colony Optimization (ACO), Honey Bee Optimization (HBO), Particle Swarm Intelligence, Genetic Algorithm (GA) and much more. Our aim is to consider the various RP techniques and compare the performance of each technique based on various parameters. To carry out this, we have considered the School Bus Routing Problem (SBRP), which ensures the on-time delivery of students to and from school by minimizing the total travel cost in terms of time and distance. The result of this research work helps to determine the best RP algorithm not only based on shortest path but also additional parameters such as pheromone level, optimum path level and distance.

Keywords Routing problem · Pheromone trail · Vehicle routing problem
Ant colony optimization · Greedy randomized approach · Honey Bee Optimization

J. Y. Lemos (✉)
St. John College of Engineering and Management, Palghar, India
e-mail: lemosjoyce01@gmail.com

A. R. Joshi
D. J. Sanghvi College of Engineering, Mumbai, India
e-mail: abhijit.joshi@djsce.ac.in

M. M. D'souza
Media.Net Corporation, Mumbai, India
e-mail: dsouzamanish@gamil.com

A. D. D'souza
Illinois Institute of Technology, Chicago, USA
e-mail: archidsouza18@gmail.com

© Springer Nature Singapore Pte Ltd. 2019
V. E. Balas et al. (eds.), *Data Management, Analytics and Innovation*,
Advances in Intelligent Systems and Computing 839,
https://doi.org/10.1007/978-981-13-1274-8_23

297

1 Introduction

The Routing Problem (RP) also referred to as Vehicle Routing Problem (VRP) is a combinatorial optimization problem, which deals with transportation of goods from service point to customers [1]. The VRP problem is solved in two parts, first, by determining the set of stops to be visited and second, finding the set of minimal number of routes between source and destination by traversing each route in routing network. The VRP problem considers the various parameters to do so, such as vehicle capacity, driver's riding time, time window of product delivery and minimum transportation cost.

The RP was first implemented at the end of 50s to solve the problem of transportation of gasoline from service station to customer point. At the same time, some researchers were set the mathematical programming formulation and algorithmic approach to solve the same problem. The definition of RP states that initially the set of vehicles 'm' is located at a depot in order to deliver the fixed quantities of goods to a set of customers 'n' [2, 3]. The RP problem is represented by determining the set of optimal routes when serving the set of customers by set of vehicles. The objective of RP is to minimize the overall transportation cost in terms of time and distance. The result of the RP problem is a set of routes starting from source to destination (which all begin and end in the depot), and which satisfies the constraint that all the customers are served only once. The transportation cost is depending on total travel distance and the required number of vehicles. The majority of the real world problems are often much more complex than the classical RP. Therefore in practice, the classical VRP problem considers the constraints such as vehicle capacity and time interval between the deliveries of services by ensuring that each customer has to be served only once. RP is hard combinatorial optimization problem which works only for small instances of problem. If the set of instances is increased, the RP does not guarantee the optimality of result.

From the last 20 years, the meta-heuristic approach has given the proper direction to solve the problems from VRP family. The School Bus Routing Problem (SBRP) was proposed in 1969 [1, 2]. The SBRP was represented as efficient scheduling of school buses in order to pick the students from various stops and delivers them to the school [4].

In the existing combinatorial optimization problems, a single sub-problem or combination of them depends on set of instances and complexity of algorithm. The SBRP's main objective is to obtain the minimum set of routes to serve the students on time and to minimize the overall transportation cost as VRP. To achieve the good results, the problem is solved by applying various algorithms and determining which algorithm is best suited to solve the problem under this research work.

2 Related Work

In this section the details of literature survey of shortest path algorithms and optimization algorithms are explained to carry out work in proposed methodology.

2.1 Dijkstra Algorithm

Dijkstra algorithm is used to find the shortest path between nodes in a graph (road network). It was proposed by computer scientist Edsger W. Dijkstra in 1956 [3]. Dijkstra algorithm considers only single node as 'source' node and finds shortest paths from the source to all other nodes in the graph. Once all the paths from source to every other node are found, the shortest path is determined by the algorithm. Dijkstra algorithm is also used to find the shortest path between single source and single destination. In road network, the nodes of graphs represent the cities and cost on each pairing edges of nodes represents the distances between the cities. For this, Dijkstra algorithm can be used to find the shortest path between one city and all other cities in the road network.

2.2 Bellman–Ford Algorithm

The Bellman–Ford algorithm is also a single source algorithm that computes shortest paths from a single source node to all of the other nodes in a weighted graph. It is slower than Dijkstra algorithm for the same problem, but more versatile, because it considers the negative weight of edges while computing the shortest path. The algorithm was proposed by Richard Bellman and Lester Ford in 1958 and 1956, respectively and hence the name 'Bellman–Ford Algorithm' [5]. If a graph contains a 'negative edge value', and that edge falls in the shortest path from source to destination, then the 'negative cycle' is formed by one walk around the edge and the new path is generated by considering the cheapest positive value.

2.3 Genetic Algorithm

In GA, initially, a random or heuristic population set is generated. Then the cycles are repeated for a number of instances in population set. GA also does the re-evaluation of cycles to determine the optimum solution. GA is sometimes termed as 'Evolutionary Algorithm' (EA) in which the main strategy is to find the optimum solution by using search operators such as natural selection, mutation and recombination to the population. GA is especially useful for problems having few but not for different solutions or problems having the exact solutions. In GA, the problems may not be having the exact procedure but the procedure estimation gives the optimum solution. This helps to eliminate some solutions and to accept another solution, i.e. one of the many good solutions [6].

2.4 Particle Swarm Optimization

In PSO, the local and global optimum search is carried out on the behaviour of particles in swarm set. Particles changes and updates their position with respect to the distance between them and their neighbourhood. With this, the global search is performed to find the optimum solution. The optimal position of node on the graph is determined by updating the particle velocities. In every iteration, the fitness of each particle's position is determined by fitness measure and the velocity of each particle is updated by keeping track of two 'best' positions [7].

2.5 Ant Colony Optimization

The Ant Colony Optimization (ACO) algorithm is a probabilistic technique based on the behaviour of ants, searching a path between their colony and a food source. In the natural world, ants' searches for their food and return to their colony by laying down pheromone trail on the path. If other ants find such a path, instead of travelling randomly, they also follow the same path as pheromone is laid down on it. After some time, the pheromone starts evaporating which results in decreasing the likelihood of the path. If pheromone is not evaporated at all on the path, the paths chosen by the first ant would be selected by all following ants [8, 9]. Thus, when one ant finds a good path from the colony to a food source, other ants are more likely to follow that path, resulting in selecting the single path by all ants.

2.6 Honey Bee Optimization

The Honey Bee optimization (HBO) algorithm was proposed in 2005 based on the behavior of bees in natural world. This algorithm is divided into two categories, according to the bee's behavior in the nature, the foraging behavior and the mating behavior. From the set of solution, the candidate solution is selected based on local optimum exists between them [5]. In HBO mating process, the set chromosomes are formed by probabilistically selecting the set of sub-solutions from the best solution. Once the set of chromosomes is formed, the crossover operation is performed to generate the subset of child solutions. Mutation operation modifies the values of chromosome from its initial state. After mutation, the solution may change entirely from the previous solution.

2.7 Greedy Randomized Adaptive Search Approach

The objective of the Greedy Randomized Adaptive Search Procedure (GRASP) is to repeatedly select the samples from set of solutions, and then use a local search procedure to define them at local optimum level. During each iteration, the present solution is chosen as new solution and the best solution is determined based on the local optimum value between them. Greedy Randomized Approach is a meta-heuristic algorithm for combinatorial problems, in which iteration consists of two phases: construction and local search [10]. The construction phase builds a feasible solution by investigating all solutions until a local optimum is reached during the local phase.

3 System Methodology

The system mainly consists of two phases namely, pheromone evaporation and pheromone deposition. First, the nodes are initialized with ants (number of buses required to pick up students from stops). The routes are generated by visiting the stops one by one. Once the particular route is visited the pheromone is deposited on that route and local pheromone trail is calculated. The process is repeated until all the routes are visited and finally the global pheromone trail is calculated, which gives the optimum route.

The route map is generated, which consists of a number of stops and number of students to be picked up from stops. The route is generated by visiting each stop. When the route is visited for the first time, the pheromone is deposited in that route and pheromone rate is increased in each subsequent visit for the same route. The distance between the nodes is calculated using shortest path algorithm and the values are stored in distance matrices.

The maximum distance value from distance matrices is considered as 'dmax' and the weight of each edge between neighbouring nodes is calculated. If the distance 'd' between neighbouring nodes is less than 'dmax', then next stop is selected using pheromone deposition rate and if it is not, then next stop is selected using Euclidean distance formula. Once the stop is selected, the pheromone trail is calculated for that path. If the same path is selected in next iteration, update pheromone trail for the same. Once the local pheromone is updated, the global pheromone is calculated to find the optimum path (Fig. 1).

4 Experimental Conditions

The following parameters are considered for analysis, which are namely, number of cycles (I), number of ants (K), algorithm parameter (q_0) and beta value (β).

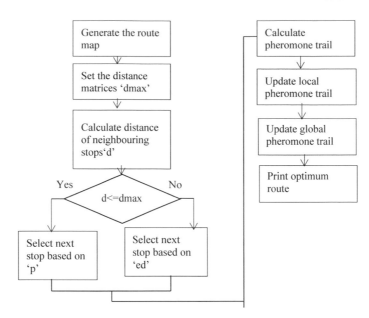

Fig. 1 Flowchart of the proposed methodology

4.1 Input Condition

Algorithm takes dataset from standard text (.txt) file. The user can provide any number of stops based on the requirement. The input to the algorithm will be number of stops followed by X-coordinate & Y-Coordinate to print the stop point on screen. The dataset can be randomly generated or taken from any standard benchmark dataset of TSP or VRP. It is also considered that on each stop single student will get picked up. It will also not accept same stop location more than once. All stops have to be distinct.

4.2 Number of Cycles (I) and Number of Ants (K)

For the sensitivity analysis, we also used 50 and 200 buses. For the analysis, we considered $I = 50$, $I = 90$, $I = 150$, $I = 240$ and $I = 300$. Each bus should visit at least 50 stops (considering total numbers of buses are 10). All buses visit 50 stops with random points allocation for the values of I. Once the random allocation of stops on graph is completed, the local value for I is determined by adding the random point values of stops as follows.

Table 1 Cycle sensitivity analysis

Bus No	Distance travel	Stops sisited
Bus 1	3333.925	50
Bus 2	3117.45605	50
Bus 3	3094.432	50
Bus 4	3283.36175	50
Bus 5	3405.294554	50
Bus 6	3258.152934	50
Bus 7	3015.280347	50
Bus 8	3261.952611	50
Bus 9	3112.165453	50
Bus 10	3616.843016	50
Total	32498.86	500

$$I_L = \sum_{i=1}^{n} R_{ij}$$

where 'n' is the total number of stops and 'R_{ij}' is random point value for stop (random point of stop on screen). When $I = 50$ then $I_L = 30.59511$ (adding the random points for each stop on screen as R_{ij}) and $I = 300$ then $I_L = 24.2742984$ for Bus 1. The same process is applied for Bus 2, Bus 3 and Bus 4. As the number of buses increases, the number of cycles required to visit the route decreases. From the local value of I, the global value for I is determined to find the total distance travelled by all buses for all the values of I. The global value for I is calculated as,

$$I_G = \frac{\left(\sum_{j=1}^{K} I_L \times n \right)}{2}$$

where 'K' is the total number of buses. For Bus 1, the value for $I_G = 3333.925$. Similarly, for Bus 2, Bus 3 and Bus 4, the global value for I is calculated (see Table 1).

4.3 Number of Ants/Buses (K)

For the analysis, we considered $K = 50$, $K = 100$, $K = 150$, and $K = 200$. To calculate the value of total travelled distance with varying values of K and I, we have to consider the global value for I (I_G), which determines the total travelled distance with all the values for K. The local value for K is calculated as

Table 2 Ant sensitivity analysis

Bus No	Distance travel	Stops visited
Bus 1	138.9135	50
Bus 2	129.8939	50
Bus 3	128.9346	50
Bus 4	136.8067	50
Bus 5	140.4039	50
Bus 6	122.3215	50
Bus 7	131.9886	50
Bus 8	125.9036	50
Bus 9	130.6015	50
Bus 10	133.9577	50
Total	1319.726	500

$$K_L = I_G \div K$$

Here 'K' value starts from 50, 100, 150 and 200. When $K = 50$, the value of K_L for Bus 1, Bus 2, Bus 3 and Bus 4 is 66.6785, 62.3491, 61.88864 and 65.6672 respectively. As the number of buses increases, the distance travelled by each bus decreases with respect to the value of I_G. That is with $K = 200$ the value of K_L for Bus 1 Bus 2, Bus 3 and Bus 4 is 16.6696, 15.5872, 15.4721 and 16.4168, respectively. Then the global value for K is calculated as

$$K_G = \sum_{i=1}^{K} K_L \times K$$

Once the global value for K is calculated, finally, the distance travelled by all buses with varying values of I is determined. The K_G value for Bus 1, Bus 2, Bus 3 and Bus 4 is 138.9135, 129.8939, 128.9346 and 136.8067, respectively. After observing Tables 1 and 2, we can say that with more number of buses the same path can be repeatedly selected so distance covered by buses is less as compared to distance coverage by more number of cycles.

4.4 Algorithm Parameter (q_0)

The parameter is also tested with values of 0.8 and 0.95. The algorithm parameter is used to find the pheromone level deposited on the visited route. It is calculated as

$$g(p) = \frac{(k_G \times q_0)}{K}$$

Table 3 Algorithm parameter sensitivity analysis

$q_0 = 0.8$			$q_0 = 0.95$		
Bus No	Distance travel	Stops visited	Bus No	Distance travel	Stops visited
Bus 1	11.11038	50	Bus 1	13.1967	50
Bus 2	10.391512	50	Bus 2	12.3399	50
Bus 3	10.314768	50	Bus 3	13.57206	50
Bus 4	10.9445	50	Bus 4	14.4007	50
Bus 5	11.6748	50	Bus 5	13.7895	50
Bus 6	11.4352	50	Bus 6	14.6757	50
Bus 7	10.5320	50	Bus 7	14.76456	50
Bus 8	11.45432	50	Bus 8	13.8789	50
Bus 9	11.60768	50	Bus 9	13.7687	50
Bus 10	10.5535	50	Bus 10	14.6678	50
Total	110.0187	500	Total	139.0545	500

where K_G is global value for K, q_0 is either 0.8 or 0.95 and K is the total number of buses. When $q_0 = 0.8$ the values are 11.11038, 10.391512, 10.314768, 10.9445 and with $q_0 = 0.95$ the values are 13.1967, 12.3399, 13.57206, 14.4007 for Bus 1, Bus 2, Bus 3 and Bus 4 respectively. With $q_0 = 0.8$, the distance travelled by each bus is less (i.e. 110.0187) than the higher value of q_0, i.e. 0.95 (i.e. 139.0545) (see Table 3). So with less value of q_0, the chances of selecting the same path are less and distance coverage is also less as compared to high value of q_0.

4.5 Beta (β)

Table 4 presents the results of two different values for β. We have tested the value for $\beta = 3$ and $\beta = 8$.

$$\rho = \frac{g(p)}{\beta}$$

where '$g(p)$' is algorithm parameter value for each bus and β is either 3 or 8. When the values are 3.7034, 3.4638, 3.4382, 3.6461 and with $\beta = 8$, the values are 1.6495, 1.5424, 1.6965, 1.8 for Bus 1, Bus 2, Bus 3 and Bus 4, respectively. With $\beta = 3$, the distance travelled by each bus is more (i.e. 35.63919) than the higher value of β, i.e. 8 (i.e. 16.32859) (see Table 4).

Table 4 Beta sensitivity analysis

$\beta = 3$			$\beta = 8$		
Bus No	Distance travel	Stops visited	Bus No	Distance travel	Stops visited
Bus 1	3.7034	50	Bus 1	1.6495	50
Bus 2	3.4638	50	Bus 2	1.5424	50
Bus 3	3.4382	50	Bus 3	1.6965	50
Bus 4	3.6481	50	Bus 4	1.8	50
Bus 5	3.7878	50	Bus 5	1.878	50
Bus 6	3.00898	50	Bus 6	1.35454	50
Bus 7	3.5251	50	Bus 7	1.4068	50
Bus 8	3.87687	50	Bus 8	1.2342	50
Bus 9	3.8654	50	Bus 9	1.8899	50
Bus 10	3.32154	50	Bus 10	1.87675	50
Total	35.63919	500	Total	16.32859	500

4.6 Pheromone Evaporation Coefficient (P)

For the analysis of p, we choose the values 0.3 and 0.7. The value of pheromone evaporation coefficient determines the evaporation speed of the pheromone trace. It is calculated as

$$P = g(p) \times \rho \times p$$

where '$g(p)$' is algorithm parameter value for each bus ρ is beta value for each bus and 'p' is either 0.3 or 0.7. When $p = 0.3$, the values 12.3438, 10.7932, 10.63926, 11.97798 and with $p = 0.7$ the values are 15.2375, 13.3231, 16.1174, 18.1448 for Bus 1, Bus 2, Bus 3 and Bus 4, respectively. With $p = 0.3$ the distance travelled by each bus is less (i.e. 117.7865) than the higher value of p, i.e. 0.7 (i.e. 158.1579) (see Table 5).

5 Analysis of ACO, HBO and Greedy Randomized Approach

The pheromone level, distance and optimum path level are three different parameters which are considered for analysis here. Pheromone level is defined as how many times the same path is selected by multiple numbers of buses. Optimum path level determines the ratio of same path selection to the number of buses. Distance parameter calculates the total distance travelled by each bus.

In ACO technique, when the numbers of buses are increased, the pheromone level obtained by buses also gets increased. In ACO, if the numbers of buses are more, then the same path can be selected by more number of buses. Also in ACO,

Table 5 Pheromone evaporation sensitivity analysis

$p = 0.3$			$p = 0.7$		
Bus No	Distance travel	Stops visited	Bus No	Distance travel	Stops visited
Bus 1	12.3438	50	Bus 1	15.2375	50
Bus 2	10.7982	50	Bus 2	13.3231	50
Bus 3	10.63926	50	Bus 3	16.1174	50
Bus 4	11.97798	50	Bus 4	18.1448	50
Bus 5	13.26654	50	Bus 5	18.1276	50
Bus 6	10.3224	50	Bus 6	13.9151	50
Bus 7	11.13789	50	Bus 7	14.7678	50
Bus 8	13.32207	50	Bus 8	15.8799	50
Bus 9	13.4607	50	Bus 9	18.76687	50
Bus 10	10.51761	50	Bus 10	13.8778	50
Total	117.7865	500	Total	158.1579	500

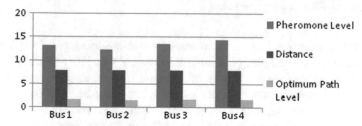

Fig. 2 ACO output with parameter values

the optimum path selection is proportional to the ratio of pheromone level to the number of buses. So when the pheromone level increases, the optimum path level also decreases. ACO mainly considers the highest value of q_0. In ACO, the pheromone level is increased with decreasing value of optimum path level. For the highest value of q_0, i.e. 0.95, the distance travelled by each bus is also more compared to the minimum value of q_0, i.e. 0.8. On the other hand, if the value of q_0 increases that is with more number of buses, the value of β (optimum path level) decreases (see Fig. 2).

When the Bus 1 starts at one particular point and its next stop is assigned randomly (as no other stops visited yet). In the next subsequent phase, the next stop can be chosen based on pheromone evaporation rate. If it is more, it means the path is selected more than once, and then the next stop is selected, which is not visited. The pheromone level is increased for that it is calculated with $q_0 = 0.95$ and the optimum path level decreases, it is calculated with $\beta = 8$. The distance parameter remains constant with changing values of pheromone level and optimum path, as it is the ratio of pheromone level to optimum path level. For Bus 1, the pheromone level is 13.167 with less optimum path value which is 1.6495. (Same for Bus 2, Bus 3 and Bus 4). But the distance value remains constant as 8 (see Table 6).

Table 6 ACO parameter values

Number of buses	Pheromone level	Distance	Optimum path level
Bus 1	13.1967	8.00042	1.6495
Bus 2	12.3399	8.00045	1.5424
Bus 3	13.57206	8	1.6965
Bus 4	14.4007	8.003	1.8

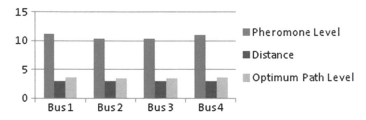

Fig. 3 HBO output with parameter values

In HBO technique, when the numbers of buses are increased, the pheromone level decreased but optimum path level increases. As in HBO when numbers of buses are decreased, the possibility of selecting the same path repeatedly gets increased. HBO mainly considers the lowest value of q_0. In HBO the pheromone level is decreased with increasing value of optimum path level. For the lowest value of q_0, i.e. 0.8, the distance travelled by each bus is also less compared to the maximum value of q_0, i.e. 0.95. On the other hand, if the value of q_0 decreases that is with less number of buses, the value of β (optimum path level) increases. It means that in HBO, the low pheromone value results in high optimum path selection (see Fig. 3).

A Bus 1 start at one particular point and its next stop is assigned randomly (as no other stops visited yet). In the next subsequent phase, the next stop can be chosen based on pheromone evaporation rate. If it is more, it means the path is selected more than once and has large visiting ratio, so the path is selected in next subsequent phase. The pheromone level is decreased for that it is calculated with $q_0 = 0.8$ and the optimum path level increases, it is calculated with $\beta = 3$. The distance parameter remains constant with changing values of pheromone level and optimum path, as it is the ratio of pheromone level to optimum path level. For Bus 1, the pheromone level is 11.11038 with more optimum path value is 3.7034. (Same for Bus 2, Bus 3 and Bus 4). But the distance value remains constant as 3 (see Table 7).

In Greedy Randomized Approach, the distance between two parameters is calculated based on Euclidian distance. The pheromone level and optimum path level are based on distance. If the distance between two stops is more, pheromone level and optimum path level get decreased otherwise it is increased. Greedy Randomized approach considers the good result obtained for q_0, β and p as 0.95, 3

Table 7 HBO parameter values

Number of buses	Pheromone level	Distance	Optimum path level
Bus 1	11.11038	3	3.7034
Bus 2	10.39115	2.99	3.4638
Bus 3	10.3147	3	3.4382
Bus 4	10.9445	3	3.6481

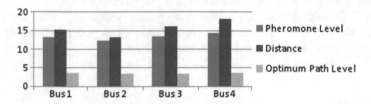

Fig. 4 Greedy randomized approach output with parameter values

Table 8 Greedy randomized approach parameter values

Number of buses	Pheromone level	Distance	Optimum path level
Bus 1	13.1967	15.2375	3.7034
Bus 2	12.3399	13.3231	3.4638
Bus 3	13.57206	16.1174	3.4382
Bus 4	14.4007	18.1448	3.6481

and 0.7. By considering the maximum value for pheromone level and optimum path level, the distance travelled by each bus is also increased (see Fig. 4).

The pheromone level is increased for that it is calculated with $q_0 = 0.95$ and the optimum path level increases, it is calculated with $\beta = 3$. The distance parameter also changes with respect to the values of q_0, β and $p = 0.7$. For Bus 1, the pheromone level is 13.1967 with more optimum path value is 3.7034 and the distance covered is also more compared to ACO and HBO i.e. 15.2375 (Same for Bus 2, Bus 3 and Bus 4) (see Table 8).

As seen in the previous section, the analysis of ACO, HBO and Greedy Randomized Approach is carried out based on pheromone level, distance and optimum path level. It is observed that the performance of all techniques is different in terms of output given by them for each parameter. From this, we can say that instead of applying one technique to solve problem in hand, one can build the hybrid model to solve the problem to a great extent.

6 Conclusion

In this research work, we have implemented the approaches to solve a real-life school bus routing problem. The problem can be solved by defining student pickup and delivery points to buses and finding the minimum distance with routing algorithm. The problem is solved using one of the routing algorithms, which gives us the shortest route as well as optimal solution based on different predefined constraints. Also, the major important task is we have done the analysis of ACO, HBO and Greedy randomized approach to decide which algorithm suits to the problem. But the observation says that none of them gives better performance for all parameters, instead all are good for different parameters.

References

1. Li, L., & Fu, Z. (2009). The school bus routing problem. *European Journal of Operational Research.* Journal of the Operational Research Society, Department of Industrial and Management Engineering, Pohang University of Science and Technology (POSTECH).
2. Riera-Ledesma, J., & Salazar-Gonza´lez, J. J. (2013). Solving school bus routing using the multiple vehicle traveling purchaser problem: A branch-and-cut approach. *Computer & Organization Reaserch*, 2013.
3. Selvi, V., & Umarani, R. (2010). Comparative analysis of ant colony and particle swarm optimization techniques. *International Journal of Computer Applications*, 2010, Tamil Nadu, India.
4. Riera-Ledesma, J., & Salazar-Gonza´lez, J. J. (2012). A column generation approach for a school bus routing problem with resource constraints. *Computer & Organization Reaserch*, 2012.
5. Rizzoli, A. E., Montemanni, R., Lucibello, E., & Gambardella, L. M. (2007). Ant colony optimization for realworld vehicle routing problems: From theory to applications, 2007.
6. Park, J., Tae, H., & Kim, B. I. (2011). A post-improvement procedure for the mixed load school bus routing problem. *European Journal of Operational Research*, 2011. Department of Industrial and Management Engineering, Pohang University of Science and Technology (POSTECH), Pohang.
7. Schittekat, P., Kinable, J., Sörensen, K., Sevaux, M., Spieksma, F., & Springael, J. (2013). A metaheuristic for the school bus routing problem with bus stop selection. *European Journal of Operational Research*, 2013.
8. Sghaier, S. B., Guedri, N. B., & Mraihi, R. (2013). Solving school bus routing problem with genetic algorithm. *High Institute of Transport and Logistics*, 2013.
9. Nha, V. T. N., Djahel, S., & Murphy Lero, J. (2012). *A comparative study of vehicles routing algorithms for route planning in smart cities.* Ireland: UCD School of Computer Science and Informatics.
10. Diaz-Parra, O., Ruiz-Vanoye, J. A., Buenabad-Arias, A., & Cocon, F. (2012). *A vertical transfer algorithm for the school bus routing problem.* Mexico: Department of Information Technology.

Recommender System for Shopping: A Design for Smart Malls

Dhvani Shah, Joel Philip, Abhijeet Panpatil, Afrid Shaikh
and Suraj Mishra

Abstract The population is growing day by day. World is moving towards advanced technologies which makes life easier and hence there is a need of smart city. Such smart city also needs smart malls which deliver the wonderful experience of shopping. Whereas many e-commerce sites give priority to each and every consumer by sending notifications about upcoming offers and sale, about the launch of new products, and predicting and showing a list of items customers may like to buy. Malls are pretty bad at advertising offers and giving personalized experience to customers and make them feel special. Our system will provide a great experience of personalized shopping which will help them to buy the product. Our system will make use of customer's previous purchase record, and current offers going on in the mall. The system will sort best offers that will be most relevant to his/her shopping style. The system will advertise using LCD's and speakers that will come along with it, which will make it much more interactive with customers using Raspberry Pi which will make it a low-cost system. Also when the customer reaches nearby the mall, he will be shown only those products advertises in which he/she would be interested, this will make him/her feel special because of giving preferences to their choices. The user will also get notified by Android app or email, whenever products or offers of their interest that would be launch in the mall. The system will keep customers connected to the mall even if they are working in an office or resting at home or roaming in the mall.

D. Shah (✉) · A. Panpatil · A. Shaikh · S. Mishra
St. John College of Engineering and Management, Palghar, India
e-mail: shahdhvani08@gmail.com

A. Panpatil
e-mail: panpatilabhijeet66@gmail.com

A. Shaikh
e-mail: afrid75@gmail.com

S. Mishra
e-mail: mi.suraj8833@gmail.com

J. Philip
Universal College of Engineering, Thane, India
e-mail: tjoelphilip@gmail.com

© Springer Nature Singapore Pte Ltd. 2019
V. E. Balas et al. (eds.), *Data Management, Analytics and Innovation*,
Advances in Intelligent Systems and Computing 839,
https://doi.org/10.1007/978-981-13-1274-8_24

311

Keywords Raspberry Pi · Smart malls · RFID · Location-based
Android application

1 Introduction

A smart city is an inner-city expansion vision to assimilate information and communication in a secure manner. ICT allows city officials to work together directly with the people and the infrastructure and to keep an eye on what is new and happening. Through the use of sensors combined with real-time systems, data are composed from general public and devices—then administered and examined. The information gathered are keys to attacking inefficiency. Shopping malls are defined as "One or more construction of the building". The smart city consists of everything which makes the human life very easy. Hectic schedules and long work hours may leave people tired of shopping. The smart city provides the solution for this. The smart city provides that facility and platforms for people with help of which they can save their time and do the work smartly.

The mall is experiencing a rapid alteration. In the upcoming times, the shopping experience in the malls will be more like shopping from an e-commerce site. Malls will be transformed to 'Smart Malls' using the technology offered by the system where it will become more and more interactive with the customer and customer-friendly. Approximately half of all offline purchases are influenced by the people living vicinity of the mall. Malls that are listed in an online search, they are listed whenever customers are watching for a precise product or brand.

Online e-commerce sites or stores need the customer's name, address, email id, credit card details or debit card details if online payment is opted for delivery of the product or downloading the e-product like movies, e-books [1]. By gaining all the details about the customers, the online stores or e-commerce sites can cleanly say based on their past purchase behaviour that what will the client choose from the variety they have to offer in the future. Thus, due to the data present and the algorithm in place, they can target the customers at the right time with the new offers and new items or products in their stores which leads to great email-to-purchase conversion rates. Malls in contrast to online stores have a very narrow system to target their customers. Their way of targeting customers is only on the basis of their products and offers in place there is no personalization/recommendation in 'offers and products for you' segment.

Smart malls will be present in the process of the online commerce. Coarsely half of the purchases from the mortar and brick malls are influenced by online commerce and social media marketing. Therefore, smart malls will be present online search, so they are found whenever consumers are looking for a specific product/brand. Additionally, smart malls store the record of shoppers and their shopping details like brands, cost range, etc. and engage shoppers with personalized communications. The Smart Malls will be different from e-commerce or online stores in a way that while online stores can reach out to anyone in the world, the Smart Malls will be able to target only that particular vicinity matching their interests and offering real-time information.

There are several ways in which the customers can be reached at home or in a mall. Customers can be reached by following ways, using email, in-app push notifications or by sending the text messages to the customers. By keeping the communication right and sending the minimum number of messages to the users, Smart Malls will be capable to attract more customers to their malls. By doing this Smart Malls will offer than just products to their customers. In fact, it offers Smart Malls' officials and data analysts a large record of shoppers that they can reach to increase the customers' base.

Smart Malls uses the profile of the customers as a data point and connects together to a single platform, so they are always up to date on offering latest offers, products by incorporating influential marketing strategy. The Shoppers dashboard will give them a look into their past shopping style and experience. This system allows the data scientist and other IT professionals to run the various algorithm on the resulting dataset, gaining insights to target the customers rightly and thus providing personalized recommendation which will eventually satisfy the customer to a greater extent [2].

Such system will independently interact with its customers inculcating one-to-one communication with them based on customers' activities and experiences. Also, it includes standards for control and child safety, and personalized recommendation for customers and lasting bi-directional message between customers and the Shopping mall during the shopping experience.

This correspondence amongst buyers and the chiefs of the shopping centre permits gives upgrades and to the shopping background, helping customers to settle on choices about what to purchase basically, consoling battle between brands, client honesty and promoting events which centre on genuine shopping inclinations and patterns (self-encounter).

2 Recommender System

Recommender systems can be usually implemented in two of the way—through collaborative and content-based filtering or the personality-based approach. Collaborative filtering develops a model from things acquired in the past. Given to those things and also comparative choices made by the different customer. This model is then used to predict items for the user may have an interest in. Content-based filtering approach uses a variety of distinct characteristics of an item and the characteristics mentioned in the user profile. That means it does a comparison between the users liking and the items/products for the sale.

2.1 Collaborative Filtering

Recommender systems has wide use of collaborative filtering. Collaborative filtering strategies depend on social affair and looking at a lot of data on clients' exercises, or inclinations and anticipating what clients will like in view of their comparability to different customer. The biggest advantage of the collaborative filtering is that it doesn't depend on machine analyzable substance and therefore and consequently it is prepared to do unequivocally suggesting complex things, for example, movies without requiring a 'understanding' of the item itself. Numerous algorithm have been utilized as a part of estimating customer likeness or item comparability in recommender systems. For instance, the k-nearest neighbour (k-NN) approach and the Pearson Correlation. Collaborative filtering depends on the fact that individuals who concurred in the past on items, offers will concur later on, and that they will like similar sorts of things. At the point when a model is worked from a client's conduct, difference is made between explicit and implicit forms of data collection. Examples of explicit information collection are to incorporate requesting that a client rate a thing on a sliding scale, requesting that a client rank a gathering of things from most loved to least loved pick, exhibiting two things to a client and soliciting to pick the better one from them, and so on. Cases of implicit information gathering incorporate recording the things that a client sees in an online store, dissecting thing/client seeing circumstances, keeping a record of the things that a client buys internet, getting a rundown of things that a client has tuned into or viewed on his/her PC, breaking down the client's interpersonal organization and finding comparable preferences and so forth [3].

3 Raspberry Pi and RFID

Raspberry Pi is a small mini-computer. It is a little PC which has all the usefulness that are conveyed by a work area PC. The Raspberry Pi has a Broadcom BCM 2836 a System on chip (SoC), 900 MHz processor with quad-core ARM Cortex A7Video Core IV GPU and with 1 GB of RAM. It utilizes a SD card for booting and information stockpiling. The RPi has an Ethernet port for organize association, USB port for interfacing outside USB gadgets, miniaturized scale USB opening for control supply, HDMI port to associate with show and General-Purpose Input Output (GPIO) pins to interface with other equipment gadgets [4]. In this research, RPi plays a very crucial role to make it a low-cost system. The latest offers ongoing in the mall as per maximum users' choice will be displayed on the screen. Once a user is identified using the RFID reader [5], the recommender system will do its tasks, publish the users' buying choice against the offers provided by a particular brand in the mall. Accordingly, a digital signage will be developed to display the offers based on the most of the users' preferences.

4 S-Mall Architecture

The proposed study comprises of technologies like: Raspberry Pi, Machine Learning, and Google Firebase (Fig. 1).

Each user is uniquely identified based on his fingerprint. After his fingerprint is captured, a temporary RFID tag will be given to the user which he will carry throughout the mall. At this point, the user's fingerprint data, his customer id and his RFID value will be saved into a central repository.

In Fig. 2 explains the S-Malls system.

The user once buys the product, purchase record containing the products purchased, amount, brand, etc. will be logged in the database. The record of every user is recorded against its customer id (customer id is unique to the user).

This data will be given to the purchase history warehouse.

Product DB, Offers DB will be maintained by the mall personnel. Product DB will contain all the data about the products/brands present in the mall and their related information against the product id. Offers DB will contain all the data about the offers related to particular product against the product id.

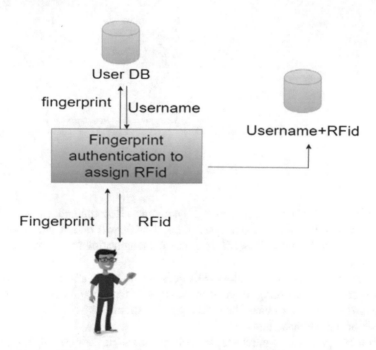

Fig. 1 *(S-Mall Register System).* User first time getting registered for the S-Mall System. In this user's data is stored in UserDB and fingerprint is used to identify the user and retrieve the username. Username is used to assign the RFid

Fig. 2 (*S-Mall System*). This
is the System Architecture for
S-Malls. The user's past
purchase data is stored in
Purchase History database
and then the products and
offers are given to the users
based on their past purchases

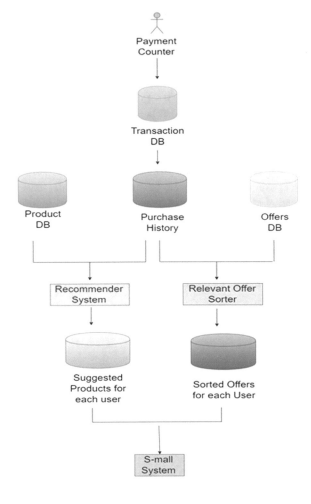

Product DB and the Purchase history will be given to Recommender system. Based
on the collaborative filtering, our system, i.e. S-Mall will output the product recommendations to the user. This will add user personalization and make our system
user-friendly.

Offers DB and the Purchase history will be given to relevant offer sorter and the
offer will be sorted according to the user. Along with the recommendations, the user
will be also being able to view the ongoing offers on his favourite band fetched on
the basis of his purchase history.

Based on his patterns of purchase we will get a rough estimate of his financial
strength. When no purchase history is available for a particular brand, S-Mall will

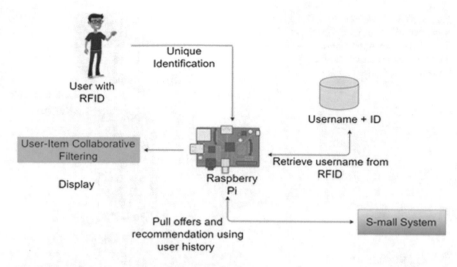

Fig. 3 *(S-Malls Tracking System)* Here the Raspberry Pi is used to retrieve the Username from Username & RFid using the username recommendation using user history

first to give him the product recommendations of brands based on his financial capability and then continue with the common offers going on if any on all brands of that product provided that brand's outlet is there in the mall.

Figure 3 represents the implementation of the following steps:

Once the user gets registered through the fingerprint, and he gets RFID tag as mentioned earlier.
At this point, now the user will be temporarily identified by this RFID value.
The customer ID and corresponding RFID value will go into our system.
Our system will provide recommendations based on his purchase history.
All the data pulled from the S-Mall system will be given to the User-Item Collaborative filtering.
User-Item Collaborative filtering will give us the suggested products/live offer on products based on past purchases of users on the display screen in the malls. User-Item Collaborative filtering outputs the products liked by the maximum people in the mall or on a particular floor of the mall.

In Fig. 4, the Location from the android application is sent to S-Malls. The Offers and Recommendation is sent to mobile application. The offers from the malls and recommendation is personalized according to individual user of application.

Fig. 4 *(S-Mall Mobile
Application)* Using the
S-Malls Mobile Application
the offers in various malls will
be given

5   Conclusion

This idea would create a revolution in the Malls and the people shopping experience. By use of this Smart Mall System user will have improved experience in shopping from the malls. Also, the use of various component makes it a system which can support large consumer base and will be a modular system. Thus, S-Mall adds a level of user personalization in the existing malls for better shopping experiences.

References

1. How the first smart shopping mall in Europe works. Available https://phys.org/news/2015-04-smart-mall-europe.html.
2. How malls become SmartMalls Using Technology to Bridge the Gap between Online and Offline to Drive Conversion from The Connected Shopper. Available: https://www.mall-connect.com/how-malls-become-smart-malls-using-technology-to-bridge-the-gap-between-online-and-offline-to-drive-conversion-from-the-connected-shopper/#_ftn1.
3. Recommender_System. Wikipedia. Available: https://en.wikipedia.org/wiki/ Recommender_system.
4. Shah, D., & Bharadi, V., Dr. (2017, March). Iot based multifunctional robot using Raspberry-Pi. *International Journal of Advances in Electronics and Computer Science, 4*(3).
5. Available. Wikipedia as Radio-frequency identification: https://en.wikipedia.org/wiki/Radio-frequency_identification.

Continuous Facial Emotion Recognition System Using PCA for Ambient Living

Anil R. Surve, Vijay R. Ghorpade and Anil S. Patthe

Abstract Nowadays, Facial Emotion Recognition is widely used and is an attractive area in affective computing especially for computer vision with healthcare applications. Facial expressions change with respect to time and person in different instances. To find out the emotions automatically by computers, facial expressions perform the most important role and also aid for human–machine interfaces. Persons can be distinguished by facial expressions easily on time but for computers, it is still a challenge. Presented work proposes the emergence-based eigenface techniques. By using PCA (Principal Component Analysis), we can extract all relevant information present in frames where human faces are detected. We know that facial expressions are conveying emotions exactly. We use PCA to reduce the dimensionality of computations. In this process we are detecting face, extracting features, reducing dimensionality using PCA, and then classifying emotions using Euclidean distance metric and after that, we apply temporal dynamics (Patthe and Anil in Temporal dynamics of continuous facial emotion recognition system, 2017) for redundant frames with emotions reduction. Eigenvectors are calculated by the set of training images, which defines the face spaces. We apply PCA for compressing eight orientations and the relevant scale of frames. In PCA, we used the database in which some frames are used for the training purpose. Rest of the frames are used for testing propose. We used training frames for emotions such as angry, disgust, happy, neutral, and surprise. We experimented on Indian Face Database. From this database, 30 frames are used for training the system and 50 frames are

A. R. Surve (✉) · A. S. Patthe (✉)
Department of Computer Science and Engineering, Walchand College
of Engineering, Sangli, Maharashtra, India
e-mail: anil.surve@walchandsangli.ac.in

A. S. Patthe
e-mail: pattheanil@gmail.com

V. R. Ghorpade (✉)
Department of Computer Science and Engineering, D. Y. Patil College
of Engineering & Technology, Kolhapur, India
e-mail: vijayghorpade@hotmail.com

© Springer Nature Singapore Pte Ltd. 2019
V. E. Balas et al. (eds.), *Data Management, Analytics and Innovation*,
Advances in Intelligent Systems and Computing 839,
https://doi.org/10.1007/978-981-13-1274-8_25

used for testing purpose. Through experimentation, we obtained a recognition rate which is 91.26%.

Keywords Affective computing · Eigenfaces · Euclidean distance
PCA · ROI · SVM · Sampling

1 Introduction

Face recognition and emotion identification are critical pattern recognition problems because of the high dimensionality of image data [1]. This scheme works based on information theory approach. Eigenfaces are used to decompose face images into the small sets of feature images; actually, they are principal components of the face images of initial training set images. For computational recognition of human faces in digital images, eigenface technique is one of the successful approaches and provides better efficiency. When we apply Principal Component Analysis (PCA) to eigenface approach, it becomes more useful for minimum dimensional representation. By projecting all face images on feature space, the system functioning span shows considerable variation in the middle of all face images. This eigenface approach provides an efficient way of finding the lowest dimensional space. The principal component is nothing but the eigenvectors so all significant features are the eigenfaces. Faces can be articulated in linear combinations of singular vectors of the faces and these singular vectors are eigenvectors of the covariance matrices.

Principal Component Analysis (PCA) increases the variances of all the extracted features of face images and because of that reconstruction error gets decreased. Also, we remove the noise present in all discarded dimensions. Face and emotion recognition technology has various commercialized and law enforcement applications [2]. For all applications which are having linear models, PCA becomes the classical technique [3].

We have organized the paper as Sect. 2 provides related work information. Section 3 provides an idea of the eigenface approach. Section 4 provides a detailed mathematical module of all procedures implicated in the eigenface approach. Section 5 presents all experimental results. Section 6 presents Conclusion.

2 Related Work

We mention here some research work already done in this domain. Accordingly, we proposed a technique for dynamically downsampling the frames for further use of emotion recognition [1]. In the first step, ROI is detected by using Voila–Jones algorithm from all the frames. After that, the video is segmented into further parts. In the next step, video is dynamically downsampled with respect to dominant frequency and then using Aviator feature extraction method, all features are extracted and classified for emotions using SVM (Support Vector Machine). Also, it was observed in earlier work that finding out emotions after all the downsampling processes are complete so the drawback of that system is some frames which have the most emotional features compared with other has the chances to be removed.

The proposed "Facial Expression Recognition System" [2] is based on Gabor feature extraction method using a Gabor filter bank by Hong Do Beng. For evaluating the performance of the Gabor filter bank, both stage feature compression methods PCA with LDA were applied for selection and compression of the Gabor features. After that for recognizing facial expressions, a minimum distance classifier was used. Rose [4] on feature vectors, tested classification performance by using the Gabor and Log-Gabor filters. They also used this process for image pixels demonstration of static face images. After that on these calculated feature vectors, PCA was performed using linear discriminate analysis and classification accuracies were compared. Khandait [3] used modified RGB, YCbCr, and HSV algorithms. These are skin detection algorithms. In this paper, they used all those algorithms as a combination of colored images. By using geometry-based method and Gabor wavelet-based method [5], features were extracted for recognizing the expressions with two-layer perceptron but they did not address face detection problem.

3 The Eigenface Approach

Based on Principal Component Analysis (PCA), Sirovich and Kirby stated that "In terms of a best coordinate system any particular face can be economically represented." This system is termed as "Eigenpictures". By performing a mathematical procedure named as Principal Component Analysis (PCA) over the larger set of standard images, a set of eigenfaces can be generated as depicting different human faces.

Eigenpictures are nothing but the eigenfunctions of all the averaged covariance of the faces. A small set of standard eigenpictures with a minimum set of its weight can be approximately represented as a collection of all face images for every standard picture (Figs. 1 and 2).

3.1 Eigenface Approach Procedures for Face Recognition

As projected by Pentland and Turk, the system is qualified with following operational steps:

1. In initial first step, all face images which are having the different classes of emotions were acquired. This set of images is nothing but the training set. This data should be annotated with emotions label.

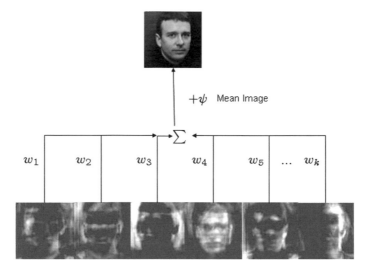

Fig. 1 Mean image extraction

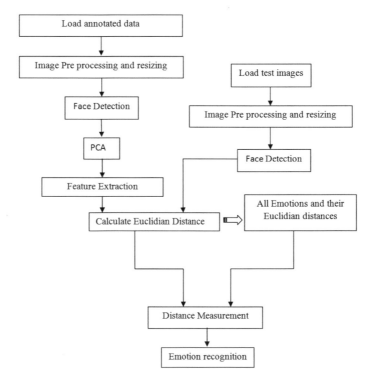

Fig. 2 Basic steps

2. Prior to all operations, all trainee data should be in one common form so that next step will be image preprocessing and image resizing. From this training dataset, we calculate the eigenfaces for each and every image. Just M eigenfaces corresponding to M largest eigenvalues are retained. These eigenfaces span the face space which constitutes the training dataset.
3. For each training image, the M eigenface weights are calculated by comparing the images onto the face spaces span by the eigenfaces. After that, every face image is represented by its M weights and tremendously squashes representation.

After the above initialization processes, the following steps were carried out to reorganize all input test images:

4. By projecting all test images onto each of the eigenfaces, the test images were obtained. In that, each test image corresponds to the set of M weights.
5. After that, we have to calculate the Euclidean distance of all training images from the neutral images present in the training dataset. All neutral images were taken as the threshold for the other emotion classes.
6. From the Euclidean distance of all trainee images, we can compare and define the class of new test images. By calculating Euclidean distance, we determine that the test image is closer or not with any training image.
7. After calculating the minimum distance of test image, the system returns the class of emotion.

4 Methodology

4.1 Create a Training Set and Load It

Trainee dataset consisting of I_i (I_0, I_1, I_2, I_3, …, I) number of face images with two ($N \times N$) dimensions. Each image has $I_{(x,y)}$; that means $x * y$ intensity values. In Image I, x and y represent pixel positions and their intensity value (Fig. 3).

4.2 Training the Recognizer

Step 1   Convert face images into a training set to face vector

We know that PCA will not work on the image directly, but it uses the image intensity matrix for processing. In our input dataset, suppose images have N by N resolution, every image generates the $N * N$ matrix. If the dataset has 100 images that time 100 matrixes of images are required but it became a crucial part of processing. So in that step, we convert each and every image into the vector form.

Fig. 3 Sample faces

If we convert each image into vector (Γ) form, it will be easy to convert all images into the single matrix (Fig. 4).

Step 2   Face Vector Normalization

Once we have converted all face images into the face vector (Γ) form, the next step is to train the Recognizer for normalization of face vector. In that normalization, we are going to remove all the common features which are shared by all images together. This common feature should be removed so that each face vector is left behind with only its unique features. For removing that commonly shared feature, we have to find out the mean of all image vectors and remove it from each image vector. Mean vector is nothing but the average face vector (Ψ). Mean vector or average vector is calculated as follows:

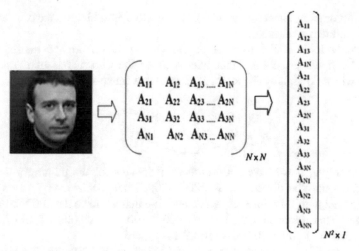

Fig. 4 Face vector

$$\Psi = \frac{1}{M} \sum_{n=1}^{M}$$

After calculating the average face vector, we have to find out normalized face vectors (Φ_i) for each and every face vector by subtracting average face vector (Ψ) from all the face-vectors (Γ_i) as follows:

$$\Phi_i = \Gamma_i - \Psi$$

(Φ_1, Φ_2, Φ_3, Φ_4, ..., Φ_i) contains the normalized vectors having different and unique features from each other.

Step 3   Reduce the dimensionality of the training set

Now, the next part of PCA algorithm is calculating the eigenvectors. For calculating the eigenvectors, we need to calculate the covariance matrix. Most important part of PCA algorithm is calculating the covariance matrix because PCA is well known for the dimensionality decomposition and this dimensionality decomposition is achieved by the covariance matrix. We simply calculate the eigenface principal component from the covariance matrix. Covariance matrix is calculated by the following formula:

$$C = A \cdot A^{\mathrm{T}}$$

Where A is nothing but all normalized face vectors as [Φ_1, Φ_2, Φ_3, Φ_4, ..., Φ_i]. Here, A is having [$N^2 \times I$] dimension. N^2 is because of every image from the dataset having $N \times N$ dimension and its vector became $N^2 \times 1$ dimension but when we combine all numbers of (I) face vectors into single matrix it forms the $[N^2 \times I]$ dimensional matrix.

But now look at the formula of calculation of covariance matrix, that is, $C = A \cdot A^T$. If we just think about what will be the dimensions of C matrix, the answer will be definitely $N^2 \times N^2$. It generates a much bigger matrix.

For example, if we have 100 numbers of images for the training purpose and each image has the dimension as 200×200. That means every image has the 40,000 pixels with 8-bit intensity value and like that we have the 100 total numbers of images. That means now we have $40,000 * 100 = 4,000,000$ values of intensity for A matrix and $100 * 40,000$ values for A^T matrix.

Now, our C will become the size of $[40,000 \times 40,000]$. So, it generates N^2 eigenvector that means 40,000 eigenvectors or principal components.

There exists a need for dimensionality reduction so that we have to find K significant eigenvectors or eigenfaces, because the principal of PCA-based recognition was to represent and train each image from the training set as a linear combination of Kth selected eigenfaces, where K must be less than or equal to the number of images (I) present in the trainee dataset. For example, if we have 100 images in our training dataset that time K will be 100 or less than 100. So now finding 100 or less than 100 eigenvectors from 40,000 eigenvectors is going to be a huge amount of calculation. Here, 40,000 eigenvectors are not the only problem for calculation but every eigenvector also has a size of $[40,000 \times 1]$. So the system may slow down terribly or run out of memory. We reacquired huge amount of memory and processing power for finding out eigenvector from the above huge calculation.

This is an essential part of PCA to find out those Kth eigenvectors but the problem is how to achieve this. This problem is solved by the main and important method of PCA, that is, dimensionality reduction. Dimensionality reduction is done by calculating eigenvectors from a covariance matrix by reducing the dimensions of the resultant matrix. But not from the previous covariance matrix but from the following covariance matrix formula:

To minimize such calculations and to minimize noise effect from eigenvectors, we have to calculate them from a covariance matrix of reduced dimensionality. From the new formula of covariance matrix with previous example data, we got covariance matrix with only [100 × 100] dimensions, because we have 100 numbers of images in our trainee dataset.

This happened only because of reversing the formula, we reversed the formula and resultant matrix dimension was also reversed.

Step 4   Calculate the eigenvector from covariance matrix C

With previous covariance matrix formula, we got the C matrix with [40,000 × 40,000] dimension. But now after reversing the formula, our resultant C matrix was converted into [100 × 100] dimension and each eigenvector was converted into [100 × 1] dimension without any feature loss. Now, it became easier to find the K number of eigenvectors which is lesser than or equal to the number of trainee images.

Step 5   Select K best eigenfaces, such that $K < I$ and it can represent the whole training dataset

We know that the final selected eigenvector must be less than or equal to the number of images present in the trainee dataset. So we can test the new test images within a short time efficiently. We have to select K eigenfaces which must be representing the whole training dataset. They must not leave any important information about the data we have referred for training purpose. These K eigenfaces were selected by the artificial intelligence procedure. But the selected K eigenfaces must be into the original dimensionality of the face vector space. So, for that purpose, we have to map it back with the previous dimensionality of data.

Step 6   Convert lower dimensional K eigenvector to original face dimensionality

By using linear algebra technique, we did this operation. And, for that, we used the following formula:

$$\mu_i = Av_i$$

where μ_i is depicting the ith eigenvector into the higher dimensional space and v_i is depicting the ith eigenvector into the lower dimensional space. From the above formula, if we multiply the matrix A with v_i it will give us the corresponding eigenvector into the higher dimensional space (Fig. 5).

Step 7   Represent each face image with a linear combination of all K eigenvectors

Once we have selected the K eigenvector faces next thing is that we need to add the mean face and weighted sum of K eigenfaces so that each face from training set can be represented. The mean image is an average image which is obtained into the normalization process. This process is needed only because we removed this average face from each image so we need to add it back with each image before all the calculations.

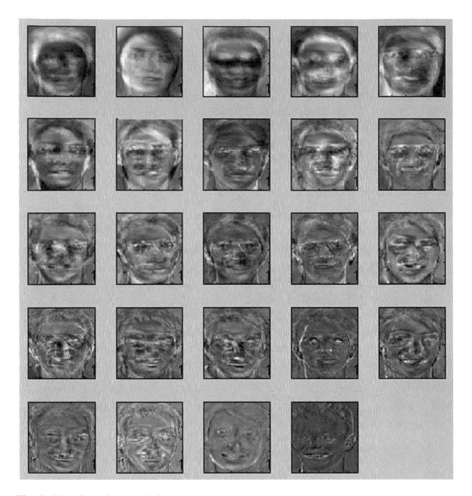

Fig. 5 Eigenfaces for sample faces

4.3 Recognizing an Unknown Face

Find out eigenfaces for all I trainee images and select a set K mainly comprising values of related eigenfaces as per $K < I$. For each image in the training set, we have to calculate and store its associated *weighted* (Euclidean distance) vector. Take input image, preprocess it, extract the face from an image, convert the image to a face vector then normalize this face vector, project this normalized face vector onto eigenface, calculate weight vector of the input image, measure and compare Euclidean distance and return the closest image vector emotion label (Fig. 6).

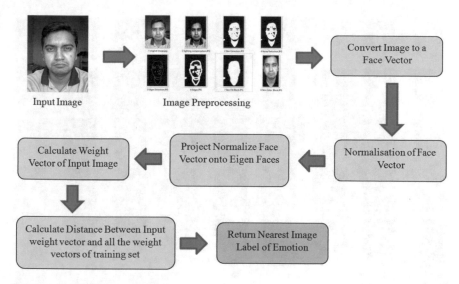

Input Image   Image Preprocessing

Fig. 6 Recognizing an unknown face

5 Experimental Results

We used face images of a single person for the experiment. In trainee dataset, we included 47 images of different emotions randomly. We included 7 images of neutral faces. Neutral faces are required in dataset because in this system we used neutral images as the threshold for other images of emotions. All 47 images are taken in the same lighting condition. We used 13 images of happy face, 11 images of disgust face, 9 images of anger face, and 7 images of sad face (Fig. 7; Table 1).

In the test dataset, we used 33 images of the same person with different emotions. These test images are also taken in the same lighting condition because PCA gives satisfactory results only when we take test images in the same lighting condition like training images. From 33 images there are 4 images of neutral face, 8 images of happy face, 10 images of disgust face, 7 images of anger face and 4 images of sad face are included (Fig. 8; Table 2).

We apply some basic operations on the trainee images as well as on test images which are shown in Fig. 9

After this image prepossessing, skin color block image is applied for PCA as input.

After all, operations are discussed in the methodology, we got the results as follows, which are shown in Table 3.

PCA gives 93.93% accuracy for this experimental setup. It gives wrong results only for two images; one from neutral images and another one from sad images because of some lack of the same lighting conditions and trainee images. For happy, disgust, and anger emotions, it gives 100% results. For 100% result of PCA, there should be

Fig. 7 Training images

Table 1 Training data

Training images						
Emotions	Neutral	Happy	Disgust	Anger	Sad	Total
No. of images	7	13	11	9	7	47

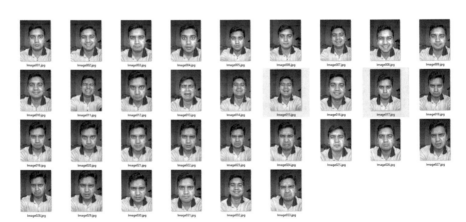

Fig. 8 Test images

Table 2 Test Data

Test images						
Emotions	Neutral	Happy	Disgust	Anger	Sad	Total
No. of images	4	8	10	7	4	33

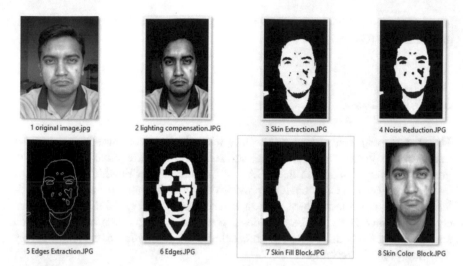

Fig. 9 Image preprocessing

Table 3 Analysis

Test images						
Emotions	Neutral	Happy	Disgust	Anger	Sad	Total
No. of images	4	8	10	7	4	33
Wrong result	1	0	0	0	1	1
Accuracy	93.93%					

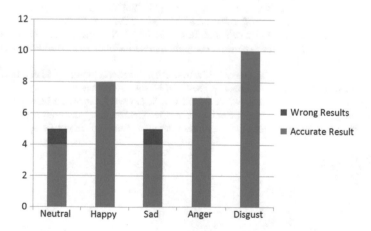

Graph 1 Analysis

enough (at least 20–25 images per emotion) images required for every emotion in the trainee set (Graph 1).

6 Conclusion

The PCA technique gives best results only when the light condition is the same for training image set as well as test image set. Also, PCA required much more training images for every emotion so that PCA can find the nearest Euclidean distance for satisfactory results. PCA technique provides simplest methods for emotion identification over other techniques. For face recognition and emotion identification, eigenface approach with PCA provides a practical solution. Face recognition and emotion identification in affective computing PCA is a quick and comparatively easy and powerful technique.

Acknowledgement We would like to thankfully acknowledge Md. Iftekhar Tanveer, whose image datasets are used for experimentation research. We are also very thankful to the college, especially to the Computer Science and Engineering department for enabling facilities and resources available for the research.

References

1. Patthe, A. S., & Surve, A. R., Dr. (2017, February). Temporal dynamics of continuous facial emotion recognition system. In *International Conference on Emerging Trends in Engineering, Technology and Architecture, iCETETA.*
2. Beng, H. D., Zin, L. W., Zen, L. X., Huang, J. C. (2005). *A new facial expression recognition method based on local gabor filter bank and PCA plus LDA* (pp. 86–96).
3. Khandait, S. P., Khandait, P. D., & Thool, R. C., Dr. (2009, November). An efficient approach to facial feature detection for expression recognition. *International Journal of Recent Trends in Engineering*, 179–182.
4. Rose, N. (2006). Facial expression classification using gabor and log-gabor filters. In *IEEE 7th International Conference on Automatic Face and Gesture Recognition.*
5. Zhang, Z., lyons, M., Schuster, M., & Akamatsu, S. (1998). Comparison between geometry based and gabor-wavelets-based facial expression recognition using multi-layer perception. In *The Third IEEE International Conference on Automatic Face & Gesture Recognition.*

Segregation of Similar and Dissimilar Live RSS News Feeds Based on Similarity Measures

Avani Sakhapara, Dipti Pawade, Hardik Chapanera,
Harshal Jani and Darpan Ramgaonkar

Abstract News in the form of text is widely available to us with a number of different sources that are available on the Internet. Having so many varied sources often makes the same news available by most of the sources. A user who prefers a thoroughgoing update on different news headlines ends up reading the same news from different sources. So we have developed a system to cluster the news based on similarity. We extract the news from the RSS link provided by the user. Using similarity measures like Edit Distance, Jaccard Similarity, Cosine Similarity and WordNet Similarity, we have implemented a system which presents the summary of identical and different news feeds from different sources and its effectiveness is measured.

Keywords RSS news feeds · Jaccard · Cosine · Edit Distance
WordNet

1 Introduction and Related Work

E-newspapers of various news agencies can be accessed over the Internet. These news websites consist of a feature of providing an instant update of incidents and events to the readers via feeds called the Really Simple Syndication (RSS) feeds. A user who prefers a thoroughgoing update on different news headlines ends up reading same news from different sources. In this paper, we are going to address the problem of news redundancy from various RSS feeds. Our objective is to design a system which will extract Live RSS news feeds from the different sources selected by the user. Then, it will analyze the headlines and will group into identical news and dissimilar news.

Many researchers have worked in this area. Burkepile and Fizzano [1] have introduced Artificial Immune based system which analyzes the content of the news

A. Sakhapara (✉) · D. Pawade · H. Chapanera · H. Jani · D. Ramgaonkar
Department of IT, K.J. Somaiya College of Engineering, Mumbai, India
e-mail: avanisakhapara@somaiya.edu

© Springer Nature Singapore Pte Ltd. 2019
V. E. Balas et al. (eds.), *Data Management, Analytics and Innovation*,
Advances in Intelligent Systems and Computing 839,
https://doi.org/10.1007/978-981-13-1274-8_26

read by a person in past. Using this data, their proposed system is analyzing the content of incoming news and finding the relevant articles. For each news feed, they have considered the title and description of article. This approach suggests that whenever a news article comes, first perform stemming and remove stop words. Then create a cell having fixed number of words. These words are selected randomly from title and description. So for a single article, N number of cells can be created. Next step is to find the relevance. For this, Jaccard Similarity coefficient is used to compare every cell in an incoming article to every cell of an article in the training set. Finally if the content is similar to that of training set, the article is said to be relevant to the particular person otherwise not. Agarwal et al. [2] used a combination of ontology and Concept Frequency-Inverse Document Frequency (CF-IDF) to segregate the news feeds according to the categories chosen by the user. They have concluded that ontology can be used as a good text classifier. Precision and recall of their approach is 0.8 and still, there is scope to improve the relevance of news feed. Davis et al. [3] have proposed a system called SociRank—which aims to identify the most discussed topic in both e-news and social media during a specific time frame. The system first stores the news feed and tweets in the database. Then this data is preprocessed. In preprocessing, keywords are extracted from the news article using TextRank algorithm proposed by Witten et al. [4]. Then tweet term extraction is carried out to create a set of all relevant and unique terms mainly those which are tagged as hashtag, noun, adjective, or verb. Based on these key sets, a graph is created having key terms as node and edges as term similarity. To achieve more precised nodes and edges, outlier detection is done. Clustering algorithm is applied on the graph to get well-defined and disjoint topic clusters. These clusters are then ranked based on relevance factor. Bergamaschi et al. [5] further extend their work of computing relevant values [6] by proposing a system which considers the syntactic, dominance, and lexical relationships of string for defining relevance value. They have developed news feed reader tool called "Relevant News". This tool groups similar news by means of data mining and clustering techniques applied to the feed titles. Karpe et al. [7] discussed spatial analysis based news feed processing using Hadoop framework. They have followed the following procedure:

- Step 1: News feeds are collected from various news agencies and social media.
- Step 2: Collected news feeds are classified into different categories using MapReduce Technique.
- Step 3: Depending upon the category, one has clicked maximum number of time, top three news from that category is provided as result. The results are presented using BI tool and it is a blend of top three results from interest domain followed by top news update and advertisement.

Setty et al. [8] have focused on the classification of Facebook feeds only. They have built a learning based classifier which segregates the feeds into two buckets, viz., liked pages posts and friends posts. They have used Binary Logistic Regression [9], Naive Bayes [10], Support Vector Machine (SVM) [11], Bayes Net [12], and J48 [13] to build the second level classifier which categorizes the segregated posts into life events posts and entertainment posts. They further apply sentiment analysis to calculate the sentiment score of the life events posts. In this step, first, tokenization is carried out to extract the terms of interest from post. Using POS, tagger part of speech is detected and post is converted into word. Using SentiWordNet dictionary words, sentiment orientation is detected, based on which post is tagged as happy, neutral, and sad post. They concluded that their proposed SVM based learning model has achieved an accuracy of 97–99% and believe that this method can be used effectively for other classification based applications.

Tabara et al. [14] discussed a framework called "ReaderBench" which provides the personalized news feeds by analyzing one's Twitter profile. Here, preprocessing includes extraction, aggregation, cleaning of tweets and news feed conversion into specific XML format which is compatible to be accepted as input by the proposed system. Thus, one's tweets and news feed are provided as input to ReaderBench framework. The collected tweets are processed using Natural Language Processing pipeline [15]. Then for topic distribution, Cohesion Network Analysis [16] is performed using semantic distances in WordNet [17], Latent Semantic Analysis (LSA) vector models [18] and Latent Dirichlet Allocation (LDA). This is basically used to extract the main essence from a subclass of Twitter text corpora. Lastly, greedy approach [19] is used to specify the relevant news category for a particular Twitter profile. This system is designed specifically for Tweeter only.

Liang et al. [20] have elaborated on demand multilingual news feed provider using Big Data. This system has two major sections: the first section deals with huge news data and other section deals with recognizing multilingual news feed based on user behavior. They maintain various databases to store data generated after each preprocessing step. In the beginning, real-time news feeds are stored in a database called original news database. After de-noising the content in original news feed database, it is stored in semantic database. On the other hand, different news event in original database are stored in event database. Event database is then analyzed to maintain the theme database. In next step, based on time, news is sorted and using theme semantic out of date news are traced. Lastly, based on the user's browsing history, personalized news feeds are provided to user. The issues related to the parallel extraction of news feeds from multiple sources are handled using Slope One feed algorithm. To improvise news correlation, correlation vector is used.

So according to our survey, some papers focus on user preferences and news categories for feed segregation. In recent years, researchers are more interested in considering the one's behavior on social networking to automatically predict the user's interest. We have observed that in most of the research papers, similarity measures have been used to find the relevance. There are many similarity measure techniques available and it becomes very difficult to choose which measure one should use. So this has motivated us to carry out our proposed work which calculates and compares the news feed relevance using different similarity measures, viz., Edit Distance (ED), Jaccard Similarity (JS), Cosine Similarity (CS), and WordNet Based Path Similarity (WPS).

Further, Sect. 2 of this paper gives the details of these four similarity measures and their comparison. In Sect. 3, the design and implementation of the proposed Live RSS news feeds segregator is discussed. In Sect. 4, analysis of the results is given and in Sect. 5, conclusion and future work are stated.

2 Similarity Measures

The RSS news feeds are extracted from the user selected sources. They are pre-processed and categorized into similar feeds using four sentence similarity algorithms namely Edit Distance, Jaccard Similarity, Cosine Similarity, and WordNet Based Path Similarity Algorithms.

2.1 Edit Distance (ED)

It counts the number of insertion, deletion, and substitution of characters required to transform one sentence $S1$ similar to another sentence $S2$.

$$ED(S1, S2) = \text{count (insertion, deletion, substitution)} \tag{1}$$

Example: $S1$: A boy is eating an apple, $S2$: A man is eating an apple.

By using Eq. (1), in $S1$, substituting "b" by "m", "o" by "a" and "y" by "n", ED $(S1, S2) = 3$. The smaller the value, the more similar the two sentences are.

2.2 Jaccard Similarity (JS)

It measures the intersection of words between two sentences $S1$ and $S2$, divided by the union of words in both the sentences.

$$JS(S1, S2) = \frac{|S1 \cap S2|}{|S1 \cup S2|}, \text{ where } 0 \leq JS(S1, S2) \leq 1 \tag{2}$$

Example: $S1$: This is a car, $S2$: This is a book.

Using Eq. (2), $|S1 \cap S2| = \{this, is, a\}$ and $|S1 \cup S2| = \{this, is, a, car, book\}$, the Jaccard Similarity is calculated as, $JS(S1, S2) = 0.6$. The higher the value, the more similar the two sentences are.

Table 1 Cosine vectors of S1 and S2

Joint set S	He	eats	bread	like	Brown
SA($S1$)	1	1	1	0	0
SB($S2$)	1	1	2	1	1

2.3 Cosine Similarity (CS)

It measures the cosine angle between the two vectors. In case of sentence similarity, a joint word set S containing all the distinct words from both the sentences $S1$ and $S2$ is used to compute vectors SA and SB for sentences $S1$ and $S2$, respectively. The vector SA for sentence $S1$ represents the number of occurrences of each word in $S1$ sentence. Similarly, a vector SB for sentence $S2$ is calculated. Then the Cosine Similarity is given as

$$CS(S1, S2) = \frac{SA \cdot SB}{\|SA\| \, \|SB\|}, \text{ where } 0 \leq CS(S1, S2) \leq 1 \tag{3}$$

Example: $S1$: He eats bread, $S2$: He eats bread like brown bread.
The vectors SA and SB for $S1$ and $S2$ are given in Table 1.
Using Eq. (3), Cosine Similarity is calculated as

$$CS(S1, S2) = \frac{(1 \cdot 1 + 1 \cdot 1 + 1 \cdot 2 + 0 \cdot 1 + 0 \cdot 1)}{\sqrt{1^2 + 1^2 + 1^2} \sqrt{1^2 + 1^2 + 2^2 + 1^2 + 1^2}} = 0.82$$

The higher the value, the more similar the two sentences are.

2.4 WordNet Based Path Similarity (WPS)

WordNet is a machine-readable lexical database for the English language [21]. In WordNet, there are over 150,000 words which are arranged in synsets and there are over 110,000 synsets. Synsets are sets of synonyms of English words. A number of relations between these synsets or their members are also stored in WordNet. The sentences "Let's have tea" and "Let's have coffee" are similar because they are beverages. WordNet uses a hierarchical approach to establish this relation. In WordNet, one word may belong to many synsets. We have used Path Similarity to get the synset which is the most similar with the other word. Path Similarity is calculated based on the following two components:

- Semantic Similarity
- Word Order Similarity

Semantic Similarity (SS) measures the Cosine Similarity between the semantic vectors SA and SB computed for each sentence $S1$ and $S2$. SA and SB are computed using a joint word set S containing all the distinct words from both the sentences $S1$ and $S2$. The semantic vector SA element is set to 1 if a word from joint word set S exists in sentence $S1$ otherwise word similarity score is calculated between the word under consideration and all the other words in sentence $S1$. Word Similarity (WS) between two words X_1 and X_2 is calculated using the formula proposed by Li, Bandar, and McLean [22].

$$WS(X_1, X_2) = e^{-\alpha l} \cdot \frac{e^{\beta h} - e^{-\beta h}}{e^{\beta h} + e^{-\beta h}} \tag{4}$$

where l = length of the shortest path between X_1 and X_2, h = depth of the subsumer, α, β = parameters representing the contribution of length and height, and $0 \leq \alpha \leq 1$ and $0 < \beta \leq 1$. For WordNet, $\alpha = 0.2$ and $\beta = 0.45$ are the optimal values.

If $WS(X_1, X_2)$ > threshold value, then semantic vector SA element is set to WS (X_1, X_2) otherwise it is set to 0. For optimal results, 0.2 is considered as threshold value.

Word Order Similarity (WOS): At times, even if the two sentences have the same words, but the order of the words is different, then the sentences yield different meaning. Thus, the computation of sentence similarity should also consider the order of the words as a measure. For computing Word Order Similarity, vectors SA and SB are computed for each sentence $S1$ and $S2$. SA and SB are computed using a joint word set S containing all the distinct words from both the sentences $S1$ and $S2$. If a word in S is present in $S1$, then SA element is set to the index of the word in $S1$ otherwise most similar word is calculated using Eq. (4). If it is greater than 0.2, then SA element is set to the index of the most similar word otherwise it is set to 0.

Example: $S1$: There is a traffic after pothole, $S2$: There is a pothole after traffic.

Any method which is based on "bag of words" will return the similarity as 1. But these sentences are not exactly similar. They are similar only to certain extent. Therefore, the order of the words also matter. The vectors SA and SB for $S1$ and $S2$ are given in Table 2.

Word Order Similarity (WOS) is calculated as

$$WOS(S1, S2) = 1 - \frac{\|SA - SB\|}{\|SA + SB\|} \tag{5}$$

Table 2 Word Order Similarity vectors of $S1$ and $S2$

Joint set S	There	is	a	traffic	after	pothole
Index	1	2	3	4	5	6
SA($S1$)	1	2	3	4	5	6
SB($S2$)	1	2	3	6	5	4

Table 3 Comparison of similarity measures

	ED	CS	JS	WPS
Context	✗	✗	✗	✓
Length	✓	✓	✓	✓
Repeated words	✗	✓	✓	✓
Joint set	✗	✓	✓	✓
Range of result	No limit	Between 0 and 1	Between 0 and 1	Between 0 and 1
Optimal result	Low	High	High	High

Since in the computation of sentence similarity, both the Semantic Similarity and Word Order Similarity play a crucial role, the overall WordNet Based Path Similarity (WPS) is calculated as

$$\text{WPS}(S1, S2) = (\delta\,(\text{SS})) - ((1 - \delta)\,(\text{WOS})), \quad \text{where } 0 \le \text{WPS} \le 1 \qquad (6)$$

where δ = the measurement of contribution of Semantic Similarity (SS) and Word Order Similarity (WOS) and $0.5 \le \delta \le 1$.

Since word order plays a less role in determining the semantics of the sentences, it is suggested that the value of δ should be greater than 0.5. For optimal results, δ is taken as 0.85. The higher the value of WPS, the more similar the two sentences are.

The different parameters considered by these similarity measures are given in Table 3.

3   Implementation Overview of Live RSS News Feeds Segregator

The working of Live RSS news feed segregator is shown in Fig. 1. It consists of the following stages:

- Extraction and Filtration of RSS Feeds
- Preprocessing
- Applying Similarity Measures

Stage 1: Extraction and Filtration of feeds

The user is presented with an interface where the user can select the sources or can input the source link from where the news feeds are to be extracted. The RSS feeds are in XML format. So the XML file of the RSS feeds is extracted. The headlines are filtered from XML tags such as <header>, <link> and <datetime> from the XML file. These filtered feeds are then stored in the database.

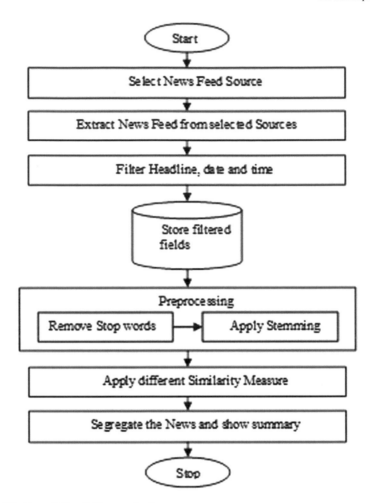

Fig. 1 Working of Live RSS news feed segregator

Stage 2: Preprocessing

Before applying the similarity algorithms, the feeds are preprocessed to remove stop words like it, as, a, the, and through stemming, the root form of the words in the feeds is obtained.

Stage 3: Applying Similarity Algorithms

The four similarity algorithms namely, Edit Distance, Jaccard Similarity, Cosine Similarity, and WordNet Based Path Similarity are applied to these processed feeds. The average of the result is taken and if it is above a threshold value than the news feeds are similar otherwise they are different. Finally, a summary of identical and different news feeds from the user chosen sources is given to the user.

4 Results and Discussions

For simulation purpose, we have extracted around 1700 Live RSS news feeds from various sources. Here, we are discussing some sample cases. For example, Live RSS news feeds extracted from the two sources, Indiatimes and The Hindu is shown in Table 4.

From Table 4, we observed that in column 1 (feeds from Indiatimes) and column 2 (feeds from The Hindu), the feeds are different. This is because we are extracting Live RSS news feeds in bulk and most of the time their sequence is jumbled. For example, in one source, news may be at feed number 5 and in another source, the same news may be at feed number 100. The RSS feeds from various sources are rated by 10 persons as "0" (zero) and "1" (one) for the similarity measurement. Here "1" indicates that the RSS feeds are similar and "0" indicates that the RSS feeds are dissimilar. The human rating of similarity measurement for sample pairs of RSS feeds is presented in Table 5.

In Table 5, it is observed that for some pair of feeds, such as I3:H5, people had different opinion on the similarity measurement due to the ambiguous sentences. For example, in feed H5, the term "tragic event" may not necessarily refer to "road

Table 4 Extracted RSS feeds

Sr. No.	Indiatimes	The Hindu
1.	I1: Mexicans burn Donald Trump effigies in Easter ritual	H1: Pakistan loses the match against India
2.	I2: England books a spot in the semifinal	H2: Modalities for Pak probe team in India to be worked out on arrival: HM
3.	I3: Two people killed in road accident	H3: Indian appointed UN adviser on human rights and businesses
4.	I4: Ready for Pak probe team: Rajnath	H4: NZ in semifinals, SA crashes out
5.	I5: India wins the match comfortably against Pakistan	H5: Two people hacked to death in a tragic event

Table 5 Similarity measure rating by humans

Pairs of RSS feeds		Similarity measure rating by 10 persons										Average (P1–P10)
Indiatimes	The Hindu	P1	P2	P3	P4	P5	P6	P7	P8	P9	P10	
I3	H5	1	0	0	0	1	1	0	0	0	0	0.3
I5	H1	1	1	1	1	1	1	1	1	1	1	1
I2	H4	0	0	0	0	0	0	0	0	0	0	0
I1	H3	0	0	0	0	0	0	0	0	0	0	0
I4	H2	1	0	1	0	1	1	1	0	1	1	0.7

accident" as stated in feed I3. "Tragic event" may refer to a building collapse or train accident, etc. Due to this ambiguity, some people have rated it as "0" and some as "1".

The same thing is applicable with automated system as well. Thus, in the proposed system, we have tried to address this issue by considering the combination of different similarity measures.

The four similarity measures are applied to the feeds and the results are shown in Table 6. From the results, it is observed that we cannot rely on any one sentence similarity method for finding identical news feeds. Edit Distance completely fails to measure the Semantic Similarity for Live RSS news feeds while Jaccard Similarity, Cosine Similarity, and WordNet Based Path Similarity succeed in identifying completely dissimilar feeds, they individually perform very poor in identifying the extent of similarity between the feeds. So the combined result of the three similarity measures, namely Jaccard Similarity, Cosine Similarity, and WordNet Based Path Similarity is considered. If the average of these three similarity measures is greater than or equal to 0.4, then the feeds are considered as Similar otherwise they are Dissimilar.

The results of Human Similarity Rating values in Table 5 is compared with the results of Similarity Measure values in Table 6. The summary of comparison is presented in Table 7. The inference of comparison of Human Similarity Rating with Similarity Measures is carried out as follows:

- If (Human Similarity Rating < 0.6 AND Similarity Measure Value $\geq$ 0.4) Then: Category = False Positive, Interpretation = Dissimilar
- If (Human Similarity Rating < 0.6 AND Similarity Measure Value < 0.4) Then: Category = True Negative, Interpretation = Dissimilar
- If (Human Similarity Rating $\geq$ 0.6 AND Similarity Measure Value $\geq$ 0.4) Then: Category = True Positive, Interpretation = Similar
- If (Human Similarity Rating $\geq$ 0.6 AND Similarity Measure Value < 0.4) Then: Category = False Negative, Interpretation = Similar

Table 6 Similarity Measure Values

Pairs of RSS feeds		Similarity measures				Average (CS, JS, WPS)	Result
Indiatimes	The Hindu	ED	CS	JS	WPS		
I3	H5	20	0.51	0.34	0.53	0.46	Similar
I5	H1	21	0.57	0.35	0.76	0.56	Similar
I2	H4	31	0.15	0.08	0.18	0.14	Dissimilar
I1	H3	58	0	0	0.02	0.01	Dissimilar
I4	H2	43	0.34	0.19	0.68	0.41	Similar

Table 7 Comparison of human similarity measurement with similarity measure values

RSS feeds		Human similarity rating	Similarity measure value	Category	Interpretation
Indiatimes	The Hindu				
I3	H5	0.3	0.46	False Positive	Dissimilar
I5	H1	1	0.56	True Positive	Similar
I2	H4	0	0.14	True Negative	Dissimilar
I1	H3	0	0.01	True Negative	Dissimilar
I4	H2	0.7	0.41	True Positive	Similar

5 Conclusion and Future Work

This paper elaborates a technique of presenting easy and faster reading of news feeds from different sources by avoiding reading same news. Since Edit Distance does not work well with the semantics, for Live RSS news feeds, the average of Jaccard Similarity, Cosine Similarity, and WordNet Based Path Similarity measures is used to present identical news feeds to the user from the user chosen sources. Individually, each of these similarity measures gives around 33% accuracy. But their combination increases the accuracy to around 67%. Further, the categorization of the feeds based on the comparison of system results with human rating is presented in the paper. The results can be improved further by considering the association of the proper nouns like people with their profession while computing the similarity measurement. This system can be extended further to compute the similarity between tweets or blog posts or Facebook posts.

References

1. Burkepile, A., & Fizzano, P. (2010). Classifying RSS feeds with an artificial immune system. In *Proceedings in IEEE 2nd International Conference on Information, Process, and Knowledge Management* (pp. 43–47). Saint Maarten, Netherlands Antilles.
2. Agarwal, S., Singhal, A., & Bedi, P. (2013). Classification of RSS feed news items using ontology. In *Proceedings in IEEE 12th International Conference on Intelligent Systems Design and Applications (ISDA)* (pp. 491–496). Kochi, India.
3. Davis, D., Figueroa, G., & Chen, Y.-S. (2017). SociRank: Identifying and ranking prevalent news topics using social media factors. *IEEE Transactions on Systems, Man, and Cybernetics: Systems, 47*(6), 979–994.
4. Witten, I. H., Paynter, G. W., Frank, E., Gutwin, C., & Nevill-Manning, C. G. (1999). KEA: Practical automatic keyphrase extraction. In *4th ACM Conference on Digital Libraries* (pp. 254–255). Berkeley, CA, USA.
5. Bergamaschi, S., Guerra, F., Vincini, M., Orsini, M., & Sartori, C. (2007). RELEVANT news: A semantic news feed aggregator. In *4th Italian Semantic Web Workshop*, Italy.
6. Bergamaschi, S., Guerra, F., Orsini, M., & Sartori, C. (2007). Extracting relevant attribute values for improved search. *IEEE Internet Computing, 11*(5), 26–35.

7. Karpe, P., Bhor, V., & Agarwal, C. (2016). News feed processing and analysis using Hadoop framework. In *Proceedings in International Journal of Computer Applications. National Conference on Advances in Computing, Communication and Networking (ACCNet-2016)* (pp. 16–18). Maharashtra, India.

8. Setty, S., Jadi, R., Shaikh, S., Mattikalli, C., & Mudenagudi, U. (2014). Classification of facebook news feeds and sentiment analysis. In *Proceedings in IEEE International Conference on Advances in Computing, Communications and Informatics (ICACCI)* (pp. 18–23). Delhi, India.

9. Cheng, Y., Zhang, K., Xie, Y., Agrawal, A., Liao, W.-K., & Choudhary, A. (2011). Learning to group web text incorporating prior information. In *11th IEEE International Conference on Data Mining Workshops* (pp. 212–219). Vancouver, Canada.

10. Chai, K. M. A., Chieu, H. L., & Ng, H. T. (2002). Bayesian online classifiers for text classification and filtering. In *25th Annual International ACM SIGIR Conference on Research and Development in Information Retrieval* (pp. 97–104). Tampere, Finland.

11. Burges, C. J. C. (1998). A tutorial on support vector machines for pattern recognition. *Data Mining and Knowledge Discovery, 2*(2), 121–167. Springer.

12. Friedman, N., & Goldszmidt, M. (1996). Learning Bayesian networks with local structure. In *12th Conference on Uncertainty in Artificial Intelligence (UAI96)* (pp. 252–262). Portland, Oregon, USA.

13. Patil, T. R., & Sherekar, S. S. (2013). Performance analysis of Naive Bayes and J48 classification for data classification. *International Journal of Computer Science and Applications (IJCA), 6*(2), 256–261.

14. Tabara, M., Dascalu, M., & Trausan-Matu, S. (2016). Building a semantic recommendation engine for news feeds based on emerging topics from tweets. In *proceedings in IEEE 15th RoEduNet Conference: Networking in Education and Research*. Bucharest, Romania.

15. Jurafsky, D., & Martin, J. H. (2008). *An introduction to natural language processing, computational linguistics, and speech recognition*. Prentice Hall Series in Artificial Intelligence.

16. Dascalu, M. (2014). Analyzing discourse and text complexity for learning and collaborating. In *Studies in Computational Intelligence 534*. Springer.

17. Budanitsky, A., & Hirst, G. (2006). Evaluating wordnet-based measures of lexical semantic relatedness. *Computational Linguistics, 32*(1), 13–47. ACM.

18. Landauer, T. K., & Dutnais, S. T. (1997). A solution to Plato's problem: The latent semantic analysis theory of acquisition, induction and representation of knowledge. *Psychological Review, 104*(2), 211–240.

19. Cormen, T. H., Leiserson, C. E., Rivest, R. L., & Stein, C. (2009). *Introduction to algorithms*. Cambridge: MIT Press.

20. Liang, Y., Guo, N., Xing, C., Zhang, Y., & Li, C. (2015). Multilingual information retrieval and smart news feed based on big data. In *Proceedings in IEEE 12th Web Information System and Application Conference* (pp. 85–88). Jinan, China.

21. Miller, G. A. (1995). WordNet: A lexical database for English. *Communications of the ACM, 38*(11), 39–41.

22. Li, Y., McLean, D., Bandar, Z. A., O'Shea, J. D., & Crockett, K. (2006). Sentence similarity based on semantic nets and corpus statistics. *IEEE Transactions on Knowledge and Data Engineering, 18*(8), 138–1150.

Machine Learning for Personality Analysis Based on Big Five Model

Joel Philip, Dhvani Shah, Shashank Nayak, Saumik Patel
and Yagnesh Devashrayee

Abstract The proposed research attempts to emulate a statistical report making system, which takes into considerations, the activities of user and their behavior online by means of their interactions on varied array of social media platforms. It is possible that youngsters may come across incidents on Internet, which probably may be inappropriate for their age group or may push them towards certain erratic psychological behaviors. This study caters to such arising needs, for various individuals—young or old, alike, so as to keep a tab upon their own online activities through browsing history which may be directly/indirectly blend into their human characteristics. On social media, people express their likes, dislikes, thoughts, opinions, and feelings which sum up to be their own personality. This data (thoughts and opinions on social platform and browsing history) can be exponentially aggregated to identify user's personality traits. It can then be used for self-monitoring, parental monitoring, or for businesses who wish to hire employees based on their personality criteria, if approved by concerned users. For this study, we have used supervised machine learning algorithms like Naïve Bayes and Support Vector Machines. We have evaluated their performance through the combinations of different feature extraction process like BOW, TF, and TF-IDF with each classifier. In conclusion, we have found that TF-IDF with SVM has the best performance.

Keywords Personality · Social media · Machine learning · Psychology
Deep learning · Data analytics · Data mining · Cognitive science
Big Five personality traits · Neural network

J. Philip · S. Nayak · S. Patel · Y. Devashrayee
Universal College of Engineering, Thane, India
e-mail: tjoelphilip@gmail.com

D. Shah (✉)
St. John College of Engineering and Management, Palghar, India
e-mail: shahdhvani08@gmail.com

© Springer Nature Singapore Pte Ltd. 2019
V. E. Balas et al. (eds.), *Data Management, Analytics and Innovation*,
Advances in Intelligent Systems and Computing 839,
https://doi.org/10.1007/978-981-13-1274-8_27

1   Introduction

A social platform is a place where a person expresses or presents themselves the way they are. The data which comes under social media is nothing but series of intercommunications among humans and post in any format. This can be in tremendous volume and is ever-increasing in size. The person opens about their likes, dislikes, thoughts, opinions, and feelings which sum up to be their own personality. Thus, this data can be collected to identify user's personality. According to the statistics [1], there were a total of 2.01 billion active users on Facebook in June itself which meant a rise of 17% since the last year and these numbers are to increase over the coming years. Also, 29.7% of users are of age 25–34, and it is the most common age demographic. Also, it comes with the responsibility of handling huge amount of data, for every 60 s: around 500,000 comments 290,000 statuses are updated, and 136,000 photos are uploaded on Facebook. All these comments, status, and photos roughly help generating a user's profile thus revealing a lot about their personality [1].

A person's personality is their characteristics and aspects of other's perception. The capability of finding connections between behavioral aspects derived from the data collected from social media can help us in bifurcating users into various groups based on their personality, this is the main objective of this research. Personality refers to individual characteristics pattern of thinking, behaving, feeling which makes them different from other individual. Personality of person is dependent on the type of situation and mood. When it comes to analyzing personality the one thing that comes into the picture is the "Big Five Model" or "Five Factor Model" which comprises of five major components such as Openness, Extraversion, Agreeableness, Conscientiousness, and Neuroticism [2].

The most famous application of machine learning which everyone is aware of is the sentimental analysis. In sentiment analysis, the machine learning model attempts to predict the emotional value based upon the input supplied to it. This research aims at doing far more than just the analysis of emotions. The aim is to label a person's personality by analyzing the textual information generated by him on social media to create a statistical report depicting the resources on which the user is investing maximum time and efforts by analyzing his browser history.

2   Related Work

In [3], the authors made use of naive Bayes and SVM models in order to classify a Tweet message into positive or negative category where positive and negative indicate the emotion based on the message. The SVM model when used correctly does work as expected but the ill-effects of it results in overfitting. In [4], the researchers have presented a method to accurately predict the user's personality by making use of data obtained from smartphones, cell phone application, Bluetooth, and SMS usage along with the use of supervised learning algorithm help constitute a data for predicting. This process of collection of data and then running it through a prediction model helps generate characteristic for the users. In [5], the personality traits of microblog users are generated. It proposes multi-task regression and incremental regression algorithms to predict the Big Five personality from online behaviors. This indicates that even online microblog can be used to accurately predict personality of users. However, no NLP method has been used as well as data set is very low, i.e., only blog of topic. In this work [6, 7], the authors have proposed a unique methodology to extract human emotions from text. It applies various machine learning algorithms to predict readers reaction.

This study aims to collect data from various resources like Twitter, Facebook, Quora, and the browsing history of a user. This study comes with two aspects: the first task is to analyze his opinions to predict his personality and the second task is to generate a report on his daily online activities to know his interest and area of research. In conclusion, people express opinions/feelings in complex ways, which makes understanding the subject of human opinions a difficult problem to solve. One of the biggest challenges of this research is data collection from various resources and categorizing the text according to the Big Five model for training purpose.

3 Proposed Idea

The basic idea here is to generate subcategories to the five factors which can help classify the personality with much better meaning. One problem faced here is that the personalities of people change according to the social networking site they use. This would mean the personalities obtained from the data provided would not be accurate, but it changes according to the social site used, e.g., A person's behavior on any social networking websites like Twitter or Facebook would be different as compared to their behavior on professional networking websites like Stack Overflow or Quora. People tend to behave differently according to the social website on which they have their online presence. If we are to follow the conventional method and try to get the personality for a person, then the output obtained would not defy correctness in their value. Our framework will be designed keeping the exact problem in mind assuming that one single user has account on social media sites as well as he is active on educational/informational sites. The framework will analyze different data differently according to the platform used by the person to check their personality. This way we can find traits related to behavioral aspects from social networking websites whereas technical qualities from professional networking websites. After we have the basic structure defined, we can start the process of determining the personality of the user.

3.1   Data Collection

The process begins with collection of information or data from the respective social networking websites.

Twitter: Twitter API provides tweets, likes, and number of followers. This helps us perform analysis on the tweets of the users to generate their personality chart.
Facebook: Graph API can help us get all the publicly available information like liked pages, liked movies, etc., can help us generate personality by going through the likes of the user.
Quora: The Quora API helps us get all the questions or answers provided by the users.

3.2   Data Cleaning

After the data collection process, comes the data cleaning process. Texts written by humans usually contain noise and we can extract features out of the texts only after the removal of the noise. Natural Language Processing helps the machines to understand the human language [8]. Techniques used for text cleaning include tokenization, POS tagging, stemming, removal of stopping words, etc. (Fig. 1).

On inspecting the dataset, we noticed that the raw data contained a lot spelling mistakes along with redundancy. To solve this problem, we ran the whole of the dataset through a spellchecker and deleted all the redundant data. Sometimes, the user writes about his feeling for a situation on his blog or writes a technical article. This sort of user data contains lots of words as compared to a simple tweet. To deal with this situation, we can use the concept of text summarization and then pass the sentences into the data cleaning process. For browsing history, all what we have is a csv file containing the URLs of the web pages visited so far. To find that where the user is investing his time, we should extract information from the URLs. To do so, we have a python library named, 'Extraction' used for extracting titles, descriptions, images and canonical URLs from web pages. Once we get the topic, we can classify it into an educational, informative, or entertainment-based content. As stated earlier, the difficult task is getting the classified historical data for the training purposes.

3.3   Features

We made use of two features in our model: a bag of words (BOW) and WordNet synset.

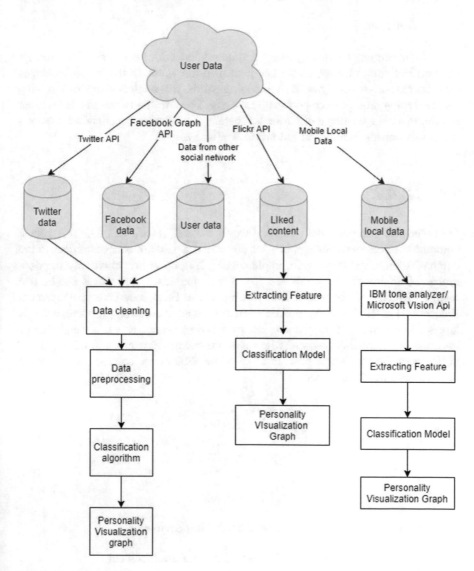

Fig. 1 System architecture

3.3.1 Bag of Words

Bag of words approach (BOW) is the most common methodology for extracting feature from sentences and documents. Here, each word counts as a feature and the histograms of words in the text are examined. We use Bag of words concept in methods of document classification where the repetition frequency of each is utilized as a feature for training a classifier.

3.3.2 WordNet

Now, for improving the quality of the feature set and reduce overfitting, the concept of WordNet comes handy, as we use it to map the words in the tweets/messages onto their synonym set (synset). By mapping words, we conclude that words having same meaning prompt same personality criteria. The quantity of features is lessened but improves the coverage of particular feature. This proves beneficial as it covers the words outside training set but are in similar synset.

3.4 TF-IDF

Observing the patterns certain function words as 'the', 'and', 'he', 'she', 'it' repeat commonly across most descriptions. As a result, it would not be a smart decision to emphasize on such words while implementing Bag of Words to classify the documents. Clipping off the words in a list of high-frequency stop words can be one method which can be used. Considering Term Frequency-Inverse Document Frequency (TF-IDF) weight of each word can be an alternative [9]. Here, more is less and less is more meaning that sometimes the less occurring words weight heavy in terms of information content whereas more occurring words can be not much useful considering information point of view. Following equation helps us to produce weights for each word:

$$\text{tf}_{i,j} = \frac{n_{i,j}}{\sum_k n_{k,j}}$$

$$\text{idf}_i = \log \frac{|D|}{|dt_i|}$$

$$\text{tfidf}_i = tf_{i,j}\text{idf}_i$$

$\text{tf}_{i,j}$: importance of term i in document j

$n_{i,j}$: number of times term i occured in document j

$\sum k\, n_{k,j}$: total number of words in document j

idf_i : general importance of term i

$|D|$: total number of documents in corpus

$|dt_i|$: number of documents where the term t_i appears

3.5 *Classifier*

For this study, we used classification algorithms like Naïve Bayes and SVM and evaluated their performance. We need to classify the text into five main categories (Openness, Extraversion, Agreeableness, Conscientiousness, and Neuroticism). Before classification, the feature extraction process takes place using the NLP techniques mentioned above. Then, the features along with the label are fed into supervised machine learning models as shown in Fig. 2. Now, the system is ready to classify the given unknown input after the training and validation process. We have used the scikit-learn machine learning library in python. Also, the feature extraction process is done using sklearn library. For this study, we have trained the data only to output only two classes, i.e., Conscientiousness (Label 1) and Neuroticism (Label 2). Label 1 contains 570 sentences whereas Label 2 contains 621 sentences for the training purpose and 20% of the data was used for the development tests. We need to provide very less amount of training data which acts as a benefit in terms of N.B and SVM classification and help decide criterion necessary for categorization [10].

4 Results/Expected Output

Till now, we have collected data for three classes, trained the data for two classes and carried out three experiments for each classifier. For each experiment, we use a "feature vector", a "classifier" and a train-test splitting strategy. For experiment 1, we used a Bag of Words (BOW) representation of each document, for second experiment we used Term Frequency and for the final experiment, we used TF-IDF along with NB classifier and then with SVM classifier.

Table 1 shows the classification report for MultinomialNB classifier whereas Table 2 shows the same for SVM model.

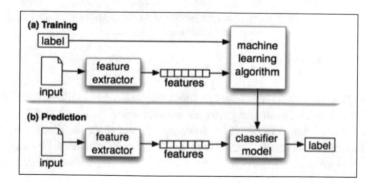

Fig. 2 System flow

352 J. Philip et al.

Table 1 Classification report NB classifier

	Precision	Recall	F1-Score	Support
Conscientiousness	0.95	0.97	0.96	577
Neuroticism	0.97	0.96	0.96	619
Avg./Total	0.96	0.96	0.96	1196

Table 2 Classification report SVM classifier

	Precision	Recall	F1-Score	Support
Conscientiousness	0.98	0.97	0.98	616
Neuroticism	0.97	0.98	0.96	585
Avg./Total	0.98	0.97	0.97	1196

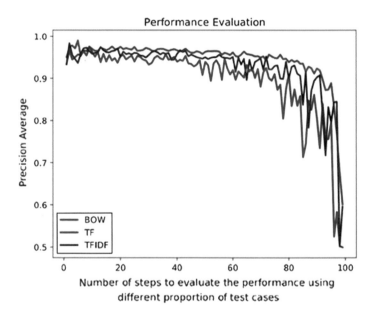

Fig. 3 Naïve Bayes classifier

Figures 3 and 4 show the average precision value for random train and test subsets for cross-validation and performance evaluation process for NB and SVM classifier models, respectively [11]. It compares the different feature extraction process like BOW, TF, and TF-IDF with BOW for each classifier. From the below results, SVM with TF-IDF outperforms Naïve Bayes Algorithm for this research study [10].

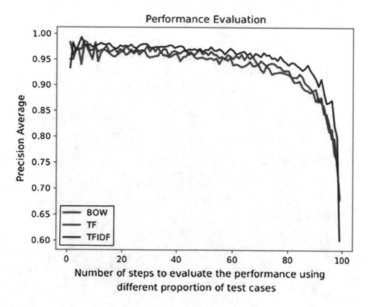

Fig. 4 Support vector machine

5 Limitation and Future Work

People with extensive use of social media platforms tend to post or upload a lot about themselves through timely updates, caption-description, images, videos, and other interests. Also, at times a user share status on ongoing topics currently happening in the real world. At such circumstances, it is required that the system should have information about that topic so that system can relate the user's status with topic and can analyze the actual meaning of status.

Figure 5 shows the expected output in the future. It will show Big Five model and the sub traits of every factor to help us know more about user. So, the system will generate multiple personality graph as system will take data from different sites. Every graph will have different values of personality traits for one user as user tends to act different on different social site. At the end, system will merge all value of personality traits and will generate one final personality analysis graph.

In Future, we can consider images, videos and voice to better understand a human's personality trait from all aspects and not only text incorporating deep learning algorithms with far better results.

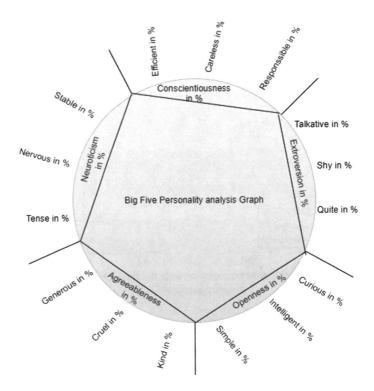

Fig. 5 Personality graph

6 Conclusion

This study comes with an efficient approach of summarizing a human's personality which is based on Big Five model using natural language processing and machine learning techniques to train the model with the personality traits and thus generating a graphical report of user's character. We found that SVM with TF-IDF feature extraction process produced better results. With more datasets and better feature extraction process like using bigrams or n-grams instead of BOW would produce better outcome. Such system will be an aid in understanding a person's characteristics which would help during recruitment process, employee satisfaction, for parental monitoring, etc.

References

1. The top 20 valuable facebook statistics. Updated September 2017. Available https://zephoria. com/top-15-valuable-facebook-statistics/. Accessed August 16, 2017.
2. McCrae, R. R., & John, O. P. (1992). An introduction to the five-factor model and its applications. *Journal of Personality, 60*(2), 175–215.
3. Golbeck, J., Robles, C., Edmondson, M., & Turner, K. (2011, October). Predicting personality from twitter. In *2011 IEEE Third International Conference on Privacy, Security, Risk and Trust (PASSAT) and 2011 IEEE Third International Conference on Social Computing (SocialCom)* (pp. 149–156). IEEE.
4. Chittaranjan, G., Blom, J., & Gatica-Perez, D. (2011, June). Who's who with big-five: Analyzing and classifying personality traits with smartphone. In *2011 15th Annual International Symposium on Wearable Computers (ISWC)* (pp. 29–36). IEEE.
5. Bai, S., Hao, B., Li, A., Yuan, S., Gao, R., & Zhu, T. (2013, November). Predicting big five personality traits of microblog users. In *2013 IEEE/WIC/ACM International Joint Conferences on Web Intelligence (WI) and Intelligent Agent Technologies (IAT)* (Vol. 1, pp. 501–508). IEEE.
6. Kalghatgi, M. P., Ramannavar, M., & Sidnal, N. S. (2015). A neural network approach to personality prediction based on the big-five model. *International Journal of Innovative Research in Advanced Engineering (IJIRAE), 2*(8), 56–63.
7. Zhang, W., Zhao, G., & Zhu, C. C. (2014). Mood detection with tweets.
8. Natural language processing. Available https://en.wikipedia.org/wiki/Natural_language_ processing. Accessed August 16, 2017.
9. Available http://www.markhneedham.com/blog/2015/02/15/pythonscikit-learn-calculating- tfidf-on-how-i-met-your-mother-transcripts/. Accessed November 2, 2017.
10. Available http://scikit-learn.org/stable/modules/feature_extraction.html. Accessed November 8, 2017.
11. Available https://www.datasciencecentral.com/profilesblogs/7-important-model-evaluation- error-metrics-everyone-should-know. Accessed November 25, 2017.

A Personalized Context-Aware Recommender System Based on User-Item Preferences

Mandheer Singh, Himanshu Sahu and Neha Sharma

Abstract In the digital world, it has become a challenging task to find items that suit users' persona and fulfill their need. The reason behind this problem is the unprecedented growth of content and product available online. Recommender System (RS) has emerged as a tool which provides personalized results to users as well as suggestion based on its behavior and past history. Collaborative Filtering (CF), the widely used technique, in the field of RS, provides useful recommendations to users based on similar users. Traditional recommendation approaches such as collaborative filtering and content-based filtering, work on two dimensions, i.e., user-item pair. In addition to this "Context used as third dimension", is also getting popular among researchers. In the present paper, a new method is proposed, i.e., Context-Aware Recommender System by utilizing both item as well as user preferences based on splitting criteria for movie recommendation applications. In this method, first single item is split into two virtual items based on contextual value and a modified dataset is created. Then, the single user is split into two virtual users based on contextual values. Splitting of any user or item is done only if there is a significant difference between two virtual items (users). Further user-based collaborative filtering is used to generate effective recommendations. The results show the effectiveness of proposed scheme in terms of various performance measure criteria using LDOSCOMODA dataset.

Keywords Collaborative filtering · Context-Aware Recommender Systems
Content-based filtering · Recommender System

M. Singh
MNNIT Allahabad, Allahabad, India
e-mail: mandheer.cse@gmail.com

H. Sahu (✉) · N. Sharma
University of Petroleum and Energy Studies, Dehradun, India
e-mail: hsahu@ddn.upes.ac.in

N. Sharma
e-mail: nehas956@gmail.com

© Springer Nature Singapore Pte Ltd. 2019
V. E. Balas et al. (eds.), *Data Management, Analytics and Innovation*,
Advances in Intelligent Systems and Computing 839,
https://doi.org/10.1007/978-981-13-1274-8_28

357

1   Introduction

The worldwide web has evolved so quick that it has become a large sea of information. The traditional (as search engine like Google, Bing, etc.) search provides the user with tons of information. It becomes a tedious task for the user in extracting relevant information that also suits her preferences. It creates a need of a tool or system which helps the user to get what they need. Although it is the ultimate choice of the user to select what he actually needs but what they need is to overcome with the bombardment of information.

Recommender System (RS) [1] solves this problem by searching through large volume of dynamically generated information to provide users with personalized content and services. RS is basically information filtering system that deals with the information overload problem by filtering vital problem fragment out of large amount of dynamically generated information according to the user's preferences, interest, or observed behavior about item [2]. The Recommender System not only helps the user but also the e-commerce company to sell their products by suggesting what a user can buy by following what he has already bought. This concept is also working for digital content company such as online music, video, or e-books stores.

Although a lot of work has been carried out in the area of Recommender System, there is still unbounded scope available due to its multifaceted application and commercial importance. There are different techniques available for recommendation system such as content-based filtering, collaborative filtering, and hybrid filtering. Collaborative filtering is one of the most successful approaches that recommends items to a user based on its similarity index with some group of users. In the present paper, context is added as a third domain to collaborative filtering which has shown a significant improvement over the existing collaborative filtering for the same set of user and items.

The remaining of the paper is organized as follows: Sect. 2 describes background and literature survey. Section 3 describes proposed work. Section 4 describes the experimental results and finally Sect. 5 gives the conclusion and future works.

2   Background and Literature Survey

This section discusses what Recommender System is and then discusses different filtering techniques. It also sets the mathematical background and describes the context in RS.

A. Recommender System

RSs collect the following information either explicitly or implicitly. These sets of data provide the basis on which recommendations take place.

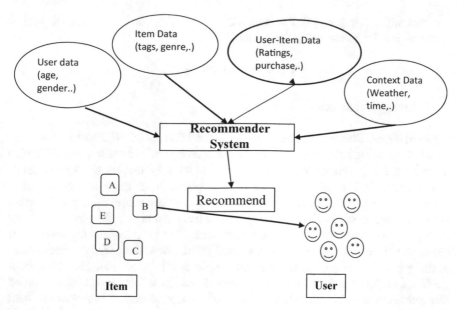

Fig. 1 Recommendation process

User data—It is the information about the user such as age, gender, address, occupation, etc. It is also called as demographic features of users [3].

Item data—Item can be movies, songs, jokes, news, travel destinations, etc. RS collects information about items available and associates them with tags and keywords [3].

Context data—It is the condition in which item is used or is to be used, like companion, season, weather, etc.

Preference data—It is a collection of information by asking user to rate the items or by tracking user's past behavior like songs heard, movies watched, and product purchased [3].

On the basis of the above information, RS provides a list of items, that user may like or interested in, by applying the recommendation techniques. Figure 1 shows an overview of recommendation process.

Recommendations can be solved as a rating estimation problem [3]. In the recommendation process, the RS's need to approximate the rating of new or unseen items using the values for the old items. After this estimation, the items with the highest rating are supposed to best suitable for the user and provided as a recommendation.

RS as a Rating Estimation Problem Let **U** be the set of users, **I** be the set of all possible items, and **R** be the set of ratings provided by user to a seen item. Rating is a nonnegative integer between certain ranges like (1–5). Where 1-signifies very bad and 5-signifies best item [4]. For each user, $u \in U$, $i' \in I$ is to be chosen in such a way that maximize the rating estimation:

$$\forall u \in U, i' = \mathbf{argmax}_{i \in I} R_{u,i} \tag{1}$$

B. Filtering Techniques

Content-based filtering In the Content-Based Filtering (CBF), user gets recommendations of the items, very similar to items user preferred earlier. In CBF rating of an item, i' is predicted based on ratings of items $i \in I$ that are similar to item i'. For example, in a music recommendation application, in order to recommend a song to user u, the CBF tries to understand the common properties (specific artist, composer, genre, etc.) among the songs user u has rated high in the past. Then song having higher similarity is going to be recommended [5]. CBF mainly focuses on the applications where item contains textual information. For example, documents, news, messages, blogs, URLs, etc. Like user profile, there can also have item profile. Let $p(i)$ be the profile of an item i, which is having set of properties characterizing item i. That profile is computed through extraction of features of item through its content. Content is mostly described in the form of keywords.

Collaborative Filtering Collaborative Filtering (CF) is the most popular approach for recommendation systems. It tries to predict the rating of item, based on ratings given by the other users, who shares the similar taste with the target user. More formally to predict rating $r_{u,i}$, first find the users $u' \in U$ such that u' have given same ratings to the other items as user u. Now rating $r_{u,i}$ is predicted based on the ratings given to i by user-u'. For example, to recommend a movie to any user, first find users, who watched and rated the movies similarly as target user, these users are called peers of user. Then aggregate the ratings of peers to predict the rating of target item [6]. CF does not require structured description or content of items. They are more often implemented then CBF [6]. There are two types of CF namely user-based collaborative filtering and item-based collaborative filtering.

User-Based Collaborative Filtering The basic idea behind user-based collaborative filtering is that in this method unrated items are ranked according to the similarity between users [7]. To find rating for item i by user u, first find the peers of user u, i.e., set of similar users then aggregate the ratings of all peers of user u. Then based on aggregated rating, the top rating can be recommended to similar set of user.

Item-Based Collaborative Filtering The basic idea behind item-based collaborative filtering is that in this method, unrated items are ranked according to the similarity between items [7]. To find rating for item i for user u, first find the items similar to i and after finding that aggregate the ratings of all items (neighbors) i to predict the rating. The similarity measures used are Pearson correlation, cosine-based similarity, vector space similarity, and so many. Pearson's correlation coefficient formula between two users' u and v is defined as

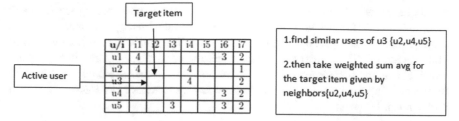

Fig. 2 User-based collaborative filtering

$$\mathbf{sim}(u, v) = \frac{\sum_{i \in S} (r_{u,i} - \bar{r}_u)(r_{v,i} - \bar{r}_v)}{\sqrt{\sum_{i \in S} (r_{u,i} - \bar{r}_u)^2 \sum_{i \in S} (r_{v,i} - \bar{r}_v)^2}} \quad (2)$$

where S is the set of co-rated items between user u and user v, i represents the current item and $\bar{r}_u$ represents mean rating of user u.

Figure 2 shows an example of user-based collaborative filtering. To predict the rating for item i_2 that user u_3 would give, first find peer or neighbor set for the user u_3. After finding peers, take weighted sum average of ratings to item given by peers $(u_1 u_2 u_4)$.

Simple Resnick's formula [8] is used to predict rating.

$$r(u, i) = \bar{r}_u + \frac{\sum_{v \in N} \mathbf{sim}(u, v) \cdot (r_{v,i} - \bar{r}_v)}{\left| \sum_{v \in N} \mathbf{sim}(u, v) \right|} \quad (3)$$

where $\bar{r}_u$ means average rating of user u and N is the set of neighbors for user u.

Hybrid Filtering Several RSs have implemented a combination of both filtering approaches, i.e., (collaborative and content-based) which is called as a hybrid filtering approach. The hybrid filtering approach helps in mitigating the limitations associated with content-based as well as collaborative filtering systems [9]. These hybrid approaches can be classified as following [4]:

1. **Combination of predictions**: First implement both methods separately and then combine their results to get the final prediction
2. **Content added in Collaborative filtering**: Add the content-based flavor to collaborative approach,
3. **Collaboration added in Content-based filtering**: Adding some collaborative characteristics into a content-based approach, and
4. **Unified Model**: A unified model that provide a combination of both type of filtering method in full fledge mode.

Context-Aware Recommender Systems Contexts are very important for recommending items correctly. Traditional approaches of Recommender System like collaborative and content-based method work on ratings provided by the user and they do not consider any situation which may also affect user preferences. But it is found that user may prefer one item in one situation and a very different item in other situation [10]. Context can be any condition or situation of user, item or event which affects the preference of user. Traditional approaches are based on user and item data. They can be called as two-dimensional approaches [11]. Recommendation problem can be viewed as

$$R : \text{Users} \times \text{Items} \rightarrow \text{Ratings}$$

Context-Aware Recommender System (CARS) [12, 13] is smart enough and works in three dimensions. CARS is quite similar to the traditional approaches but they also consider situation of user or item when the user uses the item. In CARS, the recommendation problem is different.

$$R : \text{Users} \times \text{Items} \times \text{Contexts} \rightarrow \text{Ratings}$$

The major challenge of CARS was how to incorporate the contextual information into existing two-dimensional system [10]. There are mainly three approaches for incorporating the contextual information into existing system proposed by researchers. Contextual pre-filtering, post-filtering, and contextual modeling are the approaches that are used in CARS [14].

Contextual Pre-filtering In contextual pre-filtering, the contextual information is used to discard the irrelevant data, after discarding same two-dimensional data set is achieved. Hence, existing algorithms can be used to final recommendations [14].

Post-filtering In post-filtering method, contextual information is not considered at first. The recommendation is calculated using two-dimensional approaches then and after that filtering is applied using contextual information to discard the irrelevant recommendation [14].

Contextual Modeling In this approach, data set (including contextual information) is used to fit a model. Various techniques can be used to make better user models such as regression and tensor factorization [14].

C. Literature Review

Context-Based RS

Adomavicius and Tuzhilin [15] proposed that user preferences can be influenced by contextual conditions and not only by user-item preferences. Definition of context is not very much clear but widely accepted definition is given by [11]:

> Context can be any information that can be used to characterize the situation of any entity. An entity is a person, place, or object that is considered relevant to the interaction between a user and an application, including the user and applications themselves.

Verbert et al. [16] classified the contexts into three categories computing contexts (n/w connectivity, bandwidth, etc.), user contexts (location, social situation, etc.), and physical context (noise level, traffic conditions, etc.). Contextual information can be collected in two ways either explicitly (by asking user) or implicitly (by using GPS) [10]. Ricci et al. [10] explains how contextual information can help in provinding accurate recommendations. Divides the contextual factors into three categories based on the knowledge of the system, which are fully observable, partially observable, and unobservable. Contextual information can be useful in several application domains (movies, traveling, holidays etc.).

Filters in CARS

Various techniques proposed for incorporating contextual information in existing RS. So far three main categories have been defined, i.e., pre-filtering, post-filtering, and modeling [1, 10, 14]. Ricci et al. [10] proposed pre-filtering method that discards irrelevant data using contextual values and then apply existing approach. Anand et al. [17] divided dataset into fixed number of segments by combining some contexts in one group and extracts "contextual cues" instead of using fixed contextual factors. Baltrunas et al. [18] proposed two post-filtering approaches, one is "Weight", which reorders the recommendations based on the probability of relevance in that particular context, and other is "Filter" method which discards data with small probability. Our work is inherited from the work done in [18]. They proposed item splitting in which one item is divided into two virtual items based on contextual conditions. Their work is extended and suggested to split user also and combine both (item splitting and user splitting) in linear manner.

3 Proposed Work

This proposed method is used to recommend movies considering various contextual values. It is not only the explicit ratings which affect the user preferences but contextual values can also play an important role as well. For example, a beach (holiday destination) is more preferable in summer than in winter. So it is obvious that beach will get more ratings in summer. If ratings are recorded separately for summer and winter, then there would be significant difference between them. In this work, L-DOSCOMODA context-aware dataset [19] for movies was used to provide recommendations. The method that has been proposed is applicable in general where rating data is associated with the context.

A. Overview

In this experiment below mentioned, contextual factors are used to evaluate the performance of the method. Figure 3 shows different values of context.

Contextual factors have finite number of values such as time has four contextual values Morning, Afternoon, Evening, and Night which are stored in the database as follows: Morning = 1, Afternoon = 2, Evening = 3, Night = 4.

S.No	Context	Value
1	Time	Morning, Afternoon, Evening, Night
2	Day type	Working day, Weekend, Holiday
3	Season	Spring, Summer, Autumn, Winter
4	Location	Home, Public place, Friend's house
5	Weather	Sunny/clear, Rainy, Stormy, Snowy, Cloudy
6	Social	Alone, My partner, Friends, Colleagues, Parents, Public, My family
7	EndEmo	Sad, Happy, Scared, Surprised, Angry, Disgusted, Neutral
8	DominantEmo	Sad, Happy, Scared, Surprised, Angry, Disgusted, Neutral
9	Mood	Positive, Neutral, Negative
10	Physical	Healthy, Ill

Fig. 3 Different values of context

The basic idea of item and user splitting is to split the rating set based on contextual value into two parts and check, if there is a significant difference between them, apply the changes to the dataset [18]. User may have different choices in different contextual condition. Hence, one user (or item) is considered as two virtual users (or items) associated with different contextual values. For example, a user u has given rating to an item in morning and given rating to other items in the evening, then it will be treated as two virtual users.

The proposed methodology can be divided into three phases. Each phase is described in detail in the following subsections. Figure 4 shows an overview of the proposed method. Let U be the set of all users, I be the set of all items, and C be the set of all contextual factors and corresponding contextual values. R is the set of ratings given by user. As shown in Fig. 4, after apply item splitting over I on the initial input dataset, I_{new} is obtained. On the changed dataset user splitting (over U) is applied and the new user's set U_{new} is created. Finally, in the modified data set standard two-dimensional approaches can be applied to get the final recommendation. In the proposed method, user-based collaborative filtering is used.

B. Phase I: Item Splitting

In phase-1, item splitting [18] has been performed to get new item set $I = I_{new}$. Figure 5 shows an example of item splitting performed over one item. i is an item which receives ratings from five users in different contextual values. Split item i into two virtual items, i.e., I_1 and I_2, where I_1 represents ratings given by user in one contextual value and I_2 represents ratings given by users for alternative contextual values. The algorithm for item splitting is given below.

Fig. 4 Proposed
methodology

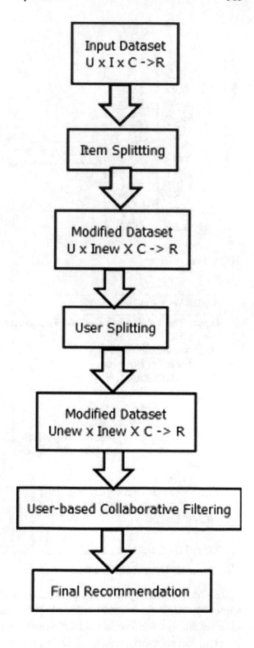

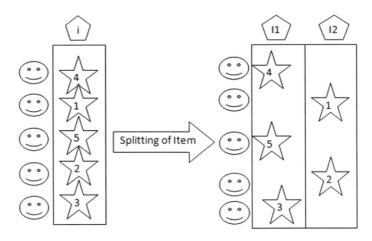

Fig. 5 Item splitting

Algorithm 1 Item splitting

Input: Input dataset, Threshold T, Splitting criteria S_{cr}
Output: Modified dataset
      for $k = 1$ to n_i**do**
         take $item$ i_k
      for$k = 1$ to n_{cf}**do**
            take $c_f \in C$
            for $cv \in c_j$**do**
                generate RI_{cv} and $RI_{cv'}$
                compute $diff(i_k, cv)$
end for
      end for
$CV_{max} \leftarrow aggr\ maxcv[diff(i_k, cv)]$
      If $diff(i_k, cv) \geq T$**then**
Modify dataset
end if
end for

C. Phase II: User Splitting

In phase-2 user splitting is performed to get new user set $U \rightarrow U_{new}$. Figure 6 shows an example of user splitting performed over one user. u is a user which receives ratings for five items in different contextual values. A single user u is split into two virtual items, i.e., U_1 and U_2, where U_1 represents ratings given by user in one contextual value and U_2 represents ratings given by users for alternative contextual values.

Fig. 6 User splitting

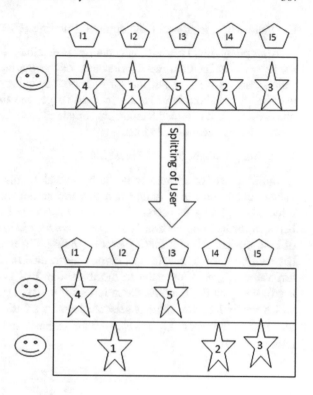

The algorithm of user splitting is shown below

Algorithm 2 User splitting

Input: Input dataset from phase I, Threshold
T, Splitting criteria S_{cr}
Output: Modified dataset
　　　　for $k = 1$ to n_u **do**
　　　　take useru_k
　　　　for$k = 1$ to n_{cf} **do**
　　　　　　take $c_f \in C$
　　　　　　for $cv \in c_j$ **do**
　generate UI_{cv} and $UI_{cv\prime}$
　　　　　　　　compute $diff(u_k, cv)$
end for
　　　　end for
　$CV_{max} \leftarrow aggr\ maxcv[diff(u_k, cv)]$
　　　　If $diff(u_k, cv) \geq T$**then**
Modify dataset
end if
end for

D. Phase-III: User-Based Collaborative Filtering (UBCF)

   After phase-1 and phase-2, new dataset is obtained with item set as I_{new} and user set as U_{new}. After this, pre-filtering any of the traditional recommendation algorithms can be used to get the final recommendation. Present results have used user-based collaborative filtering method to get the final recommendation. After phase-3, final recommendations are obtained and those recommendations are context-aware recommendations.

E. Spilting Criteria S_{cr} and Threshold T

   Splitting user or an item could be beneficial if rating set in both the conditions have enough number of ratings and that too homogeneous [18]. It is also conjectured that splitting is useful if two sets of ratings have a significant difference between the splitting criteria of set [18]. One way is to consider the average rating of both the set. For considering average rating, diff is used as the difference value between both set. The split operation is performed to any user or item, only if the diff value is greater or equal to the threshold T. T is predefined value ($T = 1$ is used). If any $cv \in C_j$ divides the rating set into two, i.e., RU_{cv} and $RU_{cv'}$. The split which maximizes diff value is selected for user splitting. Let $m_{u_{cv}}$ and $m_{u_{cv'}}$ be the mean rating for user u. $s_{u_{cv}}$ and $s_{u_{cv'}}$ are the variance and $n_{u_{cv}}$ and $n_{u_{cv'}}$ are the size of rating set [17].

$$\text{diff} = \frac{m_{u_{cv}} - m_{u_{cv'}}}{\sqrt{\frac{s_{u_{cv}}}{n_{u_{cv}}} + \frac{s_{u_{cv'}}}{n_{u_{cv'}}}}} \tag{4}$$

4   Results and Analysis

This section provides the experimental results and the analysis which have been performed over LDOSCOMODA dataset and three methods (i.e., UBCF, Item splitting and proposed method) are compared on the basis of MAE and coverage as performance measure criteria.

4.1   Performance Measure Criteria for RS

The matrix is divided into two sets, i.e., training and testing. The training matrix is used for model fitting and testing matrix is used for evaluation. Several metrics are used to evaluate Recommender Systems. The most popular metrics are mean absolute error (MAE), root mean square error (RMSE), precision and recall ratio. MAE is defined as the average absolute difference between predicted and actual ratings [20]. MAE is a measure of deviation of recommendations from real user-rated ratings, and it is most commonly used and very easy to interpret. Let n be the number of rating prediction pair then and $r_{u,i}$ represent rating of user u on item i and $\hat{r}_{u,i}$ represent predicted rating, then

$$\text{MAE} = \sum_n \frac{r_{u,i} - \hat{r}_{u,i}}{n} \tag{5}$$

If MAE is lower means RS is accurate. Thus smaller MAE is better. RMSE is defined as root of the square of the difference between actual and predicted rating [20].

$$\text{RMSE} = \sqrt{\frac{\sum_n \left(r_{u,i} - \hat{r}_{u,i} \right)^2}{n}} \tag{6}$$

Coverage measures how much percentage of items for which Recommender System is capable of making predictions.

4.2 Dataset Used

The LDOS–CoMoDa dataset [19] is used to perform the experiments, which is a context-rich movie rating database.

4.3 Experimental Setup

The experiment has been performed on the abovementioned dataset and the three methods have been compared on the basis of MAE and Coverage. A new parameter **ttr** train/test ratio is used whose range is 0–1. The proposed method has applied to the training set and to get the results. The results have been examined against testing set. Random data from the dataset is extracted to implement these methods. The next parameter which is used is k, which represents the value of neighborhood. These experiments are performed by varying the value of k and check its impact on the performance.

4.4 Results

Figure 7a–e gives the result on different runs. The result is varying due to random selection of training set and testing set. The **ttr** value is set to 0.8 and K is varied from 40 to 60. The graph shows that the MAE of the proposed method is lower than

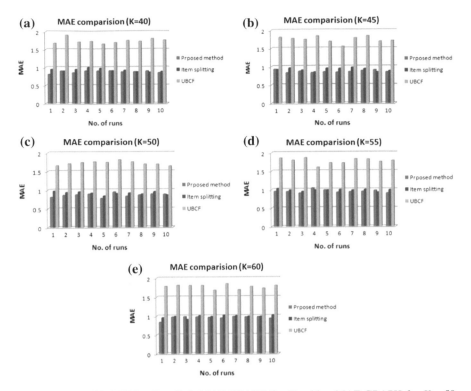

Fig. 7 **a** MAE GRAPH for $K = 40$. **b** MAE GRAPH for $K = 45$. **c** MAE GRAPH for $K = 50$. **d** MAE GRAPH for $K = 55$. **e** MAE GRAPH for $K = 60$

Table 1 Average MAE value on different run

K	Proposed method	Item splitting	UBCF
40	0.8814	0.9263	1.7464
45	0.8682	0.9192	1.7408
50	0.8609	0.913	1.7149
55	0.9497	0.982	1.7745
60	0.9517	0.9838	1.771
AVG	0.9024	0.9449	1.7495

of the two methods. So the proposed method is better than the other two methods. Table 1 shows that the average value of MAE for all values of k is 0.9024 for the proposed method. While for item splitting and UBCF, the average MAE value is 0.9449 and 1.7495 respectively.

Table 1 shows the average value of MAE which clearly shows improvement of the proposed method and the MAE value is lower as compared to other two methods.

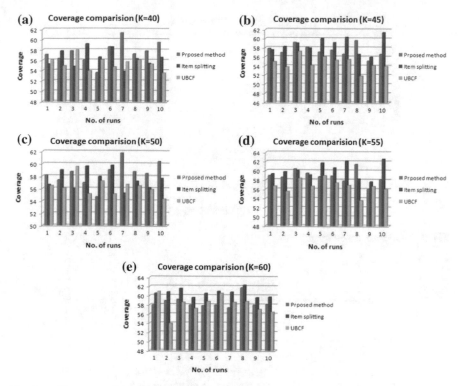

Fig. 8 **a** Coverage values for $K = 40$. **b** Coverage values for $K = 45$. **c** Coverage values for $K = 50$. **d** Coverage values for $K = 55$. **e** Coverage values for $K = 60$

Table 2 Average value of comparison

K	Proposed method	Item splitting	UBCF
40	57.5301	56.4414	55.4335
45	57.4698	58.6336	54.6820
50	58.4186	57.5225	56.2813
55	58.8403	60.1351	56.6666
60	58.5090	60.6456	58.0732
AVG	58.1536	58.6756	56.2273

Figure 8a–e shows different experimental graphs for coverage comparisons. It is clear from Fig. 8a–e that coverage of proposed method is better than UBCF and but it is quite equivalent to the coverage of item splitting.

Table 2 shows that the average value of coverage for all values of k is 58.15361 for the proposed method. While for item splitting and UBCF, the average coverage value is 58.675 and 56.2273, respectively.

Figure 9 shows the graph which compares all three method's average MAE and impact of varying value of neighborhood. Increasing the value of k decreases the

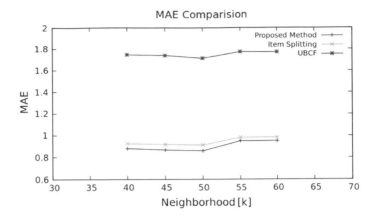

Fig. 9 MAE comparison graph

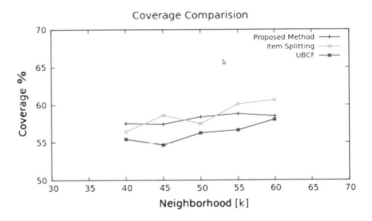

Fig. 10 Coverage comparison graph

MAE up to $k = 50$ after that it is increased with increasing value of k. As shown in graph, our proposed method is very good in comparison with UBCF and slight improvement is also there from item splitting.

Figure 10 shows the graph which compares all three methods mean coverage and impact of varying value of neighborhood. Increasing the value of k increases the coverage after $k = 50$ for UBCF and item splitting, it shows a random pattern for the proposed method. As shown in graph our proposed method is very good in comparison with UBCF and almost equivalent to item splitting.

5 Conclusion and Future Work

5.1 Conclusion

Over the last two decades, an extensive work has been done in both industry and academia on developing new approaches to RS. Many commercial RSs are widely adopting the CF technique. However, it has been argued that considering contextual information is beneficial for accuracy improvement of its recommendation ability. In this work, a new approach has been presented for CARS, which is a combination of item splitting and user splitting. Proposed method aims to enhance the accuracy of CARS where item as well as user is split into two virtual items or users based on contextual information. Here contextual pre-filtering is used to discard irrelevant data using contextual information and then used CF framework to reach out final recommendations. Experimental results demonstrated the effectiveness of our proposed system.

5.2 Future Work

In future, this work can be enhanced in many ways. In this experiment, average rating is used as splitting criteria, but other criteria like chi-square, proportion, etc., can also be used to splitting [17]. Moreover, item splitting and user splitting can be combined using other combinations or hybridization could also be a promising direction for future work.

References

1. Bobadilla, J., Ortega, F., Hernando, A., & Gutiérrez, A. (2013). Recommender systems survey. *Knowledge-based Systems, 46,* 109–132.
2. Resnick, P., & Varian, H. R. (1997). Recommender systems. *Communications of the ACM, 40*(3), 56–58.
3. Anderson, C., & Hiralall, M. (2009). Recommender systems for e-shops. Business Mathematics and Informatics paper.
4. Adomavicius, G., & Tuzhilin, A. (2005). Toward the next generation of recommender systems: A survey of the state-of-the-art and possible extensions. *IEEE Transactions on Knowledge and Data Engineering, 17*(6), 734–749.
5. Lops, P., De Gemmis, M., & Semeraro, G. (2011). Content-based recommender systems: State of the art and trends. In *Recommender systems handbook* (pp. 73–105). US: Springer.
6. Su, X., & Khoshgoftaar, T. M. (2009). A survey of collaborative filtering techniques. *Advances in Artificial Intelligence, 2009,* 4.
7. Sarwar, B., Karypis, G., Konstan, J., & Riedl, J. (2001, April). Item-based collaborative filtering recommendation algorithms. In *Proceedings of the 10th International Conference on World Wide Web* (pp. 285–295). ACM.

8. Resnick, P., Iacovou, N., Suchak, M., Bergstrom, P., & Riedl, J. (1994, October). GroupLens: An open architecture for collaborative filtering of netnews. In *Proceedings of the 1994 ACM Conference on Computer Supported Cooperative Work* (pp. 175–186). ACM.

9. Burke, R. (2002). Hybrid recommender systems: Survey and experiments. *User Modeling and User-Adapted Interaction, 12*(4), 331–370.

10. Ricci, F., Rokach, L., & Shapira, B. (2011). Introduction to recommender systems handbook. In *Recommender Systems Handbook* (pp. 1–35). US: Springer.

11. Dey, A. K. (2001). Understanding and using context. *Personal and Ubiquitous Computing, 5*(1), 4–7.

12. Adomavicius, G., & Tuzhilin, A. (2011). Context-aware recommender systems. In *Recommender Systems Handbook* (pp. 217–253). US: Springer.

13. Adomavicius, G., Sankaranarayanan, R., Sen, S., & Tuzhilin, A. (2005). Incorporating contextual information in recommender systems using a multidimensional approach. *ACM Transactions on Information Systems (TOIS), 23*(1), 103–145.

14. Panniello, U., Tuzhilin, A., & Gorgoglione, M. (2014). Comparing context-aware recommender systems in terms of accuracy and diversity. *User Modeling and User-Adapted Interaction, 24*(1–2), 35–65.

15. Adomavicius, G., & Tuzhilin, A. (2001). Extending recommender systems: A multidimensional approach. In *Proceedings of the International Joint Conference on Artificial Intelligence (IJCAI-01), Workshop on Intelligent Techniques for Web Personalization (ITWP2001)* (pp. 4–6). Seattle, Washington: Citeseer.

16. Verbert, K., Manouselis, N., Ochoa, X., Wolpers, M., Drachsler, H., Bosnic, I., et al. (2012). Context-aware recommender systems for learning: a survey and future challenges. *IEEE Transactions on Learning Technologies, 5*(4), 318–335.

17. Anand, S. S., & Mobasher, B. (2006, September). Contextual recommendation. In *Workshop on Web Mining* (pp. 142–160). Berlin: Springer.

18. Baltrunas, L., &Ricci, F. (2014). Experimental evaluation of context-dependent collaborative filtering using item splitting. *User Modeling and User-Adapted Interaction, 24*(1–2), 7–34.

19. Wu, H., Liu, X., Pei, Y., & Li, B. (2014, October). Enhancing context-aware recommendation via a unified graph model. In *International Conference on Identification, Information and Knowledge in the Internet of Things (IIKI), 2014* (pp. 76–79). IEEE.

20. Herlocker, J. L., Konstan, J. A., Terveen, L. G., & Riedl, J. T. (2004). Evaluating collaborative filtering recommender systems. *ACM Transactions on Information Systems (TOIS), 22*(1), 5–53.

Plant Classification Using Image Processing and Neural Network

Manisha M. Amlekar and Ashok T. Gaikwad

Abstract This paper is presenting here the plant classification method using imaging technology which is useful for classifying the plants by providing the leaf image as an input. The proposed method performs classification by automatically extracting shape patterns and features performing the image processing techniques and neural network model. This method gets the leaf image as an input, performs the leaf image processing tasks, and automatically extracts the leaf shape pattern and leaf shape features. It performs the classification with the help of leaf shape features using neural network techniques. This method presents plant classification using the leaf shape features and feed forward back propagation neural network model. This method results in up to 99% accuracy of classification. This method extracts the leaf shape features and patterns automatically using image processing techniques.

Keywords Feature extraction · Feed forward neural network · Leaf image processing · Shape extraction

1 Introduction

Environmental imbalance is due to deforesting. Because of this imbalance, we face the problem of drought. Plants play a very important role in the balancing in environment. Plants are useful as food. They are important for human being for health. Their extracts are used as medicines. For various industries, plants are useful in many ways. Trained taxonomist and botanist perform the task of classification of plants. In this process, plants are grouped together based on some identifiable

M. M. Amlekar (✉) · A. T. Gaikwad
Dr. Babasaheb Ambedkar Marathwada University, Aurangabad 431001,
Maharashtra, India
e-mail: manishaak2012@gmail.com

A. T. Gaikwad
e-mail: drashokgaikwad@gmail.com

© Springer Nature Singapore Pte Ltd. 2019   375
V. E. Balas et al. (eds.), *Data Management, Analytics and Innovation*,
Advances in Intelligent Systems and Computing 839,
https://doi.org/10.1007/978-981-13-1274-8_29

characteristics which are common by some criterion. Plant classification manually is time-consuming process and required experts for identifying plants based on their characteristics with visible components. They need to follow set of task and various methods. These methods manually need to take more efforts and found time consuming. Therefore, performing this task using computerized techniques becomes very much demanding research area for the use of taxonomist and botanist, as well as in agricultural requirement. Classification of plants using computerized techniques are found to be more effective and time-saving with artificial intelligence. Leaves of every plant have differentiable features that provide effective mechanism for classification of the plants. Plant classification is performed by following various methods like morphological anatomy, cell biology, and molecular biological approach [1]. Image processing techniques presented here allow to automatically extract the features from the leaf images and accordingly classify the plants based on these features. Many researchers are attempted to perform the task of classification of plants using the various features of the plant leaves. That is a great challenge to extract features and pattern recognition automatically in this field.

2 Survey of Literature

In the review, it is found that researchers proposed the plant classification based on the leaf shape features. 10-fold cross-validation technique proposed by Hossain [7] got 91% accuracy for identifying plant species using morphological features of the plants. K-NN and neural network classification techniques are found to be more effective to provide better performance.

k-NN—This classification method is k-nearest neighbor classification technique used for classifying the plants using k distance measure of Euclidean distance between samples. The researcher Gu [6] proposed k-NN where $k = 1$ with 93% of the result of the classification. Here Gu [6] also proposed the k-NN method with k=5 and got 86% accuracy for classification of plants. Du [4] has proposed classification using k-NN with $k = 4$ got 92% classification result.

Neural Network—This is model-based architecture performs classification using training the model by learning with the leaf features as input and train according to the supervised information of predefined classes of the plant species. Then the model is tested for the leaf samples of testing set. Du [5] has proposed multilayer perceptron learning model for classification based on leaf color features and got 94% accuracy. Wang [12] proposed neural network with back propagation model with accuracy of 92%. Gu [6] used neural network with radial basis function got 91% of the result of classification. Wu [13] probabilistic neural network model has been given 90% of accuracy for classification. Deokar [3] proposed feature point extraction method to extract 28–60 feature points classified using feed forward back propagation method of neural network got 80% accuracy of the classification. Beghin [2] contour-based shape feature extraction method got up to 69.2% classification accuracy. Suman [11] proposed the method of classification of plants using color coherence vector and Haralick features. Lee [10] got up to 98% accuracy with leaf shape features and deep learning. Jiazhi [9] proposed a method of classification using morphological features with accuracy 80%.

3 Methodology

This paper presents the performance of the neural network for classification of the plant species. Figure 1 shows the methodology followed.

Methodology has the following steps:

(3.1) Dataset Collection
(3.2) Preprocessing
(3.3) Shape Extraction
(3.4) Feature Extraction
(3.5) Classification

3.1 Dataset Collection

Leaf image samples are collected from ICL dataset [8], which contain various groups of plant species. Among these groups, five groups of samples are collected. From each group, 40 by 5 samples are selected from the online licensed database provided by Intelligent Computing Laboratory, China. Figure 2 shows few samples from the dataset.

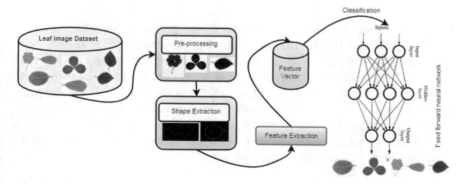

Fig. 1 Methodology

(a)   (b)   (c)   (d)   (e)

Fig. 2 Plant leaf image samples. **a** Pink wood sorrel. **b** White clover. **c** Paper mulberry. **d** Spindle tree. **e** Christmas berry

3.2 Preprocessing

Leaf image samples of these plants are preprocessed to imply uniformity in the image samples for further processing and effectiveness of classification. The leaf image samples are not in uniform color because they are in various state of the growth. For a class of plant that varies in color. These various states color forms affect the further process steps in the methodology of proposed plant classification method. Therefore, leaf image samples are transformed to the uniform gray color code form.

Leaf image with uniform gray code is enhanced to find the leaf shape. This enhancement process finds and maps the intensity of the leaf image and increases the contrast one percent at low and high intensity values. This enhancement is required for improvement in performance of leaf image segmentation for leaf shape extraction.

3.3 Shape Extraction

Preprocessed leaf image samples undergo in the process of features extraction. Shape features are found to be more effective to differentiate the plant leaf image samples. Shape extraction step extracts shape from the leaf image samples. Next step is to extract features from the leaf image samples for the plants which are useful for classification, type of features extracted are shape features using the algorithm shown in Fig. 3. This algorithm uses canny method for leaf image segmentation for extracting the shape features. The algorithm results in the shape patterns for each of the group samples. Finally, from the pattern, we extracted the shape feature to generate the feature vector.

Algorithm for shape extraction:

Step1: Begin

Step2: Original leaf image in true color is read from the database.

Step3: This original leaf image is processed pixel by pixel to map in gray
level format by combining weighted red, green and blue color code of
every pixel, as given below.

GC = 0.2989 * R + 0.5870 * G + 0.1140 * B

Step4: Enhance the leaf image by improving the image intensity low and high
contrast the leaf image by one percent.

Step5: Perform leaf image segmentation for boundary detection by using double
thresholding method Canny.

High Threshold =0.3 and

Low Threshold=0.3*0.4=0.12

4.1. Smoothing image to remove noise for boundary detection

4.2 Find large magnitude component for finding gradient of image

4.3 Find local maxima

4.4 Double thresholding for potential component

4.5 Suppress the weak components that are not strong boundary
components

Step6: End

Fig. 3 Algorithm

3.4 Feature Extraction

Feature extraction performs the automatic features extraction from the shape
extracted for the leaf image samples. Shape morphological features are extracted for
each sample of all classes of the plants [7]. Figure 7 shows the shape features for
the plant groups.

3.5 Classification

Classification task perform the clasification of plants according to the shape features
extracted for all classes. This task take the input of the feature vector extracted from
leaf image samples for each group. This is a supervised classification method, it
takes prior knowledge of the groups according to the shape features. This classical
model of the neural network takes 70% samples of leaf image features randomly for
training and 30% for testing. This method adjusts the weight for the input using LM
method of optimization for better classifying the plant groups

4 Results and Discussion

We evaluated this method on ICL Chinese dataset which we get by taking license to test our method, which contains various groups of plants that are available; among these, we selected five groups, each contains 40 samples. The preprocessing of the plant leaf samples is done by making the samples uniform in color and enhancement of the samples by increasing the contrast. The results of the preprocessing and enhancement are shown in Fig. 4.

Plant leaves samples are having various color features during the steps of their growth. That may affect the further process. Every leaf image is processed to get the uniform format for processing further, thus, implying the uniformity for color features of the plant leaf sample for each group. The shape of leaf samples shows the various features that make classification more effective. Therefore, preprocessed samples are processed further using morphological operations. Figure 5 shows the feature extraction result for few leaf image samples of the plant groups.

After finding the leaf shape for the plant groups, feature extraction method is performed that automatically extracts the features for the group; Fig. 6 shows the features for the plants. These eighteen features include leaf shape features, which are dependent on the direction and variant to the leaf image samples.

This method classifies the plants based on the leaf shape features. The biometric features of the plant leaf samples are given as input for the neural network model designed as shown in the following diagram. 18 features classification of the samples of the plants is performed. This method designs neural network classification method which adjusts the weight for the input using LM method of optimization for better classification of the plants. Feed forward neural network classification model is designed to perform the plant classification based on leaf shape features with classical architecture of 70–30% of the data set input.

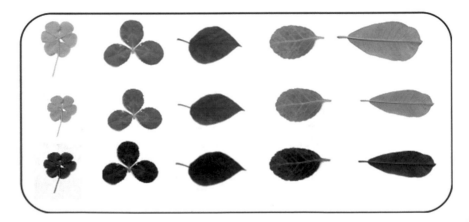

Fig. 4 Preprocessing steps

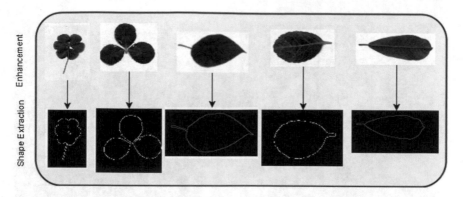

Fig. 5 Feature extraction

Sr.no.	Features ↓ plant classes →	Class1	Class2	Class3	Class4	Class5
1	Area	25.59	38.4	34.72	18.98	34.54
2	Eccentricity	0.07	0.09	0.03	0.05	0.1
3	Perimeter	40.36	64.32	40.44	24.73	53.64
4	Major Length	11.79	15.74	15.26	9.14	23.25
5	Minor Length	3.86	6.01	7.24	4.96	4.91
6	Convex Area	526.5	1007.67	1767.59	572.86	1020.18
7	Solidity	0.03	0.05	0.01	0.03	0.05
8	Distance Between Foci	0.98	1.6	0.49	0.49	2.29
9	Orientation	-33.9	-3.12	40.65	4.53	-16.24
10	Circularity	0.03	0.01	0.03	0.04	0.02
11	Equivalent diameter	1.31	1.77	1.12	0.98	1.63
12	Elongation	0.49	0.46	0.48	0.57	0.23
13	Aspect Ratio	3.28	2.62	2.11	1.85	4.73
14	Rectangularity	1.67	2.51	3.19	2.36	3.26
15	Narrowness	0.14	0.13	0.08	0.11	0.07
16	Roundness	0.45	0.33	0.2	0.33	0.1
17	C_x	36.23	81.22	24.76	41.55	107.57
18	C_y	66.36	112.62	44.11	38	60.77

Fig. 6 Shape features of plant groups

Evaluation of the method shows that this neural network model for classification given the accuracy of training the model as 99 and 96% for testing the plant leaf image samples. In total, 70% of the samples are trained and 30% samples of all five plant classes are tested with this model; confusion matrix given below shows the correctly classified samples of the plants in training and testing of the neural network model. Confusion matrix of training and testing is shown in Fig. 7. This shows one leaf image sample of class 4 and one of class 5 are misclassified in

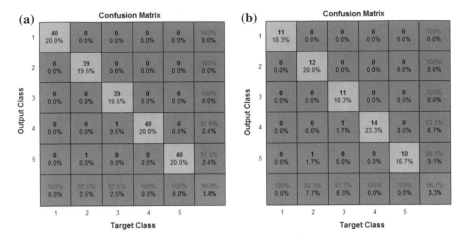

Fig. 7 **a** Confusion matrix for training. **b** Confusion matrix for testing

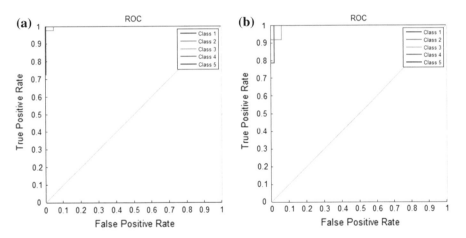

Fig. 8 **a** ROC of training set. **b** ROC for testing set

training. In testing, Class4 and Class5 samples are not classified correctly. For Class1, Class2, and Class3 leaf image samples are correctly classified with 100% accuracy.

Figure 8 shows the receiver operating curve resulted after training and testing the plants randomly selected as training set and testing set. This analyzes the result of the classification.

Figure 9a further shows the result of the classification analyzed using performance of training with blue color and testing with red color. The slope of the plot is similar for both training and testing the classification of plant groups. Figure 9b shows the error histogram.

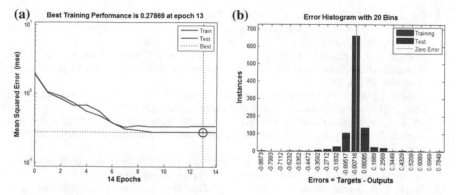

Fig. 9 **a** Performance chart. **b** Error histogram

This model takes input, as leaf shape features of samples of five classes of the plants. This classification result shows that feed forward network model has given 99% accuracy for training and 96% accuracy for testing of the classification model.

5 Conclusion

This method of plant classification is more efficient as there is no any manual intervention to extract the patterns and features of leaf shape. As it is based on new imaging technology with neural network technique, it is found more time-saving. Here, plants are classified using the neural network classifier; this method has designed an artificial neural network model in the form of feed forward with back propagation. This model results in an accuracy of 99% for training plants samples and 96% accuracy for testing remaining samples of the plants.

References

1. Amlekar, M., Gaikwad, A., Yannawar, P., & Manza, R. (2015). Leaf shape extraction for plant classification. In IEEE International Conference on Pervasive Computing (ICPC).
2. Beghin, T., Cope, J. S., Remagnino, P., & Barman, S. (2010). Shape and texture based plant leaf classification. *ACIVS, 2,* 345–353.
3. Deokar S. R., Zope, P. H., & Suralkar, S. R. (2013, January) Leaf recognition using feature point extraction and artificial neural network. *International Journal of Engineering Research & Technology (IJERT),* 2(1).
4. Du, J. X., Wang X. F., & Zhang, G. J. (2007). Leaf shape based plant species recognition. *Applied Mathematics and Computation, 185.*
5. Du, J., Huang, D., Wang, X., & Gu, X. (2005). Shape recognition based on radial basis probabilistic neural network and application to plant species identification. In *Proceedings of 2005 International Symposium of Neural Networks,* ser. LNCS 3497. Berlin: Springer.

6. Gu, X., Du, J. X., & Wang, X. F. (2005). Leaf recognition based on the combination of wavelet transform and gaussian interpolation. In *Proceedings of International Conference on Intelligent Computing 2005*, ser. LNCS3644. Berlin: Springer.
7. Hossain, J., & Amin, M. (2010). Leaf shape identification based plant biometrics. In *13th International Conference on Computer and Information Technology (ICCIT)* (pp. 458–463), December 23–25, 2010.
8. Intelligent Computing Laboratory, Chinese Academy of Sciences Homepage http://www. intelengine.cn/dataset/index.html, Plant leaf image dataset, 2010–2012.
9. Jiazhi, P, & Yong, H. (2008). Recognition of plant by leaves digital image and neural network. In *International Conference on Computer Science and Software Engineering* (Vol. 4, pp. 906–910), December 2008.
10. Lee S. H., Chang, C. S., Mayo, S. J., & Remagnino, P. (2017). How deep learning extracts and learns leaf features for plant classification. *Pattern Recognition, 71*, 1–13.
11. Suman, S. G, & Deshpande, B. K. (2017, May). Plant leaf classification using artificial neural network classifier. *International Journal of Innovative Research in Computer and Communication Engineering (IJIRCCE), 5*(5).
12. Wang, X. F., Du, J. X., & Zhang, G. J. (2005). Recognition of leaf images based on shape features using a hypersphere classifier. In *Proceedings of International Conference on Intelligent Computing 2005*, ser. LNCS 3644. Berlin: Springer.
13. Wu, S. G., Bao, F. S., Xu, E. Y., Wang, Y., Chang, Y., & Xiang, Q. (2007). A leaf recognition algorithm for plant classification using probabilistic neural network. arXiv-0707.4289V1[CS.AI].

Robot Soccer Strategy Reduction by Representatives

Václav Svatoň, Jan Martinovič, Kateřina Slaninová and Václav Snášel

Abstract The robot soccer game introduces a variable and dynamic environment for cooperating agents. Coverage of areas such as multi-agent systems, robot control, optimal path planning, real-time image processing and machine learning makes this domain very attractive. This article presents our approach to strategy description of the robot soccer game and a method of real-time strategy adaptation performed during the game. The real-time strategy adaptation method improves the strategy by adding new rules to it. During this process many new rules can be added to the original strategy, thus making it more robust but more difficult to manage. Therefore, this article presents our method for strategy reduction using representatives, in terms of the number of rules within the strategy, while preserving the quality of the adapted strategy. Strategy, as we defined it, describes a space from the real world in which we know the physical coordinates of objects located in it. Therefore, the methods we developed for strategy planning can be applied to it.

Keywords Robot soccer · Strategy planning · Strategy adaptation
Time-series · Clustering · Representatives

V. Svatoň · J. Martinovič · K. Slaninová (✉)
IT4Innovations, VŠB—Technical University of Ostrava, 17. Listopadu 15/2172,
708 33 Ostrava, Czech Republic
e-mail: katerina.slaninova@vsb.cz

V. Svatoň
e-mail: vaclav.svaton@vsb.cz

V. Snášel
VŠB - Technical University of Ostrava, 17. Listopadu 15/2172, 708 33 Ostrava
Czech Republic

© Springer Nature Singapore Pte Ltd. 2019
V. E. Balas et al. (eds.), *Data Management, Analytics and Innovation*,
Advances in Intelligent Systems and Computing 839,
https://doi.org/10.1007/978-981-13-1274-8_30

385

1   Introduction

In robot soccer games, the robots' and the ball's positions typically provide necessary information and data for reading the game situation on the playground. Utilizing the near real-time information extracted from this dynamically changing game situation, it is necessary for the robot soccer game system to constantly assign a particular action to each team robot, which is then directed by the system to perform it.

In the context of game theory [1, 2], a strategy is defined as a complete list of all available moves for every player to achieve the desired objective in any given situation. This approach is called strategy planning and it is widely used in the real world in a number of different thematic areas [3]. In general, games can serve as a representation of any situation occurring in nature. Game theory is applicable to problems that deal with the resource scheduling or resource allocation. Our approach is using strategies for describing a space and objects in it. In the following step, this strategy can be used for searching the optimal path or to relocate the objects within the given space in a specific way so as the desired goals are achieved.

Like in real soccer, the one and only objective of the robot soccer game is as simple as scoring a higher number of goals to win the game over an opponent. In order to achieve this, the team players should not only cooperate in the best possible way but also adapt to the opponent's actual strategy.

There are numerous different types of robot soccer games using various architectures. Therefore, heavy dependency of the proposed methods for strategy description, rule selection, and strategy adaptation on the applied robot soccer game architecture is to be understood. As mentioned above in the Abstract, the robot soccer game presents a dynamic and fast changing environment. Thus, the system should be capable of evolving and having the flexibility to adapt to the strategy of the opponent in order to correctly perform number of various tasks leading towards achievement of the given objective.

In the work of Huang et al., the use of the fuzzy decision making system Huang and Liang [4] as well as of bio-inspired algorithms, such as reproduction, mutation, recombination, and selection for approximation of the solution to the strategy selection and adaptation problem was demonstrated. Evolutionary method to acquire team strategy represented by a chromosome of action rules of the RoboCup players was proposed in the work of Nakashima et al. [5]. In a similar use cases Self-Organizing map was used by Tominaga et al. [6], swarm intelligence was applied by Shengbing et al. [7], evolutionary algorithms were used as a tool to adapt the strategy in the work of Larik and Haider [8] and the work of Akiama et al. [9] describes the utilization of decision-making approaches for the strategy optimization problems.

The common problem with the above mentioned methods is the different level of granularity in the definition of a strategy. For some, the term strategy is understood as a high-level view at a game situation on the game field describing all objects in it and for others as a set of moves used to control each separate robot on the game field. The paper proposes a real-time strategy adaptation method during the game with the reduction of the rules by representatives, while preserving the quality of the adapted strategy.

This paper is organized as follows: Strategy definition in our robot soccer architecture and the method used to select the rule from the strategy is introduced in Sect. 2. The Sect. 3 describes the real-time strategy adaptation method used to improve our strategy during the game. The proposed approach for strategy reduction using representatives will be presented in Sect. 4 together with the performed experiments and the evaluation of their results. At the end, the conclusion of this article and the future work in this domain will be outlined.

2 Robot Soccer Strategy

We have created our own robot soccer architecture to separate the robot soccer game into two parts; physical abstract part and logical strategy part Martinovič et al. [10]. Physical part represents the detailed coordinate system of the game field used for the robots' control while the logical part also called strategy grid is used for the description of robot soccer strategies. See Fig. 1 for the illustration of this game separation.

2.1 Strategy Description

Thanks to the grid coordinate system which is used within the strategy, we can heavily reduce the number of strategy rules that is needed to describe some specific game situations. This allows us to map very accurate physical coordinate system used for the robot control to a strategy grid coordinate system with much lower resolution and vice versa.

Strategy consists of rules that describe some specific game situations on the game field. Each rule contains the information about our robots, opponent's robots and the ball in the form of a grid coordinates. Each rule also contains destination grid coordinates of where our robots should move in the next game iteration.

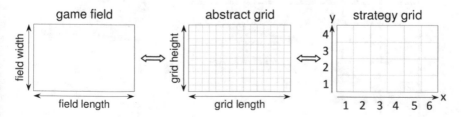

Fig. 1 Game field representation

2.2 Rule Selection

Depending on the current situation on the game field, the most similar rule for the next game step is selected from the strategy. Mechanism responsible for the rule selection method compares the current positions of all robots and the ball to the situations described by the rules in the strategy. The most similar rule in this comparison is considered as a winner and the destination coordinates are extracted from it to be used for the subsequent robot control.

There is no robot identification method (using robot's id, assigned role, etc.) implemented within our robot soccer architecture because we consider the robots to be mutually interchangeable objects. Therefore, we have developed the method for a fast and effective selection of rules from the strategy Svaton et al. [11] based on the graph description and Z-order space filling curve (see Fig. 2). Z-order represents a function which maps the two-dimensional space into the one dimension. This function also preserves the locality of objects in this space which is especially useful for our rule selection mechanism. It enables us to transform the two-dimensional game field into one-dimensional sequence of robot's coordinates. To optimize the rule selection process, the graph containing the similarity of rules from the strategy is precomputed before the actual game starts. Each rule from the strategy is represented as a node of the graph and the similarity between two nodes is represented as an edge. This similarity is computed from the two sorted sequences of robot's coordinates (applying Z-Order on the current situation on the game field and on the strategy rule) using the Euclidean distance.

Using this rule selection mechanism, we had also devised the way for how to include the description of a number of specific situations into the strategy. Thus, the concept of Substrategy was introduced. Substrategies allow the author to design a strategy with the respect to the intended game situations. Using the substrategies, the author is able to create a strategy with substrategies that represent some specific game situations such as offence on the right wing or defense in the middle. This approach results in a faster and more continuous execution of actions with respect to the predefined strategy.

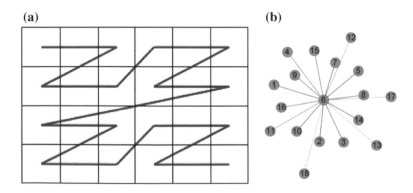

Fig. 2 **a** Z-order mapping and **b** rule graph

3 Strategy Adaptation

The basic idea of the strategy adaptation process is an ability to detect the weak parts in a definition of a strategy and adapt it accordingly. This could be done post game or real-time during the game with an opponent. The adapted strategy should be able to achieve better results than the original strategy before the adaptation process.

We have devised two versions of a strategy adaptation process. The first illustrated method is adapting the strategy based on the logs from the previously played games. The second one is designed for a real-time adaptation process which is executed during the game with the opponent. For the purpose of this article, only the real-time adaptation method was used during the experiments.

The strategy adaptation process implemented in our system consists of the following steps:

1. Information extraction from the logs of a previous games or real-time during the game.
2. Detection of a game situations which will be used for the adaptation process.
3. Extraction and analysis of rules that preceded these situations.
4. Creation of aggregated rules or anti-rules for these situations.
5. Extension of the original strategy by the new rules.

The adaptation process described above can be used for the defensive and also the offensive game situations. During the defense the goal scored by an opponent is simply identifiable during the game and thus it is considered as a relevant game situation.

During the offensive part of the game the unsuccessful attempt to score a goal is considered as a relevant game situation. This situation can be detected based on the location of the ball in relation to the opponent's gate. By changing the move-to coordinates of a selected robots to relocate them closer to this gate we are trying to put more pressure on the opponent's goalkeeper, thus increasing a change of scoring a goal.

3.1 Real-Time Strategy Adaptation

The log file generated from a standard game of 5 versus 5 robots was used for the strategy adaptation process. The game state is logged every 20 ms during the 2 min long game. The resulting log contains 6,000 records describing the progress of the entire game. The log file consists of coordinates of robots from both teams, coordinates of the ball, game score and game time. Therefore, the strategy adaptation process is able to find the relevant game situations in this log, extract the selected strategy rules and appropriate robots' coordinates.

Due to the 20 ms game step, the robots do not have enough time to move far enough on the game field. Therefore it is necessary to analyze long enough time interval for the adaptation process to be able to affect the detected game situation. 150 log entries equals to 3 s of game play, which is a sufficient time to adapt to a detected situation.

Detailed real-time strategy adaptation process:

1. Extraction of 150 log entries (3 s) for every detected relevant situation.
2. Average rule computation for each set of 50 log entries resulting in 3 new adapted rules.
3. Update of the adapted rules' move-to coordinates for selected 2 robots.
4. Insertion of unique adapted rules into the original strategy.

To create the adapted strategy, the game between the original left team strategy and reference strategy of the right team was played. The original strategy to be adapted consists of 11 rules divided into the four basic substrategies: offensive middle, offensive left, defensive middle and defensive right. The right team's reference strategy contains 31 rules that cover the main game situations which can occur during the game. Simply put, the weak left team strategy is used against the strong right team strategy. The overview of the results of the 10 games played between these two strategies can be seen in Table 1. The results show that the original left team strategy was able to win the match over the opponent's strategy in 3 cases, achieved 2 draws and lost 5 times. The sum score was 5:9 for the opponent's team, thus confirming the usage of a weak strategy for the left team.

Ten iterations of the game were used for the left team strategy adaptation process. Each scored goal for the opponent's team was detected as the relevant game situation and was used to create new defensive anti-rules that were included into the original strategy. The unsuccessful goal attempts were used in the same manner for the generation of the new offensive rules. The results of these games together with the number of adapted rules that were added in each iteration is illustrated in Table 2.

Table 1 Original left team strategy versus reference strategy

Game No.	Left score	Right score	Result
1	0	1	Loss
2	1	0	Win
3	0	2	Loss
4	0	2	Loss
5	1	0	Win
6	1	1	Draw
7	1	0	Win
8	0	1	Loss
9	1	1	Draw
10	0	1	Loss
Sum	5	9	

Table 2 Real-time strategy adaptation, 10 iterations

Game No.	Left score	Right score	Defense adapt rules	Offense adapt rules
1	0	2	4	12
2	2	0	4	19
3	1	1	7	21
4	0	1	10	24
5	1	0	10	28
6	0	1	10	28
7	2	2	13	35
8	0	0	13	43
9	1	0	13	51
10	1	0	16	54
Sum	8	7	16	54

Table 3 Fully adapted strategy versus reference strategy

Game No.	Left score	Right score	Result
1	1	0	Win
2	1	0	Win
3	0	0	Draw
4	0	0	Draw
5	3	0	Win
6	0	0	Draw
7	0	3	Loss
8	0	0	Draw
9	0	0	Draw
10	1	0	Win
Sum	6	3	

During the adaptation process, 70 new rules (16 defense rules, 54 offense rules) were added to the original strategy. To test the adapted strategy, ten games were played between fully adapted left team strategy, now consisting of 81 rules (10th iteration), and the reference strategy of the right team. The results show (see Table 3) that the adapted strategy is now much more able to compete with the reference strategy. The adapted strategy was able to win 4 matches, achieved 5 draws and lost only in 1 case.

4 Strategy Reduction

During the strategy adaptation process, the newly created rules are added to the original strategy rule set. This adaptation process detects the relevant game situations and predicts the rules that should be able to prevent the occurrence of these situations in the future. Number of adapted rules are created during the real-time adaptation process, but not all of them might be used during the actual game. The big number of rules makes the strategy more robust but also makes it harder to manage in terms of defined substrategies (game situations).

The following approach is used to detect the representatives of the game situations that occurred during the game and to use these representatives to reduce the number of rules within the adapted strategy while preserving its defensive and offensive abilities. Game situations are represented as clusters of similar sequences acquired from the game log after the game. Each sequence denotes the strategy rules that were selected during the game until the ownership of the ball changed. Using this approach, the rules that were not used during the actual game will not occur in the extracted sequences and therefore will not be present in the final reduced strategy.

4.1 Sequence Extraction

A sequence extraction method is inspired by a method widely used in the domain of social networks and adapted to the domain of robot soccer games. The game profiles described as sequences can be extracted from the log of a played game which contains the list of rules that were selected from the strategy during the game.

The definition of a game profile is as follows:

Let $U = \{u_1, u_2, \ldots, u_n\}$, be a set of games, where n is a number of games u_i. Then, sequences of strategy rules $\sigma_{ij} = \langle e_{ij1}, e_{ij2}, \ldots, e_{ijm_j} \rangle$ are sequences of strategy rules executed during a game u_i in the simulator, where $j = 1, 2, \ldots, p_i$ is number of that sequences, and m_j is a length of j-th sequence. Thus, a set $S_i = \{\sigma_{i1}, \sigma_{i2}, \ldots, \sigma_{ip_i}\}$ is a set of all sequences executed during a game u_i in the system, and p_i is a number of that sequences.

Sequences σ_{ij} extracted with relation to certain game u_i are mapped to a set of sequences $\sigma_l \in S$ without this relation to games: $\sigma_{ij} = \langle e_{ij1}, e_{ij2}, \ldots, e_{ijm_j} \rangle \rightarrow \sigma_l = \langle e_1, e_2, \ldots, e_{m_l} \rangle$, where $e_{ij1} = e_1$, $e_{ij2} = e_2$, $e_{ijm_j} = e_{m_l}$.

Define matrix $B \in N^{|U| \times |S|}$ where

$$B_{ij} = \begin{cases} \text{frequency of sequence } \sigma_l \in S \text{ for game } u_i \text{ if } \sigma_l \in S_i \\ 0 \text{ else} \end{cases}$$

A **base game profile** of the games $u_i \in U$ is a vector $b_i \in N^{|S|}$ represented by row i from matrix B.

The possession of the ball was used as a sequence classifier for the sequence extraction. Sequence labeled with number 0 denotes the free ball, 1 for the ownership of the left team and 2 for the ownership of the right team. The rules selected by the left team before the change of the ownership are part of the same sequence. Methods for sequence comparison are applied to the extracted sequences.

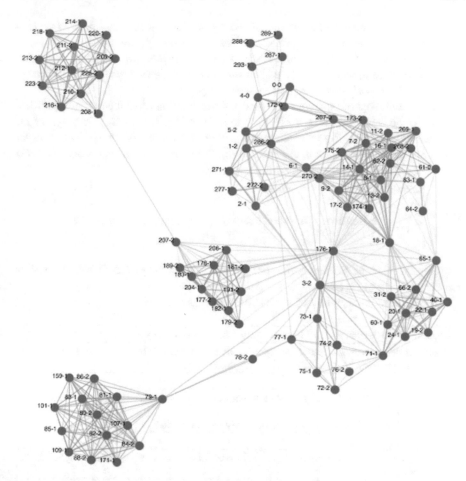

Fig. 3 Adapted strategy, 10th iteration—clusters of similar sequences

The resulting clusters of similar sequences can be easily visualized in graph. See Fig. 3 for the clusters of similar sequences extracted from the log of a 10th iteration game during the adaptation process.

4.2 Clustering and Representatives

The rules created during the real-time strategy adaptation (during the game) are sometimes numerous, and the management of the strategies/substrategies is then very difficult. Due to this reason, we propose the approach to reduce the number of the adapted rules while preserving the quality of the adapted strategy. The set of adapted rules (sequences σ_l) is clustered [12–14] and the representative for each cluster is find by the following way:

*Let C be a set of **clusters** consisted of similar event sequences. Let $R = \{\rho_1, \rho_2, \ldots, \rho_r\}$ be a set of **representatives** of clusters ρ_k, where $k = 1, 2, \ldots, r$. Then, a representative $\rho_k \in R$ of a cluster C_k is a sequence, which describes similar event sequences $\sigma_l = \langle e_1, e_2, \ldots, e_{m_l} \rangle$ in the cluster C_k (it has the maximal average similarity to the all sequences in the cluster C_k).*

Representatives of clusters were used for the creation of reduced game profiles:

Let U be a set of users u_i Thus, a set $\Pi_i = \{\pi_1, \pi_2, \ldots, \pi_{s_i}\}$ is a set of all representatives $\pi_k \in R$ for game u_i to which is mapped a set of sequences S_i executed during the game u_i A sequence σ_{ij} is mapped into a representative of a cluster π_k to which the corresponding sequence σ_{ij} .

Define matrix $P \in N^{|U| \times |R|}$ where

$$P_{ij} = \begin{cases} \text{frequency of representative } \rho_j \text{ for user } u_i \in R & \text{if } \rho_j \in \Pi_i \\ 0 & \text{else} \end{cases}$$

*A **reduced game profile** of the user $u_i \in U$ is a vector $p_i \in N^{|R|}$ represented by the row i from matrix P.*

Finding the representatives is described in Algorithm 1.

Algorithm 1 Determination of Representative in Cluster of Sequences

Goal: The determination of sequence representatives ρ which will be used for a description of the sequence clusters C, for selected similarity threshold θ.

Input:
- A cluster C, where a similarity between the objects inside the cluster is given by values $Sim(c_l, c_k)$.
- A selected threshold θ for a sequence similarity w_{seq} in clusters.

Output:
- A set of representatives $R = \{\rho_1, \rho_2, \ldots, \rho_r\}$ and a set of clusters C_ρ which correspond to the appropriate representatives.

1. For each object, belonging into a cluster C, determine its average similarity to other objects inside that cluster. Similarity between the objects c_l and c_k is given by the condition $Sim(c_l, c_k) > \theta; \forall c_l, c_k \in C$, where θ is the selected threshold for the object similarity.
2. Select the object with the maximum average similarity counted in Step 1. The selected object is labelled ρ.
3. Create a new cluster C_ρ, which contains the objects from cluster C, which meets the condition $Sim(c_l, c_k) > \theta$ and contains the representative ρ.
4. Add the representative ρ into the set R and record the appropriate cluster C_ρ, which is created in Step 5.
5. $C = C - C_\rho$.
6. If $C \neq \emptyset$, then repeat to Step 1 with a new cluster C.

4.3 Representatives Extraction

The same sequences that were acquired from the game log of the fully adapted strategy against the reference strategy, were used for the extraction of the representatives. The fully adapted strategy consists of 81 rules (11 original, 70 adapted) and 120 sequences were extracted from the game log.

These sequences were compared using the longest common subsequence method (LCSS) with the clustering coefficient ranging from 0 to 1 with the step of 0.1. The resulting number of found clusters is illustrated in Table 4.

Representative for each cluster can be found using the approach mentioned in Sect. 4.2. This representative is in a fact a sequence from this cluster. A sequence consists of the rules that were selected during the game. Therefore, it is possible to extract these rules from each of the cluster's representative and construct an aggregated strategy from these rules.

This strategy is in fact a reduced strategy, as only the rules that were selected during the game are within the extracted sequences and only the rules from the found representatives will appear in the final reduced strategy.

The reduced strategy was created for each of the clustering method illustrated in Table 4 and the created strategies are listed in Table 5. Clustering variants for the values from 0.5 to 1 obtained the same strategy. Therefore, there is only one common strategy listed in the table for these 6 variants. This was possible due to the fact, that even if there is different number of clusters in each variant, the extracted representatives contain the same strategy rules in its sequence.

The overall goal is to find the sufficiently strict clustering to reduce the number of rules within the adapted strategy while preserving its defensive and offensive abilities. The visualization of the representatives found in the clustering variant 0.1 (26 representatives) and 0.9 (73 representatives) show, that 'soft' clustering (variant 0.9) results in a high number of found representatives, that could be once again compared and visualized as the clusters of similar representatives, as shown in

Table 4 Clustering methods and the resulting clusters

Clustering method	Clusters
LCSS 0	20
LCSS 0.1	26
LCSS 0.2	28
LCSS 0.3	33
LCSS 0.4	39
LCSS 0.5	46
LCSS 0.6	53
LCSS 0.7	57
LCSS 0.8	67
LCSS 0.9	73
LCSS 1	79

Table 5 Number of rules within reduced strategies

Strategy name	Rules
0.strg	37
0.1.strg	33
0.2.strg	39
0.3.strg	41
0.4.strg	46
0.5_0.6_0.7_0.8_0.9_1.strg	48

Fig. 3. In this case, almost each of the compared sequences was labeled as the representative in its own cluster (Fig. 4).

It is important to mention that even the clustering using the parameter of value 1 returns 79 representatives from 121 processed sequences, resulting in the reduced strategy with 48 rules, instead of 81 rules as in the original fully adapted strategy. Once again, only the rules that were actually used during the game were propagated into the reduced strategy.

4.4 Reduced Strategies Results

The test the quality of the reduced strategies, the 10 games were performed for each strategy against the same reference strategy of the right team. The results from these games are illustrated in Tables 6, 7 and 8.

It is apparent that the best result was achieved by the reduced strategy 0.4. This strategy consists of 46 rules instead of 81 but the strategy was still able to win the game in 4 cases, achieved draw in 5 results and lost in only one game. It is safe to assume, that this strategy still keeps the sufficient coverage of the game field with its rules. Therefore, it is still able to quickly react to the opponent's game.

5 Conclusion

This article discussed the strategies within the robot soccer game. The description of approach for the strategy adaptation and the subsequent strategy reduction using representatives was presented. The real-time strategy adaptation method was used to improve the defensive and offensive capabilities of the robot soccer strategy.

The strategy adaptation process might result in a number of rules that are added to the original strategy. To make the adapted strategy more efficient, clear and to remove the unused or very similar rules, the approach for strategy reduction using

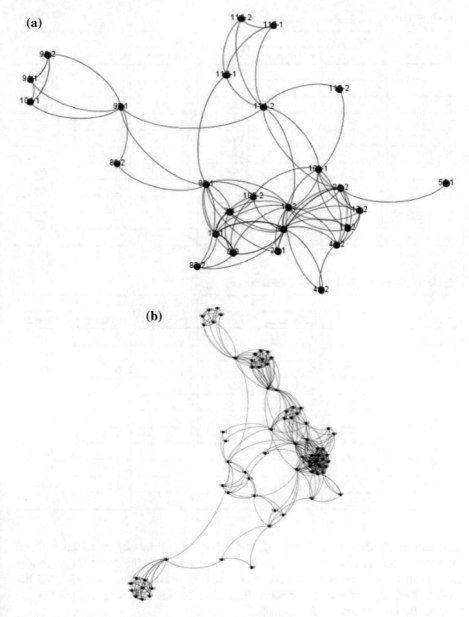

Fig. 4 Graph of representatives for variants **a** 0.1 and **b** 0.9

Table 6 0 and 0.1 strategy versus reference strategy

	0.strg			0.1.strg		
Game No.	Left score	Right score	Result	Left score	Right score	Result
1	0	0	Draw	0	1	Loss
2	1	0	Win	0	1	Loss
3	0	1	Loss	0	0	Draw
4	1	0	Win	0	1	Loss
5	0	0	Draw	0	1	Loss
6	0	1	Loss	0	0	Draw
7	0	0	Draw	1	0	Win
8	1	2	Loss	0	0	Draw
9	1	0	Win	0	0	Draw
10	0	1	Loss	1	1	Draw
Sum	4	5		2	5	

Table 7 0.2 and 0.3 strategy versus reference strategy

	0.2.strg			0.3.strg		
Game No.	Left score	Right score	Result	Left score	Right score	Result
1	2	0	Win	1	0	Win
2	1	1	Draw	0	0	Draw
3	1	1	Draw	1	0	Win
4	2	0	Win	1	2	Loss
5	0	2	Loss	0	2	Loss
6	1	0	Win	0	2	Loss
7	0	1	Loss	2	1	Win
8	0	0	Draw	1	1	Draw
9	0	1	Loss	1	2	Loss
10	0	0	Draw	2	1	Win
Sum	7	6		9	11	

cluster representatives was presented. The reduction process was tested with several variants of clustering based on the predefined clustering coefficient. With the proper setup, only the important rules are propagated into the reduced strategy and the reduced strategy is able to preserve its defensive and offensive capabilities, as shown by the performed experiments.

It is important to find a proper way of how to set up the clustering coefficient for the specific game. With the lower value of the clustering coefficient the important rules that were added during adaptation process may be lost during the reduction as only small number of representatives will be found. On the other hand, the too lenient reduction may result in a high number or representatives, therefore still too many rules within the reduced strategy.

Table 8 0.4 and 0.5_0.6_0.7_0.8_0.9_1 strategy versus reference strategy

Game No.	0.4.strg			0.5_0.6_0.7_0.8_0.9_1.strg		
	Left score	Right score	Result	Left score	Right score	Result
1	2	1	Win	0	2	Loss
2	0	1	Loss	1	0	Win
3	2	0	Win	0	0	Draw
4	0	0	Draw	3	1	Win
5	1	1	Draw	0	2	Loss
6	1	0	Win	0	1	Loss
7	0	0	Draw	0	3	Loss
8	4	0	Win	2	0	Win
9	0	0	Draw	1	1	Draw
10	1	1	Draw	0	0	Draw
Sum	11	4		7	10	

Acknowledgements This work was supported by The Ministry of Education, Youth and Sports from the National Programme of Sustainability (NPU II) project "IT4Innovations excellence in science—LQ1602".

References

1. Osborne, M. J. (2004). *An introduction to game theory*. New York Oxford: Oxford University Press.
2. Kim, J.-H., Kim, D.-H., Kim, Y.-J., & Seow, K. T. (2010). Soccer robotics, Springer tracts in advanced robotics.
3. Ontanón, S., Mishra, K., Sugandh, N., & Ram, A. (2007). Case-based planning and execution for real-time strategy games. In *Lecture Notes in Computer Science* (pp. 164–178), Vol. 4626.
4. Huang, H. P., & Liang, C. C. (2002). Strategy-based decision making of a soccer robot system using a real-time self-organizing fuzzy decision tree. *Fuzzy Sets and Systems, 127*, 1.
5. Nakashima, T., Takatani, M., Udo, M., Ishibuchi, H., & Nii, M. (2006). Performance evaluation of an evolutionary method for robocup soccer strategies. In *RoboCup 2005: Robot Soccer World Cup IX*. Berlin: Springer.
6. Tominaga, M., Takemura, Y., & Ishii, K. (2017). Strategy analysis of robocup soccer teams using self-organizing map.
7. Chen, S., Lv, G., & Wang, X. (2016). Offensive strategy in the 2D soccer simulation league using multi-group ant colony optimization. *International Journal of Advanced Robotic Systems, 13*.
8. Larik, A. S. & Haider, S. (2016). On using evolutionary computation approach for strategy optimization in robot soccer. In *2nd International Conference on Robotics and Artificial Intelligence (ICRAI)*
9. Akiyama, H., Tsuji, M., & Aramaki, S. (2016). Learning evaluation function for decision making of soccer agents using learning to rank. In *Soft Computing and Intelligent Systems (SCIS) and 17th International Symposium on Advanced Intelligent Systems, 2016 Joint 8th International Conference on*. IEEE.

10. Martinovič, J., Snášel, V., Ochodková, Zoltá, L., Wu, J., & Abraham, A. (2010). Robot soccer—Strategy description and game analysis. In *Modelling and Simulation, 24th European Conference ECMS*.
11. Svatoň, V., Martinovič, J., Slaninová, K., & Snášel, V. (2014). Improving rule selection from robot soccer strategy with substrategies. In *Computer Information Systems and Industrial Management—13th IFIP TC8 International Conference (CISIM)*.
12. Dunham, M. H. (2003). In *Data mining: Introductory and advanced topics*. New Jersey: Prentice Hall.
13. Drážkilová, P., Martinovič, J., & Slaninová, K. (2013). Spectral clustering: Left-right-oscillate algorithm for detecting communities. In *New Trends in Databases and Information Systems, Volume 185 of Advances in Intelligent Systems and Computing* (pp. 285–294). Berlin, Heidelberg: Springer.
14. Klosgen, W., & Zytkow, J. M. (2002). *Handbook of data mining and knowledge discovery*. New York, NY, USA: Oxford University Press Inc.

Part IV
Advances in Network Technologies

A Comprehensive Survey on Ransomware Attack: A Growing Havoc Cyberthreat

Aditya Tandon and Anand Nayyar

Abstract Never in the history of humanity, people all over the world are subject to exaction on a huge scale as they are today. In the recent years, the usage of PCs and the Internet has exploded and, along with this huge increase, cybercrooks have come to feed this souk, aiming acquitted consumers with a wide range of per-ware. Most of these threats are meant unswervingly or meanderingly in receiving currency from victims. Today, the ransomware appears to be one of the most unpleasant per-ware categories of the time. Several works have been published in the field of information and Internet security, various pernicious attacks, and cryptography. The objective of this research paper is to present everything with regard to latest crypto-virus trend known as ransomware. The paper explains the history, the modus operandi as well as the architecture of ransomware attack.

Keywords Ransomware · Payoff · Cryptography · Trojan · Cybercrime
Malicious code · Cybersecurity · Bitcoin

1 Introduction

Since the 50s, the world has seen the merits and the wonders of the Internet and World Wide Web (WWW). Every user today is now being connected to it at an immensely quick pace. The amount of data is now exceeding zettabytes (2^{70} bytes) since last year, and the concerns for its safety are now taking the shape of a major problem. Pernicious content and corrupt programs have been attacking and infecting various devices around the world, and the efforts for their prevention and eradication have also gained pace simultaneously. The software code written

A. Tandon (✉)
Ch. Brahm Prakash Government Engineering College, New Delhi, India
e-mail: adityat1988@outlook.com

A. Nayyar
Graduate School, Duy Tan University, Da Nang, Vietnam

© Springer Nature Singapore Pte Ltd. 2019
V. E. Balas et al. (eds.), *Data Management, Analytics and Innovation*,
Advances in Intelligent Systems and Computing 839,
https://doi.org/10.1007/978-981-13-1274-8_31

especially toward causing damage or stealing information becomes what is known as per-ware (pernicious software) or per-ware in short.

Organization of Paper

Section 2 outlines the concept of ransomware as well as elaborates the types of ransomware and provides an insight of how the ransomware attack is commenced. Section 3 goes further into the concept of the basic framework of the ransomware outbreak by explaining the detailed steps involved and stating real life and recent examples. Section 4 highlights some of the tried-and-tested preventive measures and some detection (or symptoms) before any ransomware attack. Section 5 gives detailed explanation of WannaCry—the latest ransomware scare. Section 6 concludes the paper with future scope.

1.1 History

Since the advent of the digital era, worldwide aggressors have tested the security of various companies and institutions through e-mail, phishing sites, fake antivirus, etc. The encipherment method has been implemented to ensure the exchange of information around the world. However, the concept of poly-alphabet encipherment was proposed in AD 1467 by Leon Battista Alberti, often known as the "Father of Cryptology". The need for secure and selective communication has given rise to the art of encoding messages so that only recipients may have access to the information, while unauthorized denial of information extraction, although balderdash messages fell into his hand. Art and science to hide messages to enter secret information is called cryptography. An encipherment system has been designed to implement numerous cryptographic techniques and accompanying infrastructures to ensure the security of information. A typical model of a cryptosystem (also known as cipher system) is depicted in Fig. 1.

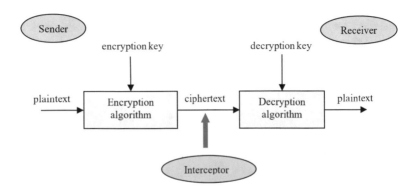

Fig. 1 A typical cipher system

This concept of applying the cipher system was a benediction to data dispensation and communications as atomic fission is to energy production. In both ways, these systems are vulnerable to attacks and, therefore, can be misused.

1.2 Concept of Ransomware

The concept of public-key cryptography (also known as symmetric-key cryptography) can also be used offensively as stated in [1]. Clearly, the authors predicted the methods used for our systems' security can be misused, also known as cryptovirology (amalgamation of public-key cryptography and Trojans or viruses). The first asymmetric ransomware prototypes were developed in the 1989 with the name Acquired Immune Deficiency Syndrome (AIDS) Trojan which was distributed through a less suspicious 5¼ floppy disk (since World Wide Web (WWW) was not famous enough) and given to the dignitaries who were attending a conference on an international level which happened to be about the AIDS disease [2]. The software enciphered file names (not the files themselves), displayed a demand for payment to a location in Panama. Ransomware is the kind of strategy that uses extortionary pernicious software to keep the computer system hostage user until a redemption is paid. Ransomware strikers often request bitcoin currency redemption due to the perceived anonymity of encipherment transactions. The per-ware blocks a user for a limited time after which the refunds or user data are destroyed [3].

Since 1990, how drastic the scare of ransomware spread throughout the globe is illustrated in Fig. 2 [4]. A typical AIDS (PC Cyborg) threat message is shown in Fig. 3.

2 Outlining Ransomware

Ransomware is a per-ware that employs asymmetric encipherment to hold a prey's information at payoff. Asymmetric (public–private) encipherment, also known as cryptography, uses a couple of keys to encode as well as decode a file. This public–private couple of keys are uniquely created by the invader for the prey, with the private key to decipher the files stored on the invader's server. The invader ensures that the private key becomes obtainable to the prey only after the payoff is paid, although that does not happen often—as observed in recent payoff software operations. Without the access to the private key, it is nearly impossible to decode the files that are being held for payoff.

It is commonly divided into two main forms—Locker ransomware and crypto-ransomware. Locker ransomware or PC locker refutes access to the PC or device. Crypto-ransomware or data locker foils access to files or data. It does not necessarily need to use encipherment to prevent users from retrieving their data, but most of the people do so. Both types of these payoff software are targeted directly at

Year		Ransomware
1989-90	-	AIDS Trojan (PC Cyborg) becomes the first known ransomware.
2005-06	-	Gpcode, TROJ.RANSOM.A, Archiveus, Krotten, Cryzip, and MayArchive. First to utilize encryption algorithms
2008	-	Gpcode.AK. Utilized 1024-bit RSA keys
2010	-	WinLock. Originated in Russia, flashed porn content on the computer screen until the user would make a $10 phone call to a premium-rate telephone number.
2011	-	Unnamed ransomware Trojan. Locked the user's computer and directed the visitor to a fake list of phone numbers which they could call to reactivate their operating system.
2012	-	Reveton ransomware would let the user know their machine has been utilized to download either copyright material or child pornography and would demand the user to pay a fine. A form of Scareware.
2013	-	CryptoLocker, the most notorious ransomware. Had increased encryption, and was extremely difficult to prevent.
	-	Locker is discovered and would demand a ransom payment of $150 in which the user had 72 hours to pay.
	-	CryptoLocker 2.0 was released and utilized Tor to increase anonymity for payment.
	-	Cryptorbit, another ransomware that utilized Tor and would encode the first 1024 bits of every file it encoded. Cryptorbit would also install a Bitcoin miner on the victim's machine to create more profit.
2014	-	CTB-Locker (Curve, Tor, Bitcoin), would leverage elliptical curve cryptography. Tor for anonymity, and Bitcoin for payment.
	-	CryptoWall, another infamous CryptoLocker clone that was responsible for infecting billions of files worldwide utilizing infected emails.
	-	Cryptoblocker didn't encrypt Windows files that were over 100MB in size. Utilized AES for encryption.
	-	SynoLocker targeted Synology NAS devices, and would encrypt all files.
2015	-	CryptoWall 2.0 used for Tor for anonymity and was delivered through multiple attack vectors.
	-	TeslaCrypt and VaultCrypt originally targeted computers that had certain games installed. Newer variants targeted non-gaming machines.
	-	CryptoWall 3.0 shared some of the same features as its predecessor but added additional features such as Anti-VM check and was delivered via exploit kits.
	-	CryptoWall 4.0 would not only encrypt the data in the files but the file names as well. It also would disable any system restore functionality and shadow volume copies.
	-	Chimera was more of a scareware ransomware that not only encrypt files but also threatened the user that it would publish them online when ransoms are not paid. Also known as doxing.
2016	-	Locky is ransomware that would not only encrypt the user's files, but would first scramble the files and then rename your file extensions to .locky.
	-	SamSam targets servers instead of end-users. The ransomware exploits vulnerabilities in JBoss application servers and compromises the server to gain shell access. SamSam then proceeds to spread to Windows machines and encrypts their files.

Fig. 2 A chronology of notable ransomware development

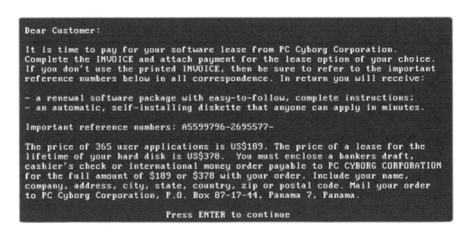

Fig. 3 PC Cyborg threat message on PC screen

our digital routine. They are created to refute us the access to something we desire and to offer to return what belongs to us in paying for a payoff. While having similar aims, the approaches adopted by any type of payoff software are very distinguishable [4].

2.1 Locker Ransomware (PC Locker)

It was created to refute access to computing reserves. This characteristically requires the method of blocking the PC or device user interface (UI) and then prompts the user to pay a fee to restore access to it. Blocked PCs are often released with restricted functionality, such as allowing only the user to interact with the per-ware and recompense the payoff [5].

Locker ransomware is specially designed to prevent access to the PC's UI, leaving largely intact the basic system and files. This means that the per-ware could possibly be removed to restore a PC to something near its default state, thus making locker ransomware less active in extracting payoffs as shown in Fig. 4. Some victims of advanced users (or superusers) can often be accessed using various tools and techniques offered by security vendors. Since this type of ransomware can be easily cleaned, it tends to use social engineering methods to pressure victims to pay or perhaps disguise themselves from a police authority by sustaining fines for alleged online users or criminal activity rumors. Those devices which have restricted choices for the users to interact with, for example, Internet of Things (IoT) devices, are at potentially greater risk than other PC systems or mobile phones.

2.2 Crypto-Per-Ware (Data Locker)

This variety of per-ware aims to find and encipher important data stored on user's PC, making the data useless, unless the user gets the decipher key. As the world

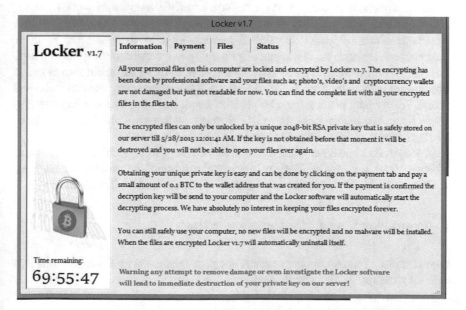

Fig. 4 A locker ransomware dialog box

becomes more and more digital, people are storing the most significant data on their PCs and devices [6].

Many users are unaware of the need to back up backups of hard drive failures or lost or stolen PCs, let alone a possible ransomware crypto attack. This may be because they have the know-how or do not realize the value of the data until it is lost. Dialog data points for these shortcomings and designers are aware of how these data are important to users, for example, memories of their loved ones, university project due to presentation, financial report for work, etc. After installation, a typical crypto-ransomware threatens and enciphers the files in silence. The goal of hacker is to remain unnoticed until it finds and enciphers all the files that might be valuable to the user. By the time the prey is presented with the threat message (the data is enciphered), the damage is already done. With data most of the cabinet infections, the infected PC continues to work normally because per-ware is not directed to critical system files and denied access to PC functionality.

2.3   Overview of a Coordinated Ransomware Attack

Ransomware is an absolute deadline for a category of per-ware that is used to digitally extradite victims in paying a precise payment. Assailants distribute these ransomwares from paid service called ransomware-as-a-service (RaaS) through numerous web servers [7]. RaaS means that the member does not need any special programming knowledge, only the will to spread ransomware (usually via botnet e-mail). The affiliate can register as an affiliate and simply download a custom binary ransomware. These customizations come from Ransom32, the world's first ransomware written in JavaScript. With the advent of the onion router (TOR) and a robust underground economy (sometimes referred to as the Dark Web), it has become significantly easier for skilled hackers to offer their services to other upcoming novice hackers who apparently do not have the skills, or the infrastructure, to deliver ransomware widely to take advantage of existing capabilities to launch a ransomware campaign. Long before the idea of RaaS came along, attackers who had amassed many prey hosts using a botnet would rent it out to anyone who wanted to launch a spam campaign or launch a DDoS attack against a target. Many varieties of the RaaS models with different offering are mention in [8].

However, these attackers pass through defined steps to perform a successful and effective attack on prey(s). Their methods under normal circumstances have been understood to follow certain series of definitive steps which are illustrated in Fig. 5.

3   Framework of a Ransomware Attack

The execution of a ransomware attack is not as sophisticated as it seems. Through a series of carefully grafted simple steps, one can easily deploy a ransomware to any such prey network or an individual. Figure 6 demonstrates how actually in the real world the ransomware attack is being executed.

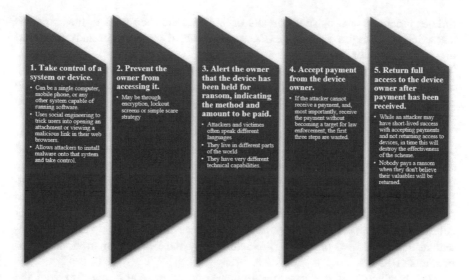

Fig. 5 Modus operandi usually followed by most attackers

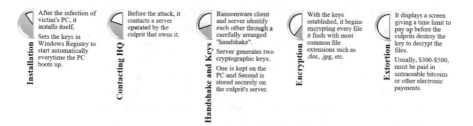

Fig. 6 Framework of a ransomware outbreak

3.1 Installation

The first step involves the installation of the infected Trojan containing the ransomware that is to be achieved first. To do that, careful *e-stalking* is executed in deploying the *infected* original files necessary for the prey's OS to download. This can be achieved either by a drive-by download, i.e., the OS implicitly downloads a portion of the per-ware or a spyware without letting the user or prey to have prior knowledge in the background. The prey is often selected based on his/her network accessibility, the web links clicked, the web searches, and dealt with popups (a novice user would click any popup showing a blinking "Your PC is infected. Run this scan now.").

Thus, the prey becomes a relatively easy catch, and the attacker can now fool the prey and easily penetrate and infect his/her PC or network. If the prey proves to be a little-experienced one, phishing emails or websites are then used to lure him/her. These luring strategies are rather extensive, untargeted spam, or specially crafted to any organization or industry [2].

This concept is like the concept of shooting the American way—fire as many bullets as you can, hoping to hit mostly on target. These emails may include attachments which might be familiar to the prey's current organization or previously worked for. These may redirect to some pernicious websites [9].

In Windows OS, the Trojan sets the keys in the Registry so that it can start automatically on every PC reboot. In case of mobile devices (especially Android devices), the unprotected phishing app stores do the job of the attacker. Some may use stolen rather valid application development certificates for iOS. The need to jailbreak or root the mobile devices has made them highly vulnerable to the ransomware attack. The necessity for side-by-side download pernicious applications substantially upsurges the danger as these devices are no longer under the protection of the fortified protocols, pre-installed by most of the smartphone manufacturers [10].

The installation of the ransomware marks the cybercrook starts to take the control over the prey's device and the parts are again divided into a flurry of processes, batch files, scripts, and other tools to avoid scanning from signature-based antivirus scanners. In another direct attack, installation techniques, obfuscation, packaging, and code operation may be more damaging to maximize rescue. Ransomware uses this initiative to spread slowly through the infected network and install it into various common opening files and systems which are obviously enciphered simultaneously when the next step instructions are sent. Most crypto-ransomware variants would first take some advantage in the form of macro-virus or an infected PDF to get onto the system or use Java or Adobe Flash. Once downloaded, the per-ware will execute its embedded code which will analyze whether the machine is real or virtual [11].

After confirming that the system is worth infecting, the per-ware disguises itself as one of the Windows processes such as *svchost.exe*. To make it more unique, the computer name's MD5 hash or any identifier like MAC address is used to ensure the culprit to know which device has been contaminated. After the phase-one dropper's achievement, the phase-two runs a flurry of scripts to certify that any inherent protections by the Windows OS are disabled (the user might observe that after a certain installation and a reboot, the Windows Security Service gets disabled automatically at system log on, and the homepage or tab pages change to rather different or weird-looking search pages). By doing this, a utility by the Windows OS environment collects and makes copies of the steady system images for backing up the currently running systems, especially the servers, without inadvertently limiting the performance and the steadiness of the services it provides, and the system recovery characteristics of this platform are turned off and any anti-per-ware processes are killed. This utility is known as the volume shadow copy service (VSS).

3.2 Contacting Headquarters (HQ)

The second stage includes contacting the source, i.e., C&C (Command and Control) or headquarters (HQ) server before the commencement of attack. This all happens in the background; the operating system (OS) keeps working fine, and there is no way that prey knows what is going on. This concept can be understood in this way—without receiving orders, it is probable that a portion of ransomware can be found lying dormant on the prey's machine right now waiting for orders. Once the pernicious code is positioned and installed, it would commence reaching out to the headquarters, searching for orders which can be numerous requests of specific types. These requests comprise mostly finding out the file types to target for encipherment, estimating the time to wait for the process to begin, and whether to continue the spread before starting the process. Some of the variants report back an important summed-up information about the prey's device, including which anti-per-ware products are installed, operating system, installed browsers, domain name, and the IP address. The information thus gathered helps the culprits to determine whether they have managed to attack a high-value target or not and ultimately defining the amount of the payoff according to the importance of target.

The communication channels can vary with different variants and categories of per-ware. In some cases, these can be as unpretentious as web-based communications (unenciphered hypertext transfer protocol (HTTP)) to complex systems involving embedded TOR services [12] to connect. More complex systems use this very concept to conceal their whereabouts and motives. Even on android platform, the embedded TOR services can still be used over TCP channels [13, 14].

3.3 Exchanging the Keys and the Concept of Handshake

Almost all the scenarios of per-ware comprised of the malevolent code being positioned on the prey's system is a client, and the control-and-command (C&C) server functioned by the pernicious foe is specifically a server. After the placement of the client, it will make sure that it is collaborating with the main culprit's server with the help of a preset handshake procedure. This handshake procedure is distinguishable from each ransomware fraternity. The CryLocker ransomware applies an exceptional method; it sends everything wrapped as an album's PNG (Portable Network Graphics) file on genuine web pages like Imgur.com and Pastee.org. Once the agreement between the client and server is done, the next step is the generation and exchange of a key. As per the complexity of the ransomware, this can range from a simple symmetric-key cipher to a rather more sophisticated RSA 4096-bit encipherment algorithm. The key exchange executes, and the private key is kept at the culprit servers while the public key is carried to the enciphering constituent of the pernicious code installed on the prey machine. However, in some cases, the use of a simple encipherment system or generation of a not-so-unique key every now and then can be easily be thwarted out using public decipherment modules and could recover the files.

3.4 Encipherment

Then, at this juncture, the key that is supposed to be applied to damage the files on the machine enciphered is now active and ready for use by the pernicious code on the prey machine. All the HQ identified files will commence the encipherment procedures by the pernicious code. This can include any file or any type of file extension ranging from all forms of MS Office applications to GIFs, JPGs, or PNGs. Some crypto-lockers not only encipher the files but also the filenames. The CryLocker ransomware enciphers the target files and changes their extension to . *cry*; the OSIRIS, a new variant of Locky ransomware, enciphers the stored data using asymmetric cryptography after infiltration using the "[8 arbitrary typescripts]-[4 arbitrary typescripts]-[4 arbitrary typescripts]-[8 arbitrary typescripts]-[12 arbitrary typescripts].osiris" pattern. It would rename a file, say, "sample.jpg" to "HL56PP89-H6B3-4T30-ER2R4O3Y-S9QT5NC0NA31.osiris".

3.5 Extortion

Following encipherment, these pernicious codes create their own unique HTML files, placing on the desktop and changing the desktop wallpaper letting the victims know that their machines have been compromised. Both the HTML file and desktop wallpaper contain the identical message stating that the files are enciphered using asymmetric encipherment algorithms and in case of OSIRIS crypto-locker, these files can only be restored using a private key only accessible when paying a payoff of 2.5 Bitcoin (currently 1 Bitcoin = $2804.23). Some ransomware variants will allow the victims to decipher only a single file for free to prove that there is key to their system. Some would delete files to scare the prey, thus enforcing them into more paying the payoff more rapidly. Upon payment, there is no assurance that the provided key by the culprits will decipher their files and moreover, there is no assurance that the per-ware would be aloof by itself.

Some ransomware lockers display a window (probably full screen) that spans the handler's entire desktop or limit the handler to just this single window by supervising the system's desktop through a background thread. The contents of these windows are generally localized (in local language) to certify that they aid, confined (local language support) content of the prey. Once the machine has been locked, the per-ware will do anything to confirm it and preserve perseverance on the machine, constituting sending signals for shutdown to the other processes, dispensing instructions to kill processes that would ultimately be used to end the per-ware executable, and generating a virtualized desktop to again ensure that the end user is incapable to get out of the simulated desktops formed by the per-ware. Some variants would lock the browser, a cross-platform mechanism by the culprits, popping-up the pernicious web pages every time the victims try to shut the browser or redirect from the affected web page [13].

4 Prevention Strategies

As more cybercrooks are looking forward to the usage of ransomware as a malicious income source, there is no surprise that the attacks are mushrooming, specifically targeting the fruitful businesses; therefore, it is the need of the hour that certain steps must be taken place to dodge these payoff bullets and put several operational systems out of the harm's way likewise in [15]. Over cloud network, the authors in [16] have proposed an enhanced ransomware prevention system CloudRPS. Authors in [14] have highlighted the formal methodologies to rescue stolen data from our smartphones over insecure networks. Thus, every business running on Windows platform, despite their Original Equipment Manufacturer (OEM) certificates, is still taken as hostages [17]. More prevention and detection strategies are mentioned in [18]. For this, we need to look at some of the measures that can be taken on diverse levels to prevent or reduce the impact of crypto-ware.

4.1 Awareness Among the Users

The first and foremost prevention measure must be taken is keeping the end users informed and aware since an infection always starts with a human fault. The risks of opening attachments, suspicious software, or the links are some of the first attacks prevalent on the basic systems. Some highly trained people may be even more prone to these attacks. It is must for the users to learn about how the ransomware works, how are they spread (like a disease), and what are they made of (actually). Moreover, the techniques that the per-ware uses needs an understanding up to a quite-a-good level, for example, the spam emails which involve the social engineering tricks. In the end, the aforementioned methods will surely benefit the Internet community to recognize and dodge further attacks.

4.2 Backing Up on a Regular Basis

Performing systematic backups of all crucial information to limit the impact of data or system loss will surely help the end users and the administrators to expedite the recovery progression. If possible, this data should be kept on an isolated device, and backups should be stored disconnected. Disabling macros which are suspiciously infected (pernicious) is a relatively good approach first since the ransomware attacks the system and mostly enciphers Microsoft Office documents. With Office 2007 and Office 2010 being the most vulnerable versions, it is highly probable that most computer workstations in the public and private sector offices and homes have this version and most of them are pirated. Thus, the effects are more far-reaching that one can imagine.

4.3   Disabling Windows Services

A service known as volume shadow copy service (VSS) is a set of Windows management instrumentation (WMI), component object model (COM), and application programming interfaces (APIs) that implement a framework (sort of a layered structure indicating what kind of programs can or should be built and how they would interrelate) to allow drive backups to be done while the programs on the operational system linger on writing into these drives. VSS provides a persistent UI that allows the proper cooperation between the user applications which alter the data on the disk by using I/O applications known as writers and those that back up these applications by using applications known as requesters which manage shadow copies to support some other functionalities such as backup and restore procedures and disk mirroring. Since Windows Vista, Microsoft has been bundling a utility called vssadmin.exe in Windows that allows an administrator to manage the shadow volume copies that are on the computer. Unfortunately, with the rise of crypto-ransomware, this tool has become more of a problem than a benefit and everyone should disable it. The developers of these per-wares are aware of shadow volume copies and designate their attacks such that they delete all these copies when the per-ware infects the computer system. This is done purely to restrict the prey from recovering any enciphered files. Ransomware injects themselves into the processes that run as administrator (to avoid any user account control prompt) so that the CMD command *vssadmin.exe Delete Shadows/All/Quiet* which will execute this utility to quietly delete all the shadow volume copies on the computer system. One must rename this file since it is rarely used in order to prevent the per-ware to utilize it to delete the shadow volume snapshots and ultimately save the files on the system. Most of the per-ware attacks can be foiled by using this method.

Another Microsoft utility which provides scripting abilities to the computing system is Windows Script Host (WSH). These code scripts can be executed directly from the desktop environment by double-clicking the code script file, or from MS-DOS. They can also be executed from either the protected-mode wscript.exe (Windows-based host) or the real-mode cscript.exe (command shell-based host). Some per-ware may use WSH to execute certain pernicious JavaScript codes when they are opened under this environment. It is then advised to disable WSH if the end user has no intention to run VB scripts in future.

4.4 File and Mail Servers

The file sharing mechanisms must be less sophisticated for the end user and more to the culprit. To do that, the shares must be fragmented as some rights on different shares must be reduced so that they (per-ware) cannot encipher; ultimately, they cannot edit these shares. Also checking up the creation of specific extensions which are used by them must be done regularly, the attachments on the gateway must be filtered by blocking those emails which contain executables (like Trojans) and those file types which should not be emailed around, for example, .chm, .lnk, and .js.

4.5 Securing the Network

Every institution or a corporation or even a lonesome Internet user, all require the connection to the Internet through basic networking devices. The end user must employ the following techniques so that the invader may not make use of the possible vulnerabilities. The first method would be using a proxy with web filtering: Some proxies allow us to filter the traffic from the blacklisted domains. It is must to ensure that the end user has the knowledge of these bad websites. This could largely reduce the chances of infection, if the list is updated thoroughly. Second, the magnitude of network sharing can become the target of certain per-ware. If the information being shared is of the highest significance, then it is highly advisable that these networks must be fragmented and thus the number of shares being available will be sufficiently reduced.

5 Latest Ransomware—WannaCry

Year 2017, month of May, the CTU scientists inspected a prevalent and cunning per-ware operation known by the names WCry, WannaCrypt, WanaDecryptor, and WannaCry that wedged numerous machines across the globe, most of them having national as well as international significance. These scientists correlate the quick infection of the per-ware to the use of a single, lone maggot or a virus component that subjugated susceptibilities in the Windows server message block (SMB) v1 protocol. It was then addressed by Microsoft in the month of March in the same year with a security bulletin MS17-010 [19]. Apparently, it is seen that WannaCry uses the MS17-010 exploit (phishing websites for the patch download) to spread to other machines through network basic input–output system (NetBIOS) as explained under common vulnerabilities and exposures (CVE).

5.1 How Are They Delivered?

The operation uses a maggot which is known as a server message block (SMB) maggot which is again used to dispense WCry donating to the per-ware's embitterment. It then tries a hypertext transfer protocol (HTTP) linkage to www [dot] asndasnqwkhekqjnskcjnkjwnekqjw [dot] com. Upon the successful linkage, the maggot halts execution and exits. Then, the maggot leverages an SMBv1 exploit (goes by the name EternalBlue) which explicitly searches for the presence of a backdoor application named as DoublePulsar [20] on infected machines. It is a backdoor implant tool developed by the NSA equation group (part of USA Security Agency) that was oozed by a hacker group known by the name "The Shadow Brokers" in early 2017. In case this backdoor is not available, the maggot tries to infect the pawn using the SMB version 1 EternalBlue [21] malcode (as mentioned earlier). Proliferation of this maggot banks on two threads (of processes), where the primary one ordains the LAN sub-nets, the SMB scans local addresses starting of the range of IP addresses that a specific ISP or datacenter owns or assigns at will and then incrementing 1-by-1 to its very end. The second thread scans arbitrarily chosen external IP addresses.

The maggot distributes itself to the infected machine as a dynamic link library (DLL) file cargo. After the DLL execution using PlayGame (a lone-exported application), it creates (or literally writes) a copy of the actual SMB maggot to the Microsoft Security Service (*mssecsvc.exe*). The SMB maggot then drops a tributary cargo from its with reserve section to the task scheduler (*tasksche.exe*) with both the files that reside in the default core directory of MS Windows operating systems are executed one after another. According to the research, this subordinate cargo is the WCry per-ware.

5.2 How They Infect Our Systems?

After the successful encipherment of the file system (whether File Allocation Table 32-bit (FAT32) or New-Technology File System (NTFS)), WannaCry displays the payoff demand dialog box as illustrated in Fig. 7. The per-ware then continuously supervises this window to make sure it remains above all windows (when the prey tries to switch over other applications), re-running if it is closed (it is like a popup on the desktop under Windows environment). Additionally, the per-ware changes the desktop wallpaper (just as Locky does).

It is then dispersed as a *.exe* file that comprises ZIP record (password-protected) in its reserves area. This record gets unwrapped when executed in the current directory and the files are the content which is shown in Fig. 8.

During the infection, some supplementary files are also created (shown in Fig. 9).

Fig. 7 WCry payoff demand interface

- b.wnry – Bitmap image used as desktop wallpaper
- c.wnry – Configuration containing TOR command (similar modus operandi of CryptoWall 2.0) and C2 addresses (HQ or the base of operations), Bitcoin addresses and other data.
- r.wnry – Ransom demand text
- s.wnry – ZIP archive containing TOR software to be installed on the prey's system; saved in TaskData directory.
- t.wnry – Encrypted DLL containing file-encryption functionality.
- u.wnry – Main module of the WCry ransomware "decryptor"
- taskdl.exe – WNCRYT temporary file cleanup program (delete any tasks native to OS)
- taskse.exe – Program that displays decryptor window to RDP sessions.
- msg – Directory containing Rich Text Format (RTF) ransom demands in multiple language (localized attack)

Fig. 8 Content of the reserve section

- 00000000.pky – Microsoft PUBLICKEYBLOB (public key binary large object) containing the RSA-2048 public key (the culprits are presumed to hold the private key).
- 00000000.res – Reserve section which holds the Data for HQ communication.
- 00000000.eky – Prey-unique RSA private key encrypted with embedded RSA public key.
- 00000000.dky – Decrypted RSA private key transmitted to prey after payoff expense.
- f.wnry – List of randomly chosen files encrypted with an embedded RSA private key that allows WCry to demonstrate decryption to the preys.
- @WannaDecryptor@.exe – Main module of the ransomware (decryptor), identical to u.wnry.
- @Please_Read_Me@.txt – Payoff demand test, identical to r.wnry.

Fig. 9 Supplementary files for the per-ware operations

The per-ware begins executing two commands when started—one is the *attrib +h* command which hides all the files as quickly as the handler moseys through every file or folder by means of the *Windows Explorer* and the second one is the *icacls ./grant Everyone:F /T/C/Q* which is used to change the file system permissions (generally NTFS file system) across various server and client operating systems. After this, WCry executes "Run As" or "runas" command to delete all shadow copies of the volume using elevated privileges; terminates several services using taskkill.exe so that their data stores can be encoded; and generates a file (particularly a batch file) using any arbitrarily generated large integer that creates a shortcut to the per-ware executable. It is then saved under "wd" registry value under HKLM registry store; otherwise, in the HKCU store.

5.3 How They Encrypt Our Files?

RSA and AES algorithms are used by this per-ware to encipher the file system. Windows crypto API for RSA encipherment and arbitrary key generation is the preferred usage choice of per-ware's method. Prior to this, WCry tallies all the accessible disks in the system. This inventory includes hard disk or local disk drives, removable drives (USB thumb drives), and network drives. However, this per-ware does not contain functionality to search the local network. WCry targets files with extensions of multimedia formats, compressed records, databases, and their add-ons likewise Locky ransomware. Most of these methods have directly been "inherited" from the other versions of CryptoWall or Locky crypto-ransomware. The file extensions targeted are .wma, .zip, .rar, .txt, .docx, .mov, .mp4, etc., perhaps almost all the extensions. Using the EKY extension (as shown in Fig. 9), a private key couple (RSA-2048) is generated and detailed to each contagion and is then stored on the local storage media and then it is used encipher the arbitrary AES-128 key produced for each enciphered file. These files are renamed using the filename format *<arbitrary_number>*.WNCRY and are replaced with the original ones. However, it is not done entirely and thus there is a hope of a forensic retrieval of file inside conditional to the milieu.

5.4 Payment Prompts and Communication
      with Headquarters (HQ)

WCry shows a meter that countdowns to the date and time when the payoff amount increments (see Fig. 7), probably 4 days or maybe 5 and when these files will become irrevocable (probably after seven days). Different variants of WCry set the payoff either in bitcoins or dollars ($300–$600). How much the culprits claim, they do not have the capability or intent to decipher the files after the prey pays them the payoff. It is very likely for them to increase the payoff demand and thus the prey falling into their trap. The TOR framework run by WCry is run as a task host service (*taskhsvc.exe*) which then creates a SOCKS5 proxy server which is again used on the loopback address, i.e., 127.0.0.1 and afterward receives the signals on port 9050 (TCP). Ultimately, it tries to access some HQ services like transmit encipherment keys, communicate with the culprits, or check payment status as elaborated in [4].

6 Conclusion

This study is more comprehensive in nature than it looks, in writing which represents a thorough analysis of the idea of a ransomware, that how cryptovirology and extortion-based cryptographic techniques gave birth to a new kind of threat in the twenty-first century, evolution of these attacks since 2005, their modes of operation, and how can we protect our computing systems and datacenters from becoming the next prey. It is clear that the ransomware will continue to grow more sophisticated and will surely become more widespread like a pneumonic plague which engulfed the Europe in the twentieth century. We believe that this cyber-plague will cripple most of the cyber-superpowers which control the Internet flow in the world today. The security agencies, anti-per-ware enterprises, antivirus multi-national corporations (MNCs), and research centers will have to be on-the-toes in tackling these novel threats to the very existence and a pillar of our digital progress. Opportunistic ransomwares like WCry use their propagation methods which powers them to spread quickly. Before the end user even knows it, the per-ware gets the hold of the prey's precious data and information and then runs phishing and trapping other victims in the network that the current machine is connected. As discussed in Sect. 4, several prevention techniques are in place now, and that the Federal Cyber Emergency Team, India (CERT.in) has communicated guidelines and strategies to protect their home machines or office machines and obviously Microsoft's MS17-010 update/patch helped most of the machines to make themselves immune to these attacks.

Future Scope
The learning, awareness, and knowing-your-foe stratagems have enabled the digital infrastructure to be able to withstand the constant evolving threats from the cyber-culprits. How threatening this evolution may seem, there is always a persistent idea of hope that the good will prevail and a fightback will always be there in counter the spread and effect of pernicious elements. Our such efforts in this paper will make the reader feel the very concept of ransomware rather than building them the sole understanding. We hope that in near future, these frequencies of these attacks will plummet and our data on the Internet can be shared in full confidence and in utmost relief. However, it would take away the motivation and may allow the systems to become a little complacent (that can do more harm than being vulnerable).

References

1. Young, A. L., & Yung, M. (2017). Cryptovirology: The birth, neglect, and explosion of ransomware. *Communications of the ACM, 60*(7), 24–26.
2. Mercaldo, F., Nardone, V., & Santone, A. (2016, August). Ransomware inside out. In *2016 11th International Conference on Availability, Reliability and Security (ARES)* (pp. 628–637). IEEE.
3. Unit 42 Palo Alto Networks Threat Report—Ransomware: Unlocking the Lucrative Criminal Business Model (2016, May). Retrieved June 21, 2017. https://www.paloaltonetworks.com/resources/research/ransomware-report.
4. Deloitte Threat Intelligence and Analytics Report (2016). Retrieved June 21, 2017. https://www2.deloitte.com/content/dam/Deloitte/us/Documents/risk/us-aers-ransomware.pdf.
5. Symantec Security Response Whitepaper. The evolution of ransomware (2015, August). Retrieved June 19, 2017. http://www.symantec.com/content/en/us/enterprise/media/security_response/whitepapers/the-evolution-of-ransomware.pdf.
6. Orman, H. (2016). Evil offspring-ransomware and crypto technology. *IEEE Internet Computing, 20*(5), 89–94.
7. McAfee Whitepaper—Understanding Ransomware and Strategies to Defeat it (2016). Retrieved June 21, 2017. https://www.mcafee.com/in/resources/white-papers/wp-understanding-ransomware-strategies-defeat.pdf.
8. Liska, A., & Gallo, T. (2017). *Ransomware: Defending against digital extortion*. Beijing; Boston; Farnham; Sebastopol; Tokyo: OReilly.
9. CERT.be (Cyber Emergency Team, Belgium) Ransomware Whitepaper (2016). Retrieved June 21, 2017. https://www.cert.be/files/ransomware_whitepaper.pdf.
10. Moore, C. (2016, August). Detecting ransomware with honeypot techniques. In *Cybersecurity and Cyberforensics Conference (CCC)*, 2016 (pp. 77–81). IEEE.
11. Scaife, N., Carter, H., Traynor, P., & Butler, K. R. (2016, June). Cryptolock (and drop it): Stopping ransomware attacks on user data. In *2016 IEEE 36th International Conference on Distributed Computing Systems (ICDCS)* (pp. 303–312). IEEE.
12. The TOR Project. Retrieved June 21, 2017. https://www.torproject.org/.
13. Yang, T., Yang, Y., Qian, K., Lo, D. C. T., Qian, Y., & Tao, L. (2015, August). Automated detection and analysis for android ransomware. In *2015 IEEE 17th International Conference on High Performance Computing and Communications (HPCC), 2015 IEEE 7th International Symposium on Cyberspace Safety and Security (CSS), 2015 IEEE 12th International Conference on Embedded Software and Systems (ICESS)* (pp. 1338–1343). IEEE.
14. Mercaldo, F., Nardone, V., Santone, A., & Visaggio, C. A. (2016, June). Ransomware steals your phone, formal methods rescue it. In *International Conference on Formal Techniques for Distributed Objects, Components, and Systems* (pp. 212–221). Cham: Springer.
15. Luo, X., & Liao, Q. (2007). Awareness education as the key to ransomware prevention. *Information Systems Security, 16*(4), 195–202.
16. Lee, J. K., Moon, S. Y., & Park, J. H. (2017). CloudRPS: A cloud analysis based enhanced ransomware prevention system. *The Journal of Supercomputing, 73*(7), 3065–3084.
17. Mansfield-Devine, S. (2016). Ransomware: Taking businesses hostage. *Network Security, 2016*(10), 8–17.
18. Brewer, R. (2016). Ransomware attacks: Detection, prevention and cure. *Network Security, 2016*(9), 5–9.
19. Microsoft Security Bulletin MS17–010—Critical (2017, March). Retrieved June 21, 2017. https://technet.microsoft.com/en-us/library/security/ms17-010.aspx.
20. Double Pulsar NSA leaked hacks in the wild (2017, April). Retrieved June 21, 2017. https://www.wired.com/beyond-the-beyond/2017/04/double-pulsar-nsa-leaked-hacks-wild/.
21. NSA-leaking Shadow Brokers (2017, April). Retrieved June 20, 2017. https://arstechnica.com/security/2017/04/nsa-leaking-shadow-brokers-just-dumped-its-most-damaging-release-yet/.

GPU-Based Integrated Security System for Minimizing Data Loss in Big Data Transmission

Shiladitya Bhattacharjee, Midhun Chakkaravarthy and Divya Midhun Chakkaravarthy

Abstract In big data transmission, data loss occurs due to excessive data and time overheads, transmission errors, and various security attacks. However, the current literature does not suggest any combinatorial solution to protect data loss by resolving these issues. Henceforth, this research proposes a GPU-based integrated technique. It includes the new dual rounds of error control operations to address every individual error bit. An idle lossless compression has been included to reduce data overhead. Furthermore, it includes a modern audio steganography which comprises a new sample selection scheme for hiding data. Finally, it reduces execution time by involving GPU during implementation. As per the experimental result, it offers higher time and space efficiencies by producing higher compression percentage and throughput. It shows substantial resistivity against various attacks and transmission errors by producing higher SNR and entropy values. It also offers lower percentage of information loss than other existing techniques.

Keywords Data loss · GPU · Compression percentage · Perceptual similarity
Throughput · Lower percentage of information loss

S. Bhattacharjee (✉) · M. Chakkaravarthy
Computer Science and Multimedia Department, Lincoln University College, Selangor Darul Ehsan, Malaysia
e-mail: shiladityaju@gmail.com

M. Chakkaravarthy
e-mail: midhun.research@gmail.com

D. Midhun Chakkaravarthy
Faculty of Post Graduate Studies, Lincoln University College, Selangor Darul Ehsan, Malaysia
e-mail: divya.phd.research@gmail.com

© Springer Nature Singapore Pte Ltd. 2019
V. E. Balas et al. (eds.), *Data Management, Analytics and Innovation*,
Advances in Intelligent Systems and Computing 839,
https://doi.org/10.1007/978-981-13-1274-8_32

1   Introduction

Data overhead, the root cause for data loss can be controlled with the application of distinct compression techniques. Conversely, between lossy and lossless compression, lossy type degrades data quality by eliminating the redundant information permanently which is unwanted for this research. Hence, this discussion relates the lossless compression only. Amidst the lossless compressions, fixed length coding (FLC) is inefficient to produce desire compression efficiency [1]. Contrariwise, the variable length coding (VLC) is dependable on prefix codes. These prefix codes can be easily affected by transmission errors which may cause data loss [1, 2].

As per the literature, another important aspect of data loss is transmission errors. Consequently, this issue increases with the increment of file size. However, the present literature fails to present any efficient error control technique which can control more than 8-bit discrete or continuous errors [3, 4]. Only cryptographic hash function can address up to 8-bit errors. Another way to replace these errors is to transmit the original information redundantly [5]. Cryptography and steganography and pattern generation as well as matching using hash function are the mostly used security techniques to protect various security attacks, performed by distinct illicit users to steal or notch information during transmission. However, these techniques are consuming lots of time and space to complete a number of iterations during execution which are not suitable for big data transmission [6].

1.1   Current Problems

The issues regarding various security schemes to protect data loss in big data transmission are as follows:

- A number of iterations are involving in existing steganography and cryptography techniques, used for protecting various security attacks. As the consequence, time and space overheads increase and data loss may occur [6].
- The existing error control techniques are not efficient to protect more than 8-bit discrete or continuous errors. The retransmission of original information increases time and space overheads [7].
- The current literature fails to suggest any integrated technique which can resolve all these root causes of data loss in big data transmission collectively [8].

1.2   Research Objectives

The primary objective is to provide an integrated solution for data loss in big data transmission by addressing the primary issues in combinatorial way. The other objectives are as follows:

- Minimizing data overhead by introducing a new lossless compression technique.
- Minimizing time overhead using single iteration during execution and GPU.
- Enhancing robustness against various security attacks and transmission errors.

2 Literature Study

This section reviews the strengths as well as the shortcomings of these available techniques and helps to develop the proposed integrated technique by analyzing them. In article [3, 4], the authors proposed a combinatorial Ethernet solution by integrating multipathing and congestion control mechanism. The proposed technique improves network throughput because its application-layer flow differentiation can make full consumption of network bandwidth and congestion control can stop unnecessary traffic from incoming network. However, it fails to utilize the buffer properly. According to article [5], the scarce inputs in wireless body sensor network (WBSN) nodes limit their abilities to survive with massive traffic during multiple, concurrent data communications. However, this technique cannot offer efficient routability and minimize congestion and overflow avoidance proficiently. As per article [6], the huge data transmission of ultrasound data becomes a serious issue of real-time transmission for a GPU-based beamformer. Yet, in this technique, phase data were fixed as 6 bits without any compression.

According to article [7, 8], fault tolerance has been observed as serious to the real use of these GPUs. In this article, the author presented an online GPU error detection, location, and correction method to include fault tolerance into matrix multiplication. However, this technique is not efficient for large number of continuous error bits and it enhances data overhead considerably. In article [9, 10], the authors define a parallel execution of a favorable similarity search procedure for an audio fingerprinting system. With simple reforms, the authors obtained up to 4 times higher GPU performance. However, it is not highly secured and robust against various transmission errors and security attacks.

Henceforth, from the above discussion, we have seen that the existing techniques for minimizing the data loss are inefficient for big data applications. These prevailing techniques are either enhancing excessive space as well as space overheads during the execution or cannot offer adequate robustness against the various transmission errors and security attacks [11, 12]. Therefore, there is further scope to develop an integrated technique which can be functioned with low time and space requirements and which can protect data loss by minimizing various data errors during the transmission process as well as protecting the various security attacks at the same time.

3   Proposed Technique

The detailed steps of implementing the proposed integrated technique are shown in Fig. 1.

The implementation of the proposed integrated technique has been performed using parallel processing in GPU environment. At the final part, the description of retrieval information at the receiving end is shown further.

3.1   *Generation and Incorporation of Error Control Bits*

The proposed dual rounds of error control bit generation and incorporation have been presented in this subsection with the help of text file as input to make it

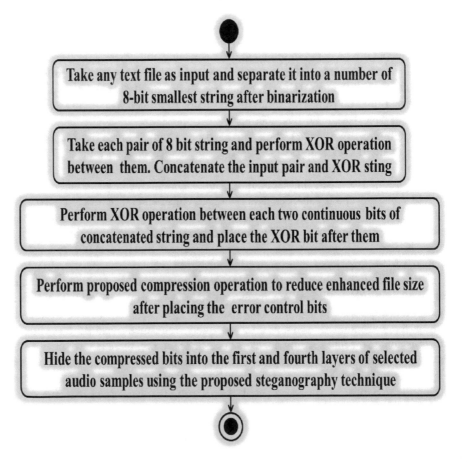

Fig. 1 Proposed integrated model for minimizing data loss using parallel processing with graphics processing unit (GPU)

simplified. Let the input text is denoted as *Text*, and after converting each character of this input text into 8-bit strings, they are stored in the string array *text*. If the total number of characters belonging to the input text is n, then the sequences of converted 8-bit strings are represented as $text_n$. Perform the first round of error control operation by

$$\left.\begin{array}{c} xorStr = text_i \oplus text_{i+1} \\ conStr = \left(text_i \times 10^{l(text_{i+1})}\right) + text_{i+1} \\ concatStr_j = \left(conStr \times 10^{l(xorStr)}\right) + xorStr \\ i = i+2, j = j+1 \end{array}\right\} \qquad (1)$$

such that

$$i, j \leq 0 \text{ and } i < (n-2), j < \left(\frac{n}{2}\right),$$

$$l(text_{i+1}) = (\log_{10}(text_{i+1}) + 1) \text{ as well as}$$

$$l(xorStr) = (\log_{10}(xorStr) + 1)$$

After completion of first round of error control bit generation and incorporation, the length of each concatenated string in array *concatStr* becomes 24 bits. Add "0" at most significant bit position of each element in array *concatStr*. Let the order of each string in *concatStr* is denoted by $concatStr_j$, where $0 \leq j < \left(\frac{n}{2}\right)$ and the bit sequence of each string in *concatStr* is signified by $concatStr_{jk}$ where $0 \leq k \leq 24$. Second round of error control bit generation and incorporation are performed to generate string array *concatStr'* of size $\left(\frac{n}{2}\right)$ as

$$\left.\begin{array}{c} xorStr' = concatStr_{pl} \oplus concatStr_{p(l+1)} \\ concatStr'_p = concatStr_{p(l+1)} \\ concatStr'_p = \left(concatStr'_p \times 10^{l(xorStr')}\right) + xorStr' \\ p = p+1 \text{ and } l = l+1 \end{array}\right\} \qquad (2)$$

such that

$$p, l \leq 0 \text{ and } p < \frac{n}{2}, l \leq 24 \text{ as well as}$$

$$l(xorStr) = (\log_{10} xorStr + 1)$$

3.2 Application of the Proposed Compression Technique

After adding error control bits, the array *concatStr'* is taken as input. Initially, create a new string array *EleRed* from *concatStr'* by truncating all redundant elements as

$$EleRed = \binom{concatStr'_m}{\frac{n}{2}} \qquad (3)$$

After creating of array *EleRed*, calculate the frequencies of each element in array *concatStr'* and put the corresponding frequencies into another array *Freq*. Sort the array *EleRed* according to higher to lower frequencies. As the input file is text and total number of ASCII characters is 128, hence, a maximum of different elements in *EleRed* will be 128. As the consequence of it, the size of *EleRed* which is m can be varied from 0 to 128. Now, let the string sequence of *EleRed* is represented by $EleRed_x$, where $0 \le x < 128$. In the next step, a string array *Comp* of size x is created to store the compressed code. If the total size of string array *EleRed* is less than or equal to 32, then generate the compressed codes and store them in *Comp* as

$$\left.\begin{array}{c} s_y = x \ modulo \ 2^y \\ t = \frac{EleRed_x}{2^i}, \ where \ 0 \le y \le 5 \\ z = z \times (10)^{l(s_y)} + s_y, \ if \ l(s_y) = log_{10}^{s_y} + 1 \\ Comp_x = z \end{array}\right\} if \ t \ne 1 \qquad (4)$$

where s, t, y, and x are integer variables where $x \ge 0$ and $x < 32$ and z is a string variable. When the size of *EleRed* > 32 and ≥ 62, then create the compressed code for zero to twenty-ninth element using the same process as

$$\left.\begin{array}{c} x = x - 30 \\ s_y = x \ modulo \ 2^y \\ t = \frac{EleRed_x}{2^i}, \ where \ 0 \le y \le 5 \\ z = z \times (10)^{l(s_y)} + s_y, \ if \ l(s_y) = log_{10}^{s_y} + 1 \end{array}\right\} if \ t \ne 1$$

Now let the 5-bit binary string, generated from 31 is used as separator (*Sep*). Concatenate separator with the compressed code as

$$z = \left(z \times 10^{l(Sep)}\right) + Sep \qquad (5)$$

where $l(Sep) = log_{10}^{Sep} + 1$.

After the concatenation operation, compressed codes are stored in array *Comp*. When the size of *EleRed* is among 63 and 90, then consider 30 and 31 as separator and repeat same steps. Similarly, when the size of *EleRed* is among 91–116, consider 29, 30, and 31 as separators. Finally, when the length of *EleRed* is more than 116, then consider 28, 29, 30, and 31 as separator and generate the compressed codes. Store these compressed codes into the *Comp*. Concatenate all strings of *concatStr* into a variable *Comp'*. Concatenate all strings of *EleRed* into a single string variable *Char*. Create the final separator *Sep'* by concatenating 100 bits into 1. Finally, concatenate *Comp'*, *Sep'* and *Char* according to generate compressed string *CompStr*.

3.3 Application of Proposed Steganography Technique

After the creation of complete *CompStr*, the .wav audio file and string *CompStr* are taken as input to create the Stego file. The creation of Stego has four parts, and they are binarization of audio file, selection of samples for incorporation compressed string, incorporation of compressed string into selected samples, and reformation of Stego audio file after incorporation of compressed string. Each part is further described in detail at the following subsections.

3.3.1 Binarization of Input Audio File

The binarization operation has been done by dividing the .wav file into few samples using sampling theory depending upon the amplitude and threshold value. The normalization is done with each sample to get the actual value. 10 is multiplied with normalized value to convert them within the range of 0–254. After multiplication, each value is converted into 8-bit binary string.

3.3.2 Selection of Samples for Incorporating Compressed String

If we hide the compressed code to each sample, then the distortion of sound will be occurred after conversion of this string to Stego sound file. So, we select the particular samples to hide the compressed code. We select the prime position of the samples and then find the sequence of number of each prime position.

3.3.3 Incorporation of Compressed String into Selected Samples

After the selection, two bits are taken in each occasion from the compressed string and incorporated into each selected sample accordingly. Let the selected sample is denoted by *Samp* and the incorporation of compressed bits into *Samp* is shown by Algorithm 1.1 and 1.2.

```
Algorithm-1.1: Audio Steganography (Part-1)
(e1)  declare selected samples as ⟨Samp⟩ string array;
(e2)   if ( Samp[3] is modified from 0 to 1)
(e3)      if (Sample[2] == 1 and Sample[4] == 1)
(e4)          for (int i = 0; i ≤ 2; i + +)
(e5)              Sample[i]=0;
(e6)          End for
(e7)      End if
(e8)      if (Sample[2] == 1 and Sample[4] == 0)
(e9)          repeat steps (e4) to (e6)
(e10)     End if
(e11)     if (Sample[2] == 0 and Sample[4] == 1)
(e12)         for (int i = 0; i ≤ 2; i + +)
(e13)             Sample[i]=1;
(e14)         End for
(e15)     End if
(e16)     if (Sample[2] == 0 and Sample[4] == 0)
(e17)         repeat steps (e12) to (e14)
(e18)         for (int j = 4; j ≤ 7; j + +)
(e19)             if (Sample[j]==1)
(e20)                 Sample[j]=0;
(e21)                 break;
(e22)             End if
(e23)             else
(e24)                 Sample[j]=1;
(e25)  End All
```

In Algorithm 1.1, line (e1) declares the 8-bit string variable *Samp*. During the incorporation, if the fourth bit of *Samp* converts from 0 to 1 and if third and fifth bits of *Samp* are 1, then lines (e1) to (e7) describe the required operations. Lines (e8) to (e10) describe the changes in *Samp* if the third bit of it remains 1 and fifth bit is 0. Similarly, lines (e11) to (e15) define the operation when the third bit of *Samp* is 0 and the fifth bit is 1. Lines (e16) to (e25) define the changes in *Samp* if the third and fifth bits are. During embedding operation, if the third bit is changed from 1 to 0, then the required operations are shown by Algorithm 1.2.

```
Algorithm-1.2: Audio Steganography(Part-2)
(e26)    if ( Samp[3] is modified from 1 to 0)
(e27)        if (Sample[2] == 0 and Sample[4] == 0)
(e28)            for (int i = 0; i ≤ 2; i + +)
(e30)                Sample[i]=1;
(e31)            End for
(e32)        End if
(e33)        if (Sample[2] == 0 and Sample[4] == 1)
(e34)            repeat steps (e30) to (e32)
(e35)        End if
(e36)        if (Sample[2] == 1 and Sample[4] == 0)
(e37)            for (int i = 0; i ≤ 2; i + +)
(e38)                Sample[i]=0;
(e39)            End for
(e40)        End if
(e41)        if (Sample[2] == 1 and Sample[4] == 1)
(e42)            repeat steps (e38) to (e40)
(e43)            for (int j = 4; j ≤ 7; j + +)
(e44)                if (Sample[j]= =0)
(e45)                    Sample[j]=1;
(e46)                    break;
(e47)                End if
(e48)                else
(e49)                    Sample[j]=0;
(e50)    End All
```

In Algorithm 1.2, we can see the required operation if the fourth bit of the *Samp* changes from 1 to 0 during the incorporation. Line (e26) tells this condition. Lines (e27) to (e31) describe the required changes of *Samp* when the third and fifth bits of *Samp* are 0. If the third bit is 0 and fifth bit is 1 of *Samp*, lines (e32) to (e34) describe the required changes of *Samp*. Lines (e35) to (e39) define the required changes of *Samp* when the third bit is 1 and fifth bit is 0. Finally, lines (e40) to (e50) define the required modification of *Samp* when the third and fifth bits of *Samp* are 1.

Reformation of Stego Audio File After Incorporation of Compressed String

After incorporation of compressed bit in the audio samples *Samp*, convert all the samples into decimal value and then divide all the samples by 10. After the division, de-normalization operation is done, depending upon the predefined threshold values. Then, combine all the samples into single file to construct the final Stego audio file.

3.4 Retrieval of Information at the Receiving End

After receiving complete Stego file, the Stego audio file is split into 8-bit binary samples. After binarization, select the sample by the sample selection technique and extract the incorporated compressed bits from the first and fourth positions of selected samples. After extracting the compressed bits from the each selected sample, they are concatenated into single string to construct final compressed string. After that, the character and compressed code strings are separated based on the separator. After separating them, compressed code array and character array are generated. From the character array, a reference code array is created. After performing first round error control operation, eliminate the entire reference error control bit. If error still exists, then generate original string with the help of reference XOR strings and replace the erroneous string with the original string. Truncate all the reference XOR strings and convert the binary strings to ASCII characters. All characters are then merged into single file to generate the original file.

4 Result Analyses

As per the research objectives, primarily we will examine the capacity of proposed compression technique to reduce file size. As per the literature, *compression percentage (CP)* is the required parameter for measuring the efficiency of any compression technique. The CP can be defined as

$$CP = \frac{\text{Actual Size} - \text{Compressed Size}}{\text{Actual Size}} \times 100 \tag{6}$$

With the help of Eq. (6), the efficiencies (in terms of CP) of different existing FLCs and VLCs have been compared with the proposed technique in Tables 1 and 2.

Table 1 Comparison of proposed technique with different VLCs

File size (GB)	Arithmetic coding	Huffman BWT	LZSS BWT	Dictionary based	Proposed technique
100	14.0	30.2	34.1	41.6	56.7
200	13.0	30.9	35.2	42.0	57.0
350	12.9	33.7	34.7	43.0	57.9
450	13.9	31.3	34.0	43.7	58.0
600	13.4	30.6	35.1	42.9	58.2
750	13.6	31.5	35.2	44.0	58.8
900	12.9	32.0	34.3	45.0	58.9
1024	12.7	29.9	36.7	44.6	59.7

Table 2 Comparison of proposed technique with different FLCs

File size (GB)	ASCII code	EBCDIC code	Run length coding	Proposed technique
35	0.0002	0.0001	−1.0055	56.7
56	0.0003	0.0002	−1.0026	57.0
90	0.0001	0.0004	0.0001	57.1
120	0.0001	0.0004	−0.998	57.9
200	0.0002	0.0001	−0.762	58.0
350	0.0003	0.0002	−0.985	58.2
450	0.0001	0.0003	−0.976	58.8
600	0.0003	0.0002	−1.112	58.9
1024	0.0002	0.0001	−1.007	59.7

Table 3 Execution speed of proposed technique in CPU and GPU

File size in GB	Throughput offered by CPU (MB/s)	Throughput offered by GPU (MB/s)
100	2.71	74.64
200	2.75	76.27
350	2.72	77.77
450	2.74	78.21
600	2.72	76.78
750	2.75	77.56
900	2.79	79.22
1024	2.76	79.67

From Tables 1 and 2, we have seen that the proposed compression technique offers higher *CP* than the other existing FLC- and VLC-based compression techniques. Hence, the proposed compression technique is efficient to compress the data than other existing which fulfills our first objective.

In the next phase, we will examine the time efficiencies of the proposed technique. As per the literature, *throughput (TP)* is the required parameter for measuring the time efficiency of any technique. *Throughput* is defined as

$$TP = \left(\frac{\text{Output file size}}{\text{Total execution time to generate output}} \right) \quad (7)$$

With the help of Eq. (7), the *throughput* produced by the proposed technique during the execution in both CPU and GPU has been calculated using distinct file of different sizes. The results have been tabulated in Table 3.

Table 3 shows that the GPU implementation offers much higher execution speed than the CPU implementation in terms of offering higher *throughputs*. As the proposed integrated technique has been implemented in GPU environment, henceforth, it is much faster than any other existing solution. Thus, we have achieved our second objective.

The third objective of this research is to examine the capacity of preventing data loss by proposed integrated technique. From the literature, we have seen that other than the time and space overheads, transmission errors and various security attacks are the main causes of data loss. The literature also shows that the capacity of any error control technique to control error can be evaluated by calculating *signal-to-noise ratio (SNR)*. The SNR is defined as

$$\text{SNR}_{\text{dB}} = 10 \times \text{Log}_{10}\left\{\sum_n x^2(n) / \sum_n \left[x^2(n) - y^2(n)\right]\right\} \tag{8}$$

In Eq. (8), $x(n)$ denotes mean amplitude values of the cover files, whereas $y(n)$ denotes mean amplitude of Stego sample file, and n is any finite number. The SNR values offered by different error control techniques and the proposed techniques are shown in Fig. 2.

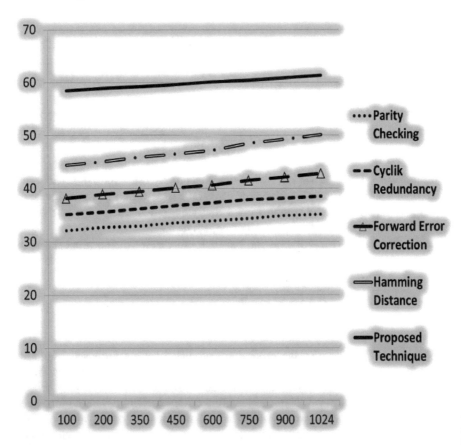

Fig. 2 SNR$_{\text{dB}}$, offered by different error control techniques

Table 4 Entropy values offered by different security techniques

Security techniques	Entropy values
Existing LSB technique	7.73
Parity coding	7.53
Phase coding	7.67
Spread spectrum	7.52
Echo hiding	7.51
Proposed technique	7.77

As per the definition, if any error control technique is effective to produce higher SNR_{dB}, then it is considered as efficient to reduce transmission errors. Figure 2 shows that proposed error control technique offers higher SNR_{dB} than others. Hence, it is more efficient to reduce transmission errors.

Consequently, the capacity to protect security attacks can be verified by calculating *entropy* values, offered by any security technique. It can be calculated as

$$H(S) = \sum_{i=0}^{2N-1} P(S_i) \log_2 \frac{1}{P(S_i)} \tag{9}$$

Here, $P(S_i)$ represents the probability of symbol S_i. With the help of Eq. (9), the *entropy* values offered by the proposed integrated and other existing technique are shown in Table 4.

From Table 4, we have seen that proposed integrated technique offers higher robustness against the various security attacks as it offers higher entropy values than others. Consequently, Fig. 3 justifies the capacity of protecting information loss by the proposed integrated technique.

Figure 3 shows that the proposed integrated technique offers lower percentage of information loss than the other exiting techniques. Hence, Fig. 2, Table 4, and Fig. 3 shows that the proposed integrated technique offers higher efficiencies to protect information loss by preventing various transmission errors and security attacks which fulfill our third objective.

5 Conclusions and Future Work

From the literature study in Sect. 2, we have seen that the existing security techniques for big data transmission are incapable of protecting data loss, which is a big challenge these days. Henceforth, this research proposed a novel integrated technique for minimizing the data loss in big data transmission and has been designed in GPU-based environment to speed up the execution process. The faster execution process will further reduce the data loss during the transmission. Apart from that, the proposed integrated technique includes a unique error control technique to protect data from various transmission errors, a novel compression technique to

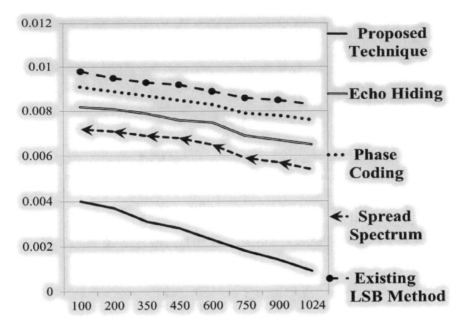

Fig. 3 Capacity of protecting information loss by different security techniques

reduce data load as well as the proposed advanced steganography technique, which helps to protect various security attacks during the transmission. In result section, from Tables 1 and 2, it can be seen that the proposed integrated technique is efficient to reduce space overheads than other existing techniques. At the same time, Table 3 shows that the proposed integrated technique reduces data loss by offering the better time efficiency in terms of providing better execution speed than other existing integrated techniques. Further, Fig. 2 shows that the proposed integrated technique can protect various transmission errors in efficient way than the other corresponding available security techniques by preventing. Table 4 shows that the proposed integrated technique is also efficient to protect various security attacks. Finally, Fig. 3 shows the efficiency of the proposed integrated technique to protect data loss with the help of different input file sizes. However, Figs. 2 and 3 show that the proposed integrated technique can be more improved that it can reduce the information loss further and it can prevent the transmission errors more accurately in the coming future. With the use of GPU environment, the overall time complexity has been reduced dramatically. However, it can be more reduced by reducing the time complexity of each of the individual parts of the proposed integrated technique and by reducing the integration process as well.

References

1. Ding, C., Karlsson, C., Liu, H., Davies, T., & Chen, Z. (2011). Matrix multiplication on GPON-line fault tolerance (pp. 311–317).
2. Lok, U.-W., & Li, P.-C. (2014). Improving performance of GPU-based software beamforming using transform-based channel data compression (pp. 2141–2144).
3. Fang, S., Yu, Y., Foh, C. H., & Aung, K. M. M., A loss-free multipathing solution for data center network using software-defined networking approach (pp. 1–8); Maxwell, J. C. (2013). A treatise on electricity and magnetism (3rd ed., Vol. 2, pp. 68–73). Oxford: Clarendon.
4. Yaakob, N., & Khalil, I. (2016). A novel congestion avoidance technique for simultaneous realtime medical data transmission. *IEEE Journal of Biomedical and Health Informatics, 20* (2), 669–681.
5. Lok, U.-W., Shih, H.-S., & Li, P.-C. (2015). Real-time channel data compression for improved software beamforming using microbeamforming with error compensation (pp. 1–4).
6. Chen, S.-n., Zheng, B.-y., & Zhou, L. (2013). A parity-based error control method for distributed compressive video sensing (pp. 105–110).
7. Xiaoliang, C., Chengshi, Z., Longhua, M., Xiaobin, C., & Xiaodong, L. (2010). Design implementation of MPEG audio layer III decoder using graphics processing units (p. 487).
8. Ouali, C., Dumouchel, P., & Gupta, V. (2015). GPU implementation of an audio fingerprints similarity search algorithm (pp. 1–6).
9. Cheung, N.-M., Au, O. C., Kung, M.-C., Wong, P. H., & Liu, C. H. (2009). Highly paralldistortion optimized intra-mode decision on multicore graphics processors. *Transactions on Circuits and Systems for Video Technology, 19*(11), 169.
10. Rahim, L. B. A., Bhattacharjee, S., & Aziz, I. B. A. (2014). An audio steganography technique to MData hiding capacity along with least modification of host. In *Proceedings of the First InterConference on Advanced Data and Information Engineering*. Singapore: Springer.
11. Bhattacharjee, S., Rahim, L. B. A., & Aziz, I. B. A. (2015). A Lossless compression technique to increase robustness in big data transmission system. *International Journal in Advances in Soft Computing and Its Application*.
12. Bhattacharjee, S., Rahim, L. B. A, & Aziz, I. B. A. (2014). A secure transmission scheme for textual data with least overhead. In *Twentieth National Conference on Communications (NCC)*. New York: IEEE.

Defending Jellyfish Attack in Mobile Ad hoc Networks via Novel Fuzzy System Rule

G. Suseendran, E. Chandrasekaran and Anand Nayyar

Abstract Security in mobile ad hoc environment is the most concerned research issue which is focused in the proposed research methodology by introducing authenticated routing based attack injection and detection framework using genetic fuzzy rule based system (AR-AIDF-GFRS). This assures both the successful detections of attacks present in the environment and secured routing by using trusted nodes. Here, initially, jellyfish attack is injected into the MANET environment. This attack is detected by genetic fuzzy based rule system which would generate rules based on which attack would be identified. And then to ensure the secured routing, trust evaluation of nodes is done by ant colony based trust evaluation method (ACTEM). This method selects the optimal nodes which are trusted in nature for establishing the route path. The overall evaluation of the proposed research method is done in NS2 simulation environment which proves that AR-AIDF-GFRS can outperform the existing research method by accurately identifying attacker nodes.

Keywords Jellyfish attack · Fuzzy rule · Secured routing · Trust value
Fuzzy rules · Quality of service

G. Suseendran (✉)
Department of Information Technology, School of Computing Sciences,
VELS Institute of Science, Technology & Advanced Studies (VISTAS),
Chennai 600117, Tamil Nadu, India
e-mail: suseendar_1234@yahoo.co.in

E. Chandrasekaran
Department of Mathematics, Veltech Dr. RR & Dr. SR University,
Chennai, India
e-mail: e_chandrasekaran@yahoo.com

A. Nayyar
Graduate School, Duy Tan University, Da Nang, Vietnam
e-mail: anandnayyar@duytan.edu.vn

© Springer Nature Singapore Pte Ltd. 2019
V. E. Balas et al. (eds.), *Data Management, Analytics and Innovation*,
Advances in Intelligent Systems and Computing 839,
https://doi.org/10.1007/978-981-13-1274-8_33

1   Introduction

The 21st century of advancements and innovations open for everybody, where there no evident limits between the usefulness of the gadgets the portable specially appointed systems administration (MANETs) assume huge part. MANETs have turned out to be a standout among the most predominant zones of research in the current years and as a result of the difficulties, it postures to the related calculations. MANET is an arrangement of remote versatile hubs that progressively self-compose in the subjective and transitory system topologies.

MANET has become a rising research area with many practical applications. Its technology provides a flexible way to set up communications in situations with geographical constraints that demand distributed networks without any centralized authority or fixed base station, such as disaster relief, emergency situations (rescue team), battlefield communications, conference rooms, and military applications [1]. Compared to the traditional wireless and wired networks, MANET is prone to varied security vulnerabilities and attacks because of its features in terms of no centralized authorities, distribution cooperation, open and shared network wireless medium, severe resource restriction, and high dynamic nature of network topology. The factors that attracted attraction of researchers around MANETs are: Self-configuration and Self-maintenance. Another unique feature of MANETs that poses security threats is its unclear defense line, i.e., no built-in security.

MANETs doesnt have committed switches, and its hubs generally work by sending the bundles to each other accordingly having no security in the correspondence, giving access to both true blue clients and aggressors [2]. For instance, hub S can speak with hub D by utilizing the most brief way S–A–BD as appeared in Fig. 1 (the dashed lines demonstrate the immediate connections between the hubs). In the event that hub A moves out of hub S range, hub A needs to locate an option course to hub D (S–C–EB–D). Accordingly, security in MANETs is the most essential worry for the fundamental usefulness of the system. The accessibility of system administrations, privacy, and trustworthiness of the information can be accomplished by guaranteeing that security issues have been met. MANETs regularly experience the ill effects of security assaults in view of its highlights like

Fig. 1 Communication between nodes in MANETs

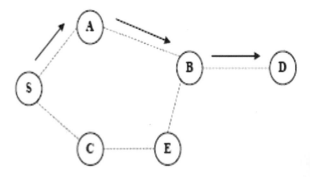

open medium, progressive topology change, absence of focal checking and administration, helpful calculations, and no unmistakable barrier system. These elements change the combat zone circumstances for MANETs against all sorts of security dangers [3].

A MANET is more open to these sorts of assaults as correspondence depends on shared trust between the hubs; there is no essential issue for arranging administration, no approval office, energetically changing topology, and restricted assets.

The main objective of the proposed method is to implement flexible network structure for MANETs environment by accurately detecting the attacks and preventing the unwanted packet dropouts. This is ensured by injecting jellyfish attack into the environment which is detected via genetic fuzzy rule based system which identifies the attack presence. The prevention of attack is done via trust-aware routing in which routing is performed via trustworthy nodes in the environment by proper means of authentication before establishing the routing path.

1.1 Structure of the paper

The rest of the paper is organized as follows: Section 2 outlines literature review of varied proposed methodologies by other researchers. Section 3 highlights the proposed research method along with suitable examples. Section 4 covers simulation based performance evaluation of proposed method and compares the novel proposed method with existing techniques. Section 5 concludes the paper with future scope.

2 Related Works

In this segment, shifting-related research strategies have been examined in detail in light of their working methodology regarding assault identification. Andel et al. [4] have characterized the undetectable hub assault and turned out to be not quite the same as the current assaults (man in the center, disguising, and wormhole) and set up its uniqueness. The authors characterized it as, in any convention that relies upon recognizable proof for any usefulness, any hub that successfully takes an interest in that convention without uncovering its personality is an imperceptible hub, and the activity and convention effect is named as INA. Considering the impacts of INA on various directing conventions, authors demonstrated it as unsolvable assault until now.

SAODV directing convention is utilized to avert against dark gap assault; however, it requires overwhelming weight encryption calculation [5]. SAR can be utilized to safeguard against dark gap assaults. In SAR, it needs intemperate encryption and decoding at each jump. ARAN can be utilized to guard against pantomime and renouncement assaults. It may not protect against validated egotistical hubs. Security convention SEAD is utilized against change assaults [6].

Di Crescenzo et al. [7] secure the administration disclosure stage by utilizing a safe different ruling set creation convention and the administration arrangement stage by utilizing a novel sort of edge signature plot. The two conventions address novel security objectives and are of free enthusiasm as they can discover applications to different ranges, most strikingly, the development of a conveyed and survivable open key framework in MANET.

Kim et al. [8] utilized the transmission control convention (TCP) to give solid information regarding transmission that has been performed with the end goal of smooth mix with the wired web. The creators propose their TCP-Vegas-impromptu convention, which is made mindful of RCs and utilizes the right BaseRTT esteems. Lee et al. [9] concluded that system coding in mix with single-bounce correspondence permits P2P document sharing frameworks in MANET to work in a more effective way and cause the frameworks to run without any hiccup. For example, dynamic topology and irregular network and in addition different issues that have been ignored in past MANET P2P explores, for example, tending to, hub/client thickness, non-helpfulness, and temperamental channel.

El Defrawy et al. [10] address various issues emerging in suspicious area-based MANET settings by planning and breaking down a protection saving and secure connection state based steering convention (Caution). Alert uses hubs' present areas to safely disperse and develop topology depictions and forward information. With the guide of cutting-edge cryptographic strategies (e.g., amass marks), caution gives both security and protection highlights, including hub confirmation, information trustworthiness, secrecy, and immovability (following protection). Zhao et al. [11] proposed a hazard mindful reaction instrument to methodically adapt to the distinguished steering assaults. Our hazard mindful approach depends on a broadened Dempster–Shafer numerical hypothesis of proof presenting a thought of significance factors [13].

3   Secured Routing with the Concern of Attacks

1. Versatile specially appointed system is one of the most regular impromptu systems with part of the issues identified with blockage and steering. It is one of the answers to secure the transmission over the system. Security angles assume a critical part in application situations given the vulnerabilities innate in remote specially appointed system administration from the very truth that radio correspondence happens (e.g., in strategic applications) to steer, man in the center and expound information infusion assaults. Security has turned into an essential worry keeping in mind the end goal to give ensured correspondence between versatile hubs in a threatening domain. The proposed framework is going to plan an interruption location framework to identify the jellyfish assault infused into the framework [14].

2. This identification framework depends on fluffy rationale. An IDS framwork is proposed in which change is made via utilization of two variables, i.e., Bundle misfortune rate and information rate. The two elements utilize fluffy rationale

which is criticial thinking control framework. Fluffy rationale gives a straight-forward approach to touch base at a clear conclusion in light of obscure, questionable, loud, or missing data. We proposed a calculation which depends on above elements. In this calculation, initially, we characterize the system with N number of hubs and we set source hub to S and goal hub D and after that, we let the currect hub as source hub. We rehash the means until the point when the current hub is not equivalent to goal hub. In this now, we discover the rundown of neighboring hubs of current hub. We distinguish the parameters of each neighbor hub, i.e., bundle misfortune and information rate.

3.1 Jellyfish Attack Injection

Jellyfish assault is one of the refusals of administration assault and furthermore a kind of inactive assault which is hard to recognize. It produces delay before the transmission and gathering of information parcels in the system. Applications, for example, HTTP, FTP, and video conferencing, are given by TCP and UDP. Jellyfish assault irritates the execution of the two conventions. Jellyfish assaults are focused against shut circle streams. TCP has surely understood vul-nerabilities to postponement, drop, and misarrange the parcels. Because of this, hubs can change the arrangement of the bundles likewise drop a portion of the information parcels. The jellyfish aggressor hubs completely obey convention rules, and henceforth this assault is called as uninvolved assault.

This assault which takes after all TCP rules has trademark in which jellyfish hub lessens the throughput, by dropping some of the bundles or deferring a few parcels or reordering a few parcels. At the point when a malignant hubs dispatch sending dismissal assaults, it likewise may follow all steering systems. A malevolent hub propelling jellyfish assaults may keep dynamic in both course finding and bundle sending with a specific end goal to keep it from identification and conclusion, yet the pernicious hub can assault the movement by means of itself by reordering parcels, dropping parcels intermittently, or expanding nerves. The jellyfish assault is particularly destructive to the TCP movement in that agreeable hubs can scarcely separate these assaults from the system blockage. It focuses on TCP's blockage control component.

As appeared in Fig. 2, hub JF is a jellyfish, and hub S begins to speak with hub D after a way by means of the jellyfish hub is built up. At that point, the dissent of administration assaults propelled by hub JF will cause bundle misfortune and sever the correspon-dences between hubs S and D at the end. In our work, jellyfish occasional assault has been infused into the MANET condition for the enhanced system execution.

Intermittent dropping is conceivable due to snidely picked period by the evil hub. This sort of occasional dropping is conceivable at hand-off hubs. Assume that blockage misfortunes drive a hub to drop $\alpha\%$ of bundles. Presently, consider that

Fig. 2 Jellyfish attack
scenario

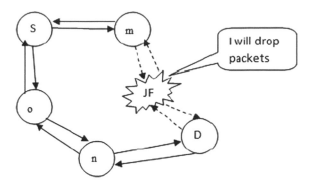

the hub drops α% of parcels intermittently, then TCP's throughput might be lessened to almost zero notwithstanding for little esteems.

3.2 Genetic Fuzzy Based Attack Detection

This technique involves the detection of attacker levels in the network layers using fuzzy logic technique. The following steps determine the fuzzy rule based interference:

- Fuzzification: This involves obtaining the crisp inputs from the selected input variables and estimating the degree to which the inputs belong to each of the suitable fuzzy sets.
- Rule evaluation: The fuzzified inputs are taken and applied to the antecedents of the fuzzy rules. It is then applied to the consequent membership function.
- Aggregation of the rule outputs: This involves merging of the output of all rules.
- Defuzzification: The merged output of the aggregate output fuzzy set is the input for the defuzzification process and a single crisp number is obtained as the output.

Initially, the fuzzy logic engine analyzes each layer, namely, the MAC layer, physical layer, and routing layer for the detection of abnormal behaviors. Then, the information gathered are stored in an attack database whose format is shown in Table 1.

Table 1 Attack database

Layer	Intrusion frequency (F)	Probability of successful attack (P)	Severity (S)
MAC Layer	F_1	P_1	S_1
Physical layer	F_2	P_2	S_2
Routing layer	F_3	P_3	S_3

Fig. 3 Fuzzy interference
system

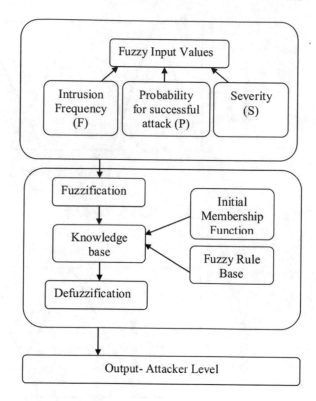

The parameters in the table are: Interruption recurrence (F). It is characterized as the assault force against the layer that is liable to checking. Its unit is in assaults/unit time.

Likelihood for effective assault (P): It portrays the strategy by which the aggressor handles to defeat the proactive controls. It varies in the range of (0–1).

Severity (S): It depicts the effect of an assault on the layer. The fluffy derivation framework is represented utilizing Fig. 3.

3.2.1 Fuzzification

This includes fuzzification of information factors, for example, interruption recurrence (F), likelihood of effective assault (P), and seriousness (S), and these data sources are given a degree to suitable fluffy sets. The fresh information sources are blend of F, P, and S. We take two potential outcomes, high and low for F, P, and S. Figures 4, 5, 6, and 7 demonstrate the participation work for the information and yield factors. Because of the computational proficiency and uncomplicated recipes, the triangulation capacities are used which are generally used for progressive applications. Likewise, a positive effect is offered by this outline of enrollment work.

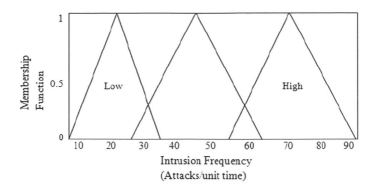

Fig. 4 Membership function of intrusion frequency

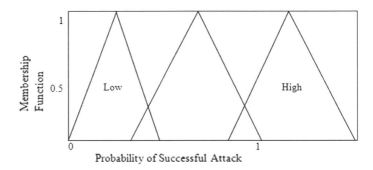

Fig. 5 Membership function of successful attack probability

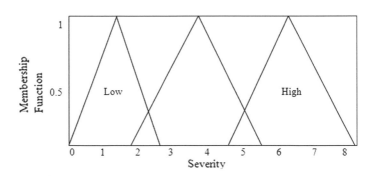

Fig. 6 Membership function of severity

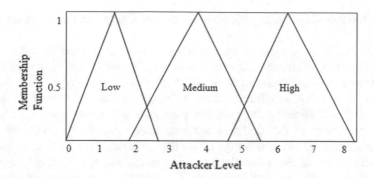

Fig. 7 Membership function of attacker level

Table 2 Fuzzy rules for the determining output

S. No	F	P	S	AL
1.	Low	Low	Low	Low
2.	Low	Low	High	Medium
3.	Low	High	Low	Medium
4.	Low	High	High	High
5.	High	Low	Low	Medium
6.	High	Low	High	High
7.	High	High	Low	High
8.	High	High	High	High

In Table 2, F, P, and S are given as sources of info and the yield speaks to the level of aggressor (AL) in every hub in the individual layer. In light of the aggressor level, the trust estimation of noxious hub is decreased (Clarified in Sect. 3.1). The eight fluffy sets are characterized with the blends introduced in Table 2.

Table 2 shows the composed fluffy induction framework. This outlines the capacity of the deduction motor and strategy by which the yields of each manage are consolidated to create the fluffy choice.

In the event that F, P, and S are low, at that point the attack level is low.
In the event that F and P are low, S is high, at that point the attack level is medium.
In the event that F and S are low, P is high, at that point the attack level is medium.
In the event that F is low, P and S are high, at that point the attack level is high.
In the event that F is high, P and S are low, at that point the attack level is medium.
In the event that F and S are high, P is low, at that point the attack level is high.
In the event that F and P are high, S is low, at that point the attack level is high.
In the event that F, P, and S are high, at that point the attack level is high.

3.2.2 Genetic-Based Rule Selection to Reduce Computation Overhead

An overabundance number of standards may not deliver great execution and it makes hard to comprehend the model conduct. To choose and tune a minimized arrangement of fluffy affiliation rules with high characterization precision from the manage base, a GA is utilized, where rules depend on the semantic two-tuple portrayal. The emblematic interpretation parameter of a semantic term is a number inside the interim $[- 0.5, 0.5)$ that communicates the area of a mark when it is moving between its two parallel names. In the event that S is set of marks speaking to a fluffy parcel, at that point, there is a couple $(S_i, \alpha i)$, $S_i \in S$, $\alpha i \in [- 0.5, 0.5)$. The CHC approach makes utilization of an interbreeding aversion instrument and a restarting procedure to energize assorted variety in the populace, rather than the notable transformation administrator. This inbreeding aversion instrument will be considered keeping in mind the end goal to apply the hybrid administrator, i.e., two guardians are crossed if their hamming separation isolated by 2 is more than a foreordained edge L. This edge esteem is instated as the most extreme conceivable separation between two people (the quantity of non-coordinating qualities in the chromosome) isolated by 4. Following the first CHC plot, L is decremented by 1 when there are no new people in the populace in one era. Keeping in mind the end goal to make this technique free of the quantity of qualities in the chromosome, for this situation, L will be decremented by $\varphi\%$ of its underlying worth (where φ controlled by the client, typically 10%). At the point when L is beneath zero, the calculation restarts the populace. Plan of this GA is as per the following:

Codification and beginning quality pool: To consolidate the govern determination with the worldwide horizontal tuning, a twofold coding plan for both run choice CS and sidelong tuning CT is utilized. For the CS part, every chromosome is a parallel vector that decides when a govern is chosen or not (alleles "1" and "0" individually).

Chromosome assessment: To assess a decided chromosome punishing countless, arrangement rate is figured and the wellness work is expanded. This capacity must be in the agreement with the system of imbalanced datasets. Along these lines, the normal of total of effectively ordered preparing designs by the standards in the chromosome part CS is utilized as wellness work.

$$\text{Fitness}(C) = \frac{\sum_{i=1}^{N_{rs}} \text{NCP}_{(R_i)}}{N_{rs}}$$

where N_{rs} is the number of rules in the rule set and NCP(Ri) is the number of correctly classified training. In the event that there is no less than one class without chose rules or if there are no secured designs, the wellness estimation of a chromosome will be punished with the quantity of classes without choosing rules and the quantity of revealed designs.

Hybrid administrator: The hybrid administrator will rely upon the chromosome part where it is connected. In the CS part, the half-uniform hybrid plan (HUX) is utilized. The HUX hybrid precisely trades the mid of the alleles that are diverse in the guardians (the qualities to be crossed are arbitrarily chosen from among those that are distinctive in the guardians). This administrator guarantees the most extreme separation of the posterity to their folks (investigation).

Restarting approach: To make tracks in an opposite direction from neighborhood optima, a restarting approach has been utilized. For this situation, the best chromosome is kept up, and the remaining are created aimlessly. The restart technique is connected when the limit esteem L is beneath zero, which implies that every one of the people existing together in the populace is fundamentally the same.

3.2.3 Defuzzification

The system by which a fresh esteems are removed from a fluffy set as a portrayal esteem is alluded to as defuzzification. The centroid of region plot is mulled over for defuzzification amid fluffy basic leadership process. The recipe (1) portrays the defuzzifier strategy.

$$
\text{Fuzzy\_cost} = \left[\sum_{\text{all rules}} z_i * \lambda(z_i) \right] / \left[\sum_{\text{all rules}} \lambda(z_i) \right]
$$

where fuzzy_cost is utilized to determine the level of basic leadership, z_i is the fluffy all tenets, and variable $\lambda(z_i)$ is its enrollment work. The yield of the fluffy cost work is adjusted to fresh an incentive according to this defuzzification technique.

3.3 Secured Routing Using Trusted Nodes

In this system, we consider swarm insight in view of insect state enhancement (ACO) method for performing confirmed directing. This procedure includes two insect operator to be specific forward subterranean insect (FA) and in reverse insect (BA) [12]. The means associated with this calculation are as per the following.

Stage 1 When source (S) needs to transmit the information parcel to goal (D), it dispatches FA with a limit put stock in esteem (T_{th}) connected with it.

Stage 2 The versatility of FA going to every N_i depends on probabilistic choice run shown in [14].

$$
P_r(N_i, S) = \begin{cases} \dfrac{a(N_i,S)^{\xi}.[b(N_i,S)]^{\sigma}}{\sum_{N_i \in N_n}[s(N_1,S)]^{\xi}.[b](N_i.S)^{\sigma}} & \text{if } r0, \text{otherwise} \\ 0, \text{otherwise} \end{cases}
$$

where $a(N_i, S)$ i represents pheromone value.
$b(N_i, S_o)$ represents the bandwidth related heuristic value.
N_R represents the receiver node.
RT (N_i) represents the routing table for N_i.

ξ and σ are the parameters that control the relative weight of the pheromone and heuristic value, respectively.

Step 3   FA travels through N_i using the control portrayed in stage 2 and confirms whether the trust estimation of the went to hub is more prominent than the trust edge esteem.

If $T_i > T_{th,}$ then
FA continues its path and keeps updating the routing table until it reaches D
Else if $T_i < T_{th}$ Then
The node is omitted from getting updated in the routing table.
End if

Step 4   Each FA deposits a quantity of pheromone ($\Delta\tau^u(r)$) in the visiting N_i as per the following equation:

$$\Delta\tau^u(r) = \frac{1}{X_s^u(r)}$$

where $X_s^u(r)$ represents the total number of N_i visited by FA during its tour at iteration r and $u = 1, 2, \ldots, n$.

Step 5   At the point when FA achieves D, BA is produced and the whole data gathered by FA is exchanged for BA.
Step 6   The BA at that point takes an indistinguishable way from that of its relating forward subterranean insect, yet the other way. It refreshes the pheromone table with the confide in estimation of the separate N_i.
Step 7   Once S gets the BA, it gathers the directing data about all N_i along every way from its refreshed pheromone table.
Step 8   From the gathered data, S picks the course with dependable hubs for information correspondence.

4   Experimental Results

This section enlists all simulation paramaters utilized for creating MANET scenario for evaluating the proposed methodology for combating Jellyfish attack. Ubuntu 17.04 is utilized as the working framework since it is easy to use which makes it simple to oversee. All the testing of the proposed method is performed on NS-2.35 simulator. In Table 3, we portray MANET parameters that are utilized as a part of this reenactment to quantify its execution and contrast it and distinctive conventions over a MANET organize. In the reproduction, we examine the connection between various MANET execution parameters regarding bundles' size.

Table 3 Parameters used in simulation

Parameter	Value
Operating system	Ubuntu 17.04
NS2-2 version	2.35
Channel type	Wireless channel
Number of nodes	100
Speed	3, 5, 7, 10, 20, 25
Data type	UDP
Simulation time	160 s
MAC protocol	MAC/802.11
Data packet size	100, 300, 500, 700, 800, 1000, 1500 and 2000
Area of simulation	700 * 700
Radio propagation model	Two-ray ground
Routing protocol	AODV/DSR

The proposed authenticated routing based attack injection and detection framework using genetic fuzzy rule based system (AR-AIDF-GFRS) is compared with the existing techniques like artificial bee colony (ABC), and memetic artificial bee colony (MABC) algorithm and performance is measured in terms of throughput, packet delivery ratio (PDR), end-to-end delay (E2E), routing efficiency, routing overhead (RO), and so on. The performance and results of the routing algorithm are as follows.

4.1 Throughput

The throughput is the number of bytes transmitted or received per second. The throughput is denoted by T,

$$\text{Throughput} = \text{received node/simulation time}$$

$$T = \frac{\sum_{i=1}^{n} N_i^r}{\sum_{i=1}^{n} N_i^s} \times 100\%$$

where N_i^r is the average receiving node for the ith application, N_i^s = average sending node for the ith application, and n = number of applications.

Figure 8 shows throughput comparison results of the attack detection algorithms such as ABC, MABC, and AR-AIDF-GFRS. From these figures, it concludes that the proposed AR-AIDF-GFRS based AD algorithm has improved throughput as compared to ABC and MABC. Based on Fig. 8, it is observed that AR-AIDF-GFRS performs better when the number of nodes increase and provides stable path from source to destination. It demonstrates that the number of Mbps transmitted from source to destination has increased when using AR-AIDF-GFRS based AD algorithm.

Fig. 8 Throughput
comparison results of AODV
protocol

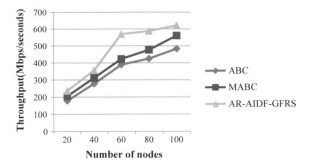

Number of nodes

No. of nodes	Throughput (Mbps/Seconds)-AODV		
	ABC	MABC	AR-AIDF-GFRS
20	178	205	236
40	278	312	356
60	389	423	569
80	425	476	587
100	483	561	621
Avg	350.6	395.4	473.8

Table 4 Throughput comparison results of AODV protocol

The values of these algorithms are tabulated in Table 4. Table 4 outlines average throughput results of AR-AIDF-GFRS based AD algorithm i.e. 473.8 Mbps/s for AODV protocol, 295.4 Mbps/s for MABC algorithm and 350.6 Mbps/s for ABC algorithm.

4.2 Packet Delivery Ratio (PDR)

It can be measured as the proportion of the received packets by the receiver node as compared to the packets transmitted by source node.

$$PDR = (\text{number of received packets} / \text{number of sent packets}) * 100$$

$$T = \frac{\sum_{i=1}^{n} \left(N_i^s - N_i^r\right)}{\sum_{i=1}^{n} N_i^s} \times 100\%$$

Figure 9 shows the packet delivery ratio of the proposed AR-AIDF-GFRS based AD algorithm, compared with existing optimization algorithms. Since the more number of the attacks has been detected, those routes are removed from original routing table. AODV increases the PDR of the proposed system and slightly decreases if the number of nodes increases. It shows that the number of packets transmitted from source to destination has increased for AR-AIDF-GFRS based AD

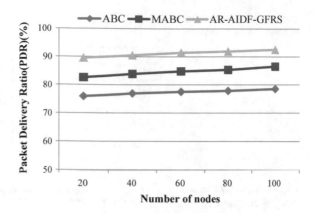

Fig. 9 PDR comparison results of AODV protocol

algorithm. The values of these algorithms are tabulated in Table 5. Table 5 shows that the proposed AR-AIDF-GFRS based AD algorithm produces average PDR results of 91.176% for AODV protocol, whereas the average PDR results of MABC and ABC are 84.61 and 77.344%, respectively.

4.3 Dropped Packets Ratio

It can be measured as the ratio of the number of packets that sent by the source node that fails to reach the destination node.

$$\text{Dropped packets} = \text{sent packets} - \text{received packets}$$

$$T = \sum_{i=1}^{n}\left(N_i^s - N_i^r\right) - \sum_{i=1}^{n} N_i^s$$

Figure 10 shows the drop packet ratio of the proposed AR-AIDF-GFRS based AD algorithm, compared with existing optimization algorithms. From the results, it concludes that the proposed AR-AIDF-GFRS based AD algorithm has less number of dropped packets as compared to other existing algorithms for both routing protocols, since the number of attacks detected in the proposed work is high. Those routes have been removed from original routing table and thereby reduces the number of dropped packets. It shows that the number of packets transmitted from source to destination has been higher. The values of these algorithms are tabulated in Table 6. Table 6 shows that the proposed AR-AIDF-GFRS based AD algorithm produces drop packets ratio results of 7.47% for 100 number of nodes in the AODV protocol, whereas the drop packets ratio results of MABC and ABC are 13.48 and 21.48%, respectively.

Table 5 PDR comparison results of AODV protocol

No. of nodes	Packet Delivery Ratio (PDR) (%)-AODV		
	ABC	MABC	AR-AIDF-GFRS
20	75.89	82.56	89.52
40	76.93	83.87	90.51
60	77.52	84.79	91.45
80	77.86	85.31	91.87
100	78.52	86.52	92.53
Avg	77.344	84.61	91.176

Fig. 10 Drop packets ratio comparison results of AODV protocol

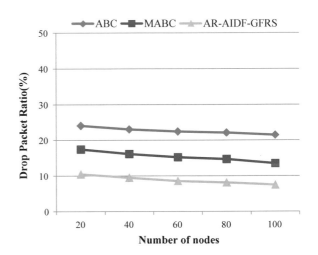

4.4 End-to-End Delay (E2E)

It represents the time required to move the packet from the source node to the destination node.

E-2-E delay [packetid] = received time [packetid] − sent time [packetid]

The average E2E can be calculated by summing the times taken by all received packets divided by its total numbers

$$D = \frac{\sum_{i=1}^{n} d_i}{n}$$

where d_i = average end-to-end delay of node of ith application and n = number of application.

Figure 11 shows the E2E delay performance comparison results of proposed AR-AIDF-GFRS based AD algorithm, compared with other existing optimization algorithms. From the results, it concludes that the proposed AR-AIDF-GFRS based

Table 6 Average drop packets ratio comparison results of AODV protocol

No. of nodes	Drop packets ratio (%)-AODV		
	ABC	MABC	AR-AIDF-GFRS
20	24.11	17.44	10.48
40	23.07	16.13	9.49
60	22.48	15.21	8.55
80	22.14	14.69	8.13
100	21.48	13.48	7.47

Fig. 11 End-to-end delay (E2E) comparison results of AODV protocol

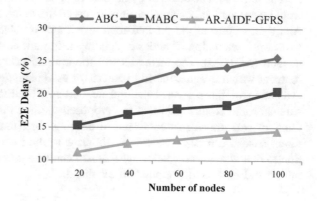

Table 7 Average E2E delay comparison results of AODV protocol

No. of nodes	E2E Delay (%)-AODV		
	ABC	MABC	AR-AIDF-GFRS
20	20.56	15.36	11.25
40	21.43	16.98	12.58
60	23.58	17.87	13.17
80	24.15	18.46	13.95
100	25.63	20.51	14.47
Avg	23.07	17.836	13.084

AD algorithm has lesser E2E when compared to existing algorithms for both routing protocols. It shows that the number of packets transmitted from source to destination has been higher. The values of these algorithms are tabulated in Table 7. Table 7 shows that the proposed AR-AIDF-GFRS based AD algorithm produces average E2E delay of 13.084% for AODV protocol, whereas the average E2E delay results of MABC and ABC are 17.836 and 23.07%, respectively.

5 Conclusion

In this work, detecting a malicious node and launching new optimization algorithm may lead to serious security concerns. The proposed scheme uses new method, namely, authenticated routing based attack injection and detection framework using genetic fuzzy rule based system (AR-AIDF-GFRS). The proposed research assures successful detection of Jellyfish attack and secured routing via trusted nodes. In this work, initially, jellyfish attack would be injected into the MANET environment. This attack would be detected by introducing genetic fuzzy based rule system which would generate various number of rules based on which attack would be identified. And then to ensure the secured routing without involvement of intruders, in this work trust evaluation of nodes is done by using ant colony based trust evaluation method (ACTEM). The method select the optimal nodes from the MANET environment which is more trusted in nature for establishing the route path. The overall evaluation of the proposed method i.e. AR-AIDF-GFRS is done using NS-2.35 simulator and results state that the proposed method outshines in every aspect in terms of performance as compared to existing algorithms. In future scenario, different attacks can be considered for improving the network performance. In addition to that attacks, prevention mechanism can be introduced to avoid the network failure and unwanted computation overhead.

References

1. Deng, H., Li, W., & Agrawal, D. P. (2002). Routing security in wireless ad hoc networks. *IEEE Communications Magazine, 40,* 70–75.
2. Yang, H., Luo, H., Ye, F., Lu, S., & Zhang, L. (2004). Security in mobile ad hoc networks: challenges and solutions. *IEEE Wireless Communications, 11,* 38–47.
3. Karpijoki, V. (2000, December). Security in ad hoc networks. In *Proceedings of the Helsinki University of Technology, Seminars on Network Security,* Helsinki, Finland.
4. Andel, T. R., & Yasinsac, A. (2007). The invisible node attack revisited. *Proceedings of IEEE SoutheastCon, 2007,* 686–691.
5. Hu, Y., Perrig, A., & Johnson, D. (2002). Ariadne: A secure on-demand routing for ad hoc networks. In *Proceedings of MobiCom 2002,* Atlanta.
6. Zhang, Y., & Lee, W. (2000). Intrusion detection in wireless ad-hoc networks. In *Proceedings of the Sixth Annual International Conference on Mobile Computing and Networking (MOBICOM),* Boston.
7. Di Crescenzo, G., Telcordia Technol, N. J., Ge, R., & Arce, G. R. (2006). Securing reliable server pooling in MANET against byzantine adversaries‖. *IEEE Journal on Selected Areas in Communications, 24,* 357–369.
8. Kim, D., Bae, H., & Toh, C. K. (2007). Improving TCP-Vegas performance over MANET routing protocols. *IEEE Transactions on Vehicular Technology, 56*(1), 372–377.
9. Lee, U., Park, J.-S., Lee, S.-H., Ro, W. W., Pau, G., & Gerla, M. (2008). Efficient peer-to-peer file sharing using network coding in MANET. *IEEE Journal on Communications and Networks, 10,* 422–429.
10. El Defrawy, K., & Tsudik, G. (2010). ALARM: Anonymous location-aided routing in suspicious MANETs. *IEEE Journal on Mobile Computing, 10,* 1345–1358.

11. Zhao, Z. Hu, H., Ahn, G.-J., & Wu, R. (2011). Risk-aware mitigation for MANET routing attacks. *IEEE Journal on Dependable and Secure Computing, 9*, 250–260. (Security Eng. for Future Comput. Lab., Arizona State Univ., Tempe, AZ, USA).
12. Kaur, M., Sarangal, M., & Nayyar, A. (2014). Simulation of jelly fish periodic attack in mobile ad hoc networks. *International Journal of Computer Trends and Technology (IJCTT), 15*.
13. Kaur, M., & Nayyar, A. (2013). A comprehensive review of mobile adhoc networks (MANETS). *International journal of emerging trends & technology in computer science (IJETTCS), 2*(6), 196–210.
14. Kaur, M., Rani, M., & Nayyar, A. (2014, September). A novel defense mechanism via Genetic Algorithm for counterfeiting and combating jelly fish attack in mobile ad-hoc networks. In *2014 5th International Conference Confluence The Next Generation Information Technology Summit (Confluence)* (pp. 359–364). IEEE.

Firefly Swarm: Metaheuristic Swarm Intelligence Technique for Mathematical Optimization

Gopi Krishna Durbhaka, Barani Selvaraj and Anand Nayyar

Abstract In this paper, we briefly reviewed the firefly algorithm fundamentals and its experimentation with diverse applications, highlighting its performance in engineering research and industrial applications in specific to machinery extracting the features to confine the deformities.

Keywords Swarm intelligence · Metaheuristics · Firefly algorithm
Optimization · Feature extraction · Prognostics and diagnostics
Fault diagnostics

1 Introduction

Optimization is an important tool in making decisions and in analyzing the systems. Optimization is regarded as finding the best solution among a set of all feasible solutions. Most of the methods developed or introduced by the mathematicians and engineers for solving the optimization problems were based on the behavior of the insects or animals, that worked together in order to solve complex problems.

Stochastic algorithms can be defined in two types, heuristic and metaheuristic. Heuristic means to learn or to discover by trial and error and hence do not guarantee the solution to be an optimal one. The performance of metaheuristics is good in comparison with a heuristic algorithm based on its search process i.e. randomization instead of local search [1]. The computational downsides of existing numerical

G. K. Durbhaka (✉) · B. Selvaraj
Department of Electronics and Control Engineering, Satyabhama University,
Chennai, India
e-mail: gopikrishna.durbhaka@gmail.com

B. Selvaraj
e-mail: baraniselvaraj77@gmail.com

A. Nayyar
Graduate School, Duy Tan University, Da Nang, Vietnam
e-mail: anandnayyar@duytan.edu.vn

© Springer Nature Singapore Pte Ltd. 2019
V. E. Balas et al. (eds.), *Data Management, Analytics and Innovation*,
Advances in Intelligent Systems and Computing 839,
https://doi.org/10.1007/978-981-13-1274-8_34

strategies have constrained scientists to depend on metaheuristic algorithms in view of simulations to tackle engineering optimization issues [2, 3].

Yang proposed Firefly algorithm and is based on fireflies flashing behavior. Firefly swarm optimization can be utilized in computer science to create an optimization solution to either incomplete or imperfect solution in an optimal manner.

There had been different types of optimization methods in light of systems and frameworks inspired by the nature. Various issues exist in our everyday life, which are hard to unravel by the traditional methods and approaches due to their restrictions. Hence, numerous analysts and researchers have moved their concentration from conventional methods for nature roused approaches to take care of these issues. The swarm algorithms have their own strategies and standards to determine issues. For this one ought to comprehend the nature standards, principles and instruments of working. The way toward outlining keen frameworks through nature motivation has the accompanying stage: understanding the nature procedure, plan examples of nature handle, ID of analogies and mechanical demonstrating for the issue.

One of the key components of nature motivated frameworks is looking through the best arrangements in enhancement space. More often than not, enhancement handle needs emphases of working sessions. Scrounging conduct in ants and bees frameworks, birds flocking, fishes, chameleon, genetic systems, fireflies, cuckoo, eagles and penguins.

Organization of the paper

Section 2 states in brief enlisting the different types of swarm intelligence algorithms along with their motivation and behavior. Section 3 states in brief about the firefly algorithm, its working principle, pseudo code, structure of firefly algorithm and its advantages. Section 4 states the different areas in specific to industrial applications where in firefly algorithm had been applied with a review of few case studies and evaluations performed till date. Section 5 state the conclusion and future scope of this review study.

2 Motivation and Behavior of Swarm Optimization Algorithms

There have been multiple different types of optimization algorithms inspired by nature inspired species and they have been listed in Table 1 [4, 5].

Firefly algorithm is compared with Particle Swarm Optimization (PSO) and Bee Colony optimization (BCO) on varied parameters [6]. Thirteen typical benchmark data sets are taken as base parameters for performing comparisons. From the results obtained, on comparison with the performance it had been concluded that for clustering purposes Firefly algorithm is highly efficient and accurate.

Table 1 Motivation and behavior of different swarm optimization algorithms

Evolutionary optimization algorithm	Motivation and behavior
Ant colony optimization	The behavior of ants wandering randomly laying the pheromone and seeking an optimal route between their colonies and food sources
Bee colony optimization	It primarily mimics the behavior of honey bees in food collection and is based on population search algorithm
Particle swarm optimization	The simulated social behavior of birds flocking together, fish schooling is the motivation here
Firefly algorithm	Fireflies operation, that tend to the light source and interact with light emitted by other brighter partner moving towards it, in proportion to the mutual distance. The main functions are: Attraction of mating partners and possible prey
Cuckoo search algorithm	The behavior of cuckoo species and their approach of eggs laying in other birds nests
Cockroach swarm optimization	The behavior of cockroaches looking for food and how they acquire food, in a scattered method and also escaping from light is the motivation here
Eagles strategy algorithm	The foraging behavior of eagles is the inspiration. An eagle flies randomly in respective territory. The eagle keeps changing its search strategy till the prey is sighted with different chasing tactics efficiently. The random search and intensive chase by locking its aim on the target are key motivations that shall be considered here
Bat algorithm	The echo location behavior of micro bats with varying pulse rates of emission and loudness had been inspired as a metaheuristic algorithm, widely applied for global optimization

3 Firefly Algorithm

3.1 Working Principle

Firefly algorithm is a type of swarm intelligence technique which has been inspired by the behavior and flashing pattern of the fireflies formed its basis. Short and rhythmic flashes for correspondence and pulling in the potential chase are utilized by the majority of the fireflies. Yang [7] introduced this firefly algorithm. Firefly takes a shot at three rhapsodize rules.

1. Fireflies are Unisex, with the prime objective that one firefly will be pulled into different fireflies irrespective of their sex.
2. Attractiveness is proportional to their brightness. Any two fireflies flashing, the less flashing firefly will be attracted towards the brighter one. The attractiveness is proportional to the brightness and they both decrease as their distance

increases. On the off chance if there is no brighter one than a specific firefly, it will move randomly.

3. Objective function determines the brightness of Firefly.

3.2 Pseudo Code of the Firefly Algorithm

Firefly algorithm [7] has been highlighted in Fig. 1.

3.3 Structure of Firefly Algorithm

As we know, the attractiveness is based on the light intensity of the firefly. The light intensity and attractiveness is inversely proportional to the particular distance from the light source; which means, as the distance increases the light and attractiveness decreases. Variation of light intensity and the formulation of the attractiveness are the two important issues in the firefly algorithm. Brightness is associated with the objective function of the optimization problem and attractiveness of a firefly is determined by the brightness. The functionality of the algorithm has been showcased it as a flowchart [8] in Fig. 2.

```
Firefly algorithm
Objective function of f (x), where x=(x1,......,xd) ^T
Generate initial population of fireflies or xi (i=1, 2...n)
Define light intensity of Ii at xi via f(xi)
Define light absorption coefficient γ
While (t< MaxGeneration)
    For i=1 to n (all n fireflies)
        For j=1 to n (all n fireflies)
            If (Ii>Ii)
                    Move firefly i towards j
                    Attractiveness varies with distance r via exp[−γr]
            end if
            Evaluate new solutions and update light intensity
        End for j
    End for i
Rank the fireflies and find the current best
End while
Post process results and visualization
```

Fig. 1 Pseudo code of the firefly algorithm

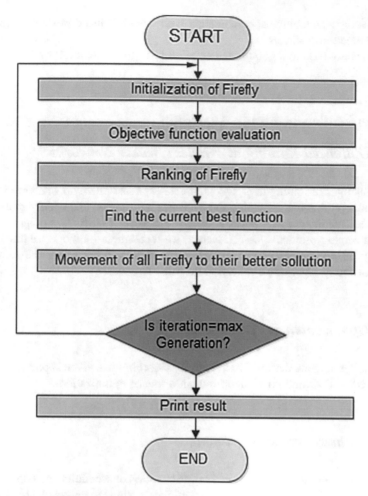

Fig. 2 Flowchart of the firefly algorithm

3.4 Advantages

a. Firefly calculation can manage profoundly nonlinear, multi-modular improvement issues normally and productively.
b. It does not utilize velocity, and there is no issue as that related with speed in PSO.
c. The speed of joining of firefly algorithm is high in the likelihood of finding the optimized solution.

d. It has the adaptability of combination with other advanced procedures to frame half breed instruments.

e. It does not require a good initial solution to begin its iteration cycle procedure.

4 Application of Firefly Algorithm

4.1 Optimized Routing in Wireless Sensor Networks

The challenges of routing in protocols in wireless sensor networks had been widely solved by optimizing their routes. One of the most common and widely applied area of swarm intelligence is "Wireless Sensor Networks (WSN)" to propose energy efficient routing protocols [9–11]. Hence, we focus more on other use cases rather than optimized routing in wireless sensor networks to recognize the scope and extent of firefly algorithm application in different zones.

4.2 Optimization in Antenna Design

Considering antenna design [12, 13], firefly algorithm has given supreme results as compared to PSO and ABC algorithms in terms of optimization.

4.3 Optimization in Scheduling

Firefly algorithm can also be applied in various sorts of scheduling and optimization problems like Traveling Salesman Problem with optimized approach [14, 15]. And further, can also be used to solve flow shop scheduling problems by discretizing firefly algorithm [16].

Manufacturing companies from various segments and sectors of the economy are waiting for solutions in numerous territories of their activities. The paper [17] highlights varied applications of firefly and cockroach algorithms to optimize two queuing systems and flow shop scheduling problems with the objective of make-span minimization.

In another research, [18] a novel distributed version of the firefly algorithm known as Firefly Colony Optimization (FCO) had been proposed which combines luminescent and foraging behavior of fireflies as well as ants. The FCO had been very efficient for solving the combinatorial and discrete optimization problems. The FCO model was applied to solve several challenges in packaging. As stated in the research, the results were very encouraging. FCO obtained optimal encouraging results with the ability of quick convergence. Henceforth, FCO was found to be more effective for discrete optimization as mentioned in this study.

4.4 Optimization of Paths in Indoor Positioning and Localization Systems

In large manufacturing factory warehouses and stock areas, the whereabouts of assets needs to be identified so as to reach them and perform respective work operations and conduct maintainence works, whenever required.

The significant drawback of GPS route is that it does not work inside, i.e. inside any building, mall, factories, and so on. With the help of Beacons, Indoor positioning can be implemented.

Estimating and determining the optimal location of the asset or human beings based on the Received Signal Strength Indication (RSSI) routing of beacons along with triangulation and trilateration approaches. Firefly algoithm helps to optimize the path navigation system to facilitate indoor localization.

4.5 Feature Extraction in Digital Image Processing

In the area of Digital image processing, swarm intelligence based techniques are applied to feature extraction. The correlation between adjacent image focuses for extraction of features from the images that has been proposed utilizing the perceptual chart. The difficult step in the processing of the image is Image segmentation i.e. clustering aims at extracting the information from the image. In this case study [19] the k-Means and firefly algorithms have been grouped to cluster image pixels into k cluster for segmentation. In general k-means clustering algorithm gets trapped in local optima during the computation of the centroid and hence here it is optimized applying the firefly algorithm.

4.6 Feature Selection and Detection of Fault Diagnostics and Localizing the Defects

In a study [20] of a tuning approach to obtain the Proportional Integral Derivative (PID) controller parameters in an Automatic Voltage Regulator system (AVR) continuous firefly algorithm had been applied. On final evaluation and comparison of both the methods PSO and CFA, the new CFA tuning results had better performance of the AVR control system in terms of time domain specifications and set-point tracking.

A fault diagnosis study of bearings for fault prediction was performed [21] wherein firefly algorithm had been deployed to optimize the SVM kernel function to improve the classification accuracy.

A contextual investigated study wherein data clustering was performed by a modified model of k-means algorithm along with the firefly algorithm in a hybridized approach known as k-FA. This k-FA had been proposed [22] to optimize the centroid of the k-Means algorithm further to improve the efficiency and accuracy of the algorithm. The experimental evaluation results of this study was performed over five UCI open source numerical data sets Iris, WDBC, Sonar, Glass and wine datasets with different models known as k-Means, PSO, k-PSO and k-FA. Finally, from the comparitive results of this study the conclusion was derived that k-FA outperformed k-means, PSO and k-PSO in accuracy and efficiency.

One of a research [23] states about the fault behaviour pattern analysis and recognition where the in-depth behaviour analysis had been performed and fault features of the rolling bearings have been extracted and grouped accordingly with a k-means algorithm. Here, the experimental evaluation of the data set had been performed and bearing faults have been categorized as inner racing, outer racing and ball racing. Another study [24] has been performed with a few more models based on a collaborative recommendation approach. Further study is in progress to optimize the computational analysis of similar models to take up the research further and avail quality results.

In order to obtain best possible power outcomes, Wind Turbines can be installed in open and unobstructed area. This study [25] attempts to solve the layout issues of Wind farms deployment via nature inspired techniques. It has been demonstrated that the firefly algorithm is a better technique for the optimization of large wind farms in comparison to genetic algorithms and by the use of spreadsheet methods such as finite difference method. This study also demonstrates the higher power derived and the lower cost of power produced by using the firefly algorithm.

5   Conclusion and Future Scope

This review paper discussed the significance of nature inspired metaheuristic model commonly known as firefly algorithm, illustrating its working flow both in pseudo code and flowchart. As per this survey, till now widely the published literature of swarm intelligence has been focussing on improving the protocols in specific to wireless sensor networks only. There is also a need to focus on the industrial application areas of how firefly algorithm can be applied herewith which have been very rarely focused upon, applied and implemented till date. There lies an enormous degree of development and advancement of efficient feature selection and capturing the behavior patterns of the industrial assets either machinery or devices to localize the defect that helps to enhance preventive and predictive maintenance.

5.1 Future Scope

Further study to optimize the computational analysis of the wind turbine diagnostics using different models such as k-means, k-Nearest Neighbour (k-NN), Support Vector machine (SVM) and neural networks using firefly algorithm is in progress. This will additionally confine the deformity more effectively by extracting the features. This detailed performance shall be evaluated showcasing the results in an effective ensemble approach.

References

1. Yang, X. S. (2010). *Engineering optimization: An introduction with metaheuristic applications*. John Wiley & Sons.
2. Blum, C., & Roli, A. (2003). Metaheuristics in combinatorial optimization: Overview and conceptual comparison. *ACM Computing Surveys, 35*(3), 268–308.
3. Yagiura, M., & Ibaraki, T. (2001). On metaheuristic algorithms for combinatorial optimization problems. *Systems and Computers in Japan, 32*(3), 33–55.
4. Yang, X. S. (2010). Firefly algorithm, stochastic test functions and design optimisation. *International Journal of Bio-Inspired Computation, 2*(2), 78. https://doi.org/10.1504/ijbic. 2010.032124.
5. Gheraibia, Y., & Moussaoui, A. (2013). Penguins search optimization algorithm (PeSOA). *Lecture Notes in Computer Science, 222–231.* https://doi.org/10.1007/978-3-642-38577-3_23.
6. Senthilnath, J., Omkar, S. N., & Mani, V. (2011). Clustering using firefly algorithm: Performance study. *Swarm and Evolutionary Computation, 1*(3), 164–171. https://doi.org/10. 1016/j.swevo.2011.06.003.
7. Yang, X. S. (2008). *Nature-inspired metaheuristic algorithms* (2nd ed.). Luniver Press.
8. Kumar, R., Talukdar, F., Dey, N., & Balas, V. (2016). Quality factor optimization of spiral inductor using firefly algorithm and its application in amplifier. *International Journal of Advanced Intelligence Paradigms*.
9. Nayyar, A., & Singh, R. (2014). A comprehensive review of ant colony optimization (ACO) based energy-efficient routing protocols for wireless sensor networks. *International Journal of Wireless Networks and Broadband Technologies (IJWNBT), 3*(3), 33–55.
10. Nayyar, A., & Singh, R. (2016). Ant colony optimization—computational swarm intelligence technique. In *2016 3rd International Conference on Computing for Sustainable Global Development (INDIACom)* (pp. 1493–1499). IEEE.
11. Nayyar, A., & Singh, R. (2017). Ant colony optimization (ACO) based routing protocols for wireless sensor networks (WSN): a survey. *International Journal of Advanced Computer Science and Applications, 8,* 148–155.
12. Basu, B., & Mahanti, G. K. (2011). Fire fly and artificial bees colony algorithm for synthesis of scanned and broadside linear array antenna. *Progress in Electromagnetics Research B, 32,* 169–190. https://doi.org/10.2528/pierb11053108.
13. Zaman, M. A., & Abdul Matin, M. (2012). Nonuniformly spaced linear antenna array design using firefly algorithm. *International Journal of Microwave Science and Technology, 2012,* 1–8. https://doi.org/10.1155/2012/256759.
14. Jati, G. K., & Suyanto. (2011). Evolutionary discrete firefly algorithm for travelling salesman problem. *Lecture Notes in Computer Science, 393–403.* https://doi.org/10.1007/978-3-642-23857-4_38.

15. Palit, S., Sinha, S. N., Molla, M. A., Khanra, A., & Kule, M. (2011). A cryptanalytic attack on the knapsack cryptosystem using binary firefly algorithm. In *2011 2nd International Conference on Computer and Communication Technology (ICCCT-2011)*. https://doi.org/10.1109/iccct.2011.6075143.
16. Sayadi, M. K., Ramezanian, R., & Ghaffari-Nasab, N. (2010). A discrete firefly meta-heuristic with local search for makespan minimization in permutation flow shop scheduling problems. *International Journal of Industrial Engineering Computations, 1*(1), 1–10. https://doi.org/10.5267/j.ijiec.2010.01.001.
17. Kwiecień, J., & Filipowicz, B. (2014). Comparison of firefly and cockroach algorithms in selected discrete and combinatorial problems. *Bulletin of the Polish Academy of Sciences Technical Sciences, 62*(4). https://doi.org/10.2478/bpasts-2014-0087.
18. Layeb, A., & Benayad, Z. (2014). A novel firefly algorithm based ant colony optimization for solving combinatorial optimization problems. *International Journal of Computer Science and Applications, 11*(2), 19–37.
19. Sharma, A., & Sehgal, S. (2016). Image segmentation using firefly algorithm. In *2016 International Conference on Information Technology (InCITe)—The Next Generation IT Summit on the Theme—Internet of Things: Connect Your Worlds*. https://doi.org/10.1109/incite.2016.7857598.
20. Bendjeghaba, O. (2014). Continuous firefly algorithm for optimal tuning of PID controller in AVR system. *Journal of Electrical Engineering, 65*(1). https://doi.org/10.2478/jee-2014-0006.
21. Thelaidjia, T., Moussaoui, A., & Chenikher, S. (2014). Support vector machine based on firefly algorithm for bearing fault diagnosis. In *International Conference of Modeling and Simulation (ICMS 14)*.
22. Hassanzadeh, T., & Meybodi, M. R. (2012). A new hybrid approach for data clustering using firefly algorithm and K-means. In *The 16th CSI International Symposium on Artificial Intelligence and Signal Processing (AISP 2012)*. https://doi.org/10.1109/aisp.2012.6313708.
23. Durbhaka, G. K., & Barani, S. (2016). Fault behaviour pattern analysis and recognition. In *2016 International Conference on Information Science (ICIS)*. https://doi.org/10.1109/infosci.2016.7845325.
24. Durbhaka, G. K., & Selvaraj, B. (2016). Predictive maintenance for wind turbine diagnostics using vibration signal analysis based on collaborative recommendation approach. In *2016 International Conference on Advances in Computing, Communications and Informatics (ICACCI)*. https://doi.org/10.1109/icacci.2016.7732316.
25. Massan, S.-R., Wagan, A. I., Shaikh, M. M., & Abro, R. (2015). Wind turbine micrositing by using the firefly algorithm. *Applied Soft Computing, 27*, 450–456. https://doi.org/10.1016/j.asoc.2014.09.048.

An Analysis of Cloud Computing Issues on Data Integrity, Privacy and Its Current Solutions

B. Mahalakshmi and G. Suseendran

Abstract Cloud computing refers to data sharing, storing of data in cloud storage and sharing resources to the client. While considering data it should be securely protected from unauthorized access. Everything regarding the data security is a top threatening issues, in this paper we discuss about the data integrity issues of the cloud. How the server maintained the data without any loss or any damage using some encryption techniques. In this paper integrity checking is considered as an issue where some unauthorized person is accessing the data. Here the authorization of the data is checked by the data owner or a TPA—third party auditing using RSA and MD5 cryptographic algorithm. The data owner is responsible for giving access permission for the data which it may be public or private for integrity checking. Here the proposed system tells about how the third party auditing checks the integrity to avoid the overwhelming workload for the data owner.

Keywords Data integrity · Data loss · Data possession · Cloud
TPA · Cryptographic algorithm

1 Introduction

Today cloud computing becomes our daily routine of using the data in the cloud, whether knowingly or unknowingly cloud becomes our part of our life. All type of organizations such as small to large is using cloud because it provides fast access

B. Mahalakshmi (✉)
Department of Computer Science, School of Computing Science,
VELS Institute of Science, Technology & Advanced Studies (VISTAS),
Chennai 600117, Tamilnadu, India
e-mail: maha.karthik921@gmail.com

G. Suseendran
Department of Information Technology, School of Computing Sciences,
VELS Institute of Science, Technology & Advanced Studies (VISTAS),
Chennai 600117, Tamilnadu, India
e-mail: suseendar_1234@yahoo.co.in

© Springer Nature Singapore Pte Ltd. 2019
V. E. Balas et al. (eds.), *Data Management, Analytics and Innovation*,
Advances in Intelligent Systems and Computing 839,
https://doi.org/10.1007/978-981-13-1274-8_35

and the cost is also reduced as it gives the option for pay for use. The computing power of cloud is also a business technique we can either buy or can be rented too. Cloud is having some delivery services which all the organizations are utilizing the services. The data is owned by some users are getting decreased because of the agility and responsiveness is getting increased. Nowadays organizations are concentrating on their business profits not on the infrastructure. Therefore the usage of cloud computing is increased and a vast development is seen on the cloud environment. So the requirements received from the organization, academics, and industrials as well as all sort of communities are developing each moment. The NIST-National Institute of Standards and Technology have a defined that cloud computing is a model for allowing ubiquitously with well-situated, network for on demand use for a shared pool of configurable computing possessions such as networks, servers, storage, service and applications which are quickly released and prerequisite with minimal service provider interaction or minimal management effort. The cloud computing is mainly having five characteristics, four deployment models and three service models. Apart from this it also provides memory, processor storage and bandwidth which are virtualized and a user can access through internet from any part virtually. There are lots of technology available such as virtualization, service oriented architecture and web 2.0, etc. Because of its massive usage there are many issues regarding the security is considered to be a concern. Cloud also provides data availability and retrieval in an efficient way. Resource optimization is also one of the responsibilities of cloud.

2 Literature Review

Chang et al. has proposed that CCAF is used for protecting the data in a multi-layered structure in real time and three layers of security such as identity management, firewall, and access control and convergent encryption method. Ethical hacking experiments are undertaken in this paper and result found too prove CCAF is highly secured when it with BPMN simulation for getting the better results. This results in the formation of integrated solution to the problem [1].

Yu et al. here the author proposed that the security against malicious server and the ID-based RDIC is done against a third party verifier. The ID based RDIC does not leak any information regarding the RDIC process. The proposed system and the implementation result analysis are proved securely that the ID based RDIC is trustworthy secured for real-time application in the cloud. Here the numerical analysis and the implementation are done effectively [2].

Ali et al. has given a detailed survey report regarding the security challenges in the cloud environment. The recent solutions for the security issues are discussed in this paper. Furthermore brief discussions about the cloud vulnerabilities are explained. In the conclusion part the detailed study about the cloud security and the future plan about which threat the further study is going to be taken is discussed there [3].

Mahalakshmi et al. has proposed that deduplication of data is eliminated in the cloud storage and the data are secured in the hybrid cloud storage. These are secured using the convergent encryption algorithm for protecting the data and the replication of data are controlled using the algorithm. By using some proposed methodology the cloud storage is secured and replication is found and the reducing the storage space. In future the encryption algorithm is changed for providing more security to the data [4].

Wang et al. explained about the data integrity in cloud storage while the data outsourcing is done. The integrity has to be checked with some special technique called verifying the Meta data that is the signatures of the data for the data owners. Hence the organization can have its own authentication method for checking the integrity and proving the PDP Meta data of its own. Here the term data privacy also is used for protecting the data from unauthorized access. The anonymous interruption of data integrity is avoided in this technique so the data owner can feel free to check the integrity whenever the outsourcing of data occurs [5].

Chen et al. provides an analysis on data security and protection of data privacy issues with cloud computing environment throughout the data life cycle. Here some current solutions are discussed. The privacy protection and data security protection are integrated together and the solution for both the problems is discussed [6].

3 Characteristics, Service Models and Deployment Models of Cloud Computing

3.1 Characteristics of Cloud

On demand self service is where the user can have the access to perform their task without the interruption of humans and with the cloud service provider. **Broad network access** is nothing but the user can access the resource from any type of platform such as laptop, mobile, etc. **Resource pooling**, resources are available in a pool like structure so that multiple users can access and share the resources. **Rapid elasticity** is means the resources are increased and decreased according to the needs. **Measured service** is used to know how much resource is consumed by the user and also for the cloud provider to know the amount of cloud storage used by the user for billing. Figure 1 shows the characteristics, service and deployment model of cloud.

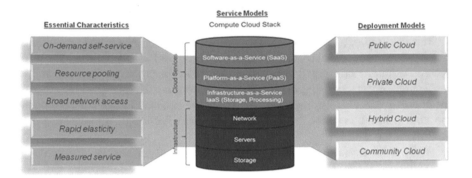

Fig. 1 Characteristic, service and deployment models

3.2 Service Models

3.2.1 Software as a Service (SaaS)

The cloud service provider provides a service to the client that is software with some cloud infrastructure and the client only can work with the software. Any changes cannot be made in the infrastructure as it is maintained by the cloud provider. All the controls over the setting of the software are under the CSP so that the client cannot modify it. Through the web browser only the client can able to access the data as a thin client.

3.2.2 Platform as a Service (PaaS)

This is same as SaaS where the platform is provided by the cloud service provider for deploying the data in the cloud environment. The client can able to install the software and deploy it using the tool provided by the CSP. Physical settings are under the control of the provider only the application side part is shared by the provider.

3.2.3 Infrastructure as a Service (IaaS)

In this cloud providers offer an infrastructure for the client to access the software and platform. It is mainly for the organization for having the own infrastructure provided by the cloud. They are having certain software and operating system to work in the cloud environment either physically or virtually this is done in the IaaS [7, 8].

3.3 Deployment Model

3.3.1 Private Cloud

The name itself implies that the cloud infrastructure is particularly owned for the organization or the individuals. The individual may be a third party provider or organization or an individual.

3.3.2 Community Cloud

This is for a group of people like an organization or some social group can able to access the resources in the cloud.

3.3.3 Public Cloud

This is a common one for all the people so anyone can access the resources available in the cloud without any permission required. This model is public so anyone can able to customize the data in the cloud storage.

3.3.4 Hybrid Cloud

The combination of all the above are said to be hybrid cloud. Here only the complication arises. Here all type of community is available do that the data should be protected for certain people and certain data is shown to all of the community.

4 Data Security Issues

Cloud computing is a pool of resource sharing and storage place for the data which is also done by the third-party provider. They are using different deployment models and service models in the cloud where all the data are not securely protected. While considering the security it is categorized into two. One is security issues faced by the cloud service providers and another is faced by the customer. But the responsibility is for the provider that is he has to ensure that the infrastructure of the cloud is safe and the application and the data of the customer are safe and protected. The confidentially, access control and data integrity are the three major concerns of the data security issues. Here we discuss about the data integrity issues [5, 7]. Figure 2 represent the security issues in cloud.

Fig. 2 Cloud security

4.1 Data Integrity Issues

In the cloud, data stored could face some security issues while outsourcing. Since the outsourced data and work based on the data are in remote location it is necessary to check whether the data and computation are secured for proving the data integrity. The data integrity is to safe the data from unauthorized user access that is modification and all. And computation integrity shows that the computational activities are analyzed and to be protected from the malware, attackers, etc., so that the program execution is maintained and the results are perfect. If any deviation occurs from normal execution of program it should be detected. Hence the computational and the data integrity should be verified at both levels. So that Data integrity leads to identify the lost data or to notify there is some exploitation occurs in data level. Below explains about the integrity issues in cloud [9].

4.2 Data Loss/Manipulation

The cloud is providing one of the services is storage as a service as there are huge numbers of data in cloud which are accessed daily or rarely. Whatever it is there is a need for protecting the data and keep as it is. The cloud is responsible for security of data which is outsourced in remote cloud but it is unreliable and unsecured. So there is a possibility for loss of data due to the unauthorized modification. In some cases the data are modified accidentally or intentionally by the attackers. Some administrative error also occurs during the attack that is recovering the incorrect data or loss of data will occur during the time of backup service. So the chance of utilizing the data from the cloud is easy for the attacker since the control over data is lost [7, 10, 11].

5 Existing System for Data Integrity

In the existing system the data owner is having the secret key to perform the task such as batch auditing of multiple tasks, multiuser modification etc. and all the other users are having the access permission from the data owner can able to read the data. In this case the data owner has to be online for every moment so that whatever changes or modification is done by the user can be upgraded only by the data owner. So apparently there is a terrific workload for the data owner in the cloud storage environment. The tie consumption and the cost of saving data everything is out of control in the existing system. So the data integrity is a difficult task to manage alone by the data owners.

6 Proposed System

While auditing the cloud server it shows the entire latest data without any interruption or corruption the server is in an ensuring state. For that type of service some schemas are discussed here. In most of the existing schemes the data owner is the only responsibility of reading and writing the data in the server. For this a signature-based public integrity auditing is proposed and a TPA is here to manage the integrity of data after some corrections are made by some public. So the scalability of the mediator along with all the schemes is aggregated for integrity check for multiple tasks. As the TPA belongs to any cloud he can able to go for public integrity auditing by having the public key provided by the data owners.

6.1 Protecting Data Integrity

The costumer of cloud environment assumes that while outsourcing the data it is encrypted before is moved to cloud and it is safe enough from attacks. Encryption is nothing but the protection of data from the malicious attackers and provides confidentiality to the user or client. It is done by the attackers and not by a configuration or bug errors. The traditionally we can prove the integrity of the outsourced data in a remote server by two ways. The truthfulness verification is done by the client or by the third party. In the first method it checks by downloading the file and then verifying hash value. Here the MAC is used and it is having two inputs one is secret key and data variable length so the output is the MAC tag and this process is done on the user side. After this the data is outsourced using the MAC value by the data owner in the cloud. The owner of the data is downloading it to check the integrity and MAC calculation is done and comparison is done before the data is outsourced. The changes will be detected by using this method and having the key the authentication of the data is confined and the user only having the key can able to check the integrity and authentication. It is overwhelming process for large files for manipulating and downloading MAC of the file. Hence it is a large file it consumes more bandwidth so the need of lighter technique is used for calculating the hash value [9].

Another method is by using the hash tree the hash value is calculated in the cloud. Here the ranking is assembled from bottom to top. The twigs are the data and the hashing occurs until the parents reached the root. The root is stored by the owner of the data. Whenever the data owner desires to check the integrity of data, the owner required the root value and does the comparison with the hash value he is having. It is also quite difficult as it is not possible for the computing huge volume of data. In case of providing service to data not for computation the file is downloaded by the user and sends it to the third party as it consumes large bandwidth. Therefore the need for data integrity and the computation power need to be checked. So remote data auditing is method in which data integrity is verified remotely [12–14].

6.2 Third Party Auditor

TPA is a person who carries out the auditing process with a skilled and effectively. TPA is mainly for data integrity checking. Even though lots of doubt is there the cloud user need to depend on the TPA. One of the proposed framework is the data owner is checking the integrity of the outsourced data so the result is favor for integrity checking and assures the data as a secured. The owner is aware of all the data integrity in cloud, so the data owner itself involving the auditing process. Initially the TPA is going with normal auditing method, after finding the modification of data the owner gets notification. The owner then checks the auditing

Fig. 3 Flow of data integrity
with TPA

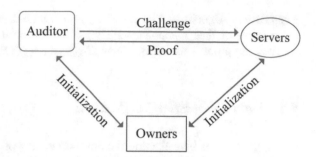

process logs for validating the changes. He can able to check if his data is unnecessarily accessed and the auditor also able to check. Therefore the proprietor is following the adaptation of their own data [2, 15, 16]. Figure 3 represent the flow of data integrity of TPA.

6.3 General Auditing Algorithm

For the general auditing algorithm Consider a file has a component of data which is to be uploaded by the data owner. The owner is responsible for the data which is to be encrypted for private data and for public there is no need of encryption. In this case every data is divided into blocks and each of its size is considered as a parameter for security. The fragmentation method is used for the above process and the security parameter is a restricted one. For every data block a tag is generated so the count of tag is less for the data.

The general auditing algorithm for storage having the following protocols:

Key Generation No input is taken instead a key is generated such as secret key and the hash key are the two security parameters of the TPA and it gives the output of public key tag $pk_t = g_2^{skt} \in G_2$.

Tag Generation For every data component it takes the secret key and the hash key as input and the output tag is created for each data $T = \{t_i\}_{i \in [1, n]}$.

Challenge Algorithm It chooses the information of data as input. It selects some data for creating a challenge set and a random number is generated for each of the chosen data block. Then a challenge is computed for each data and the output $C = (\{i, v_i\}_{i \in Q}, R)$.

Prove Algorithm It takes the data as input and also the challenge received. The result consists of the tag proof and data proof. For generating data proof the sector linear combination is used and then the output proof is $P = (TP, DP)$.

Verification Algorithm The proof P, secret hash key, public tag key and the information of data component are used for the verification of input challenge C. Then the proof is as follows $DP \cdot e(H_{\text{chal}}, pk_t) = e(TP, g_2^r)$ [9, 15].

6.4 Data Possession Verification

Data possession is investigating the appropriateness of the outsourced data in the cloud without retrieving statically in the cloud storage. In the remote server the data stored is in its possession and the original data is with the server without the retrieval process. Set of blocks are used randomly from the server for proving the possession. The RSA-based tag is used to prove the server either the user is having right to use the exact block or not. In the case of attacker the proofing data possession is failed. In some cases the PDP is checked partially and dynamically using symmetric cryptographic method. So there is a limitation for this that is it does not support auditability. Here the files from the blocks are under auditing for integrity checking and if there is need then only it the block are under corruption because of some hardware issues [8, 17, 18].

6.5 Proof of Retrievability

It is a cryptographic based protocol where the data retrieval is done by proving the data is together and without getting it back from the cloud. Using the hash key function the hash key value of the block is verified and the proof of retrievability is identified. The data owner is having the hash value of the file using the hash function. The data owner sends the file to the remote server after receiving the hash value and the key. When the data retrieval has to be checked by the data owner which sends the key the server has to reply the hash value. If both the hash values of the owner and the server are same the retrieval is done. This method is simple and easy to implement but in the case of huge data the owner has to maintain large number of keys each time [4, 19].

6.6 Proof of Ownership

The notion of proof of ownership of the client proved to the server while outsourcing the data. For some sensitive data both the POR and PDP needs to be in contact with each other for data. In such a case the need for proving the data ownership is necessary before the outsourcing of data the client and the server checks whether the file in need of secret. The owner of the file needs some proof for proving the ownership to the server [20].

7   Architecture of the Proposed Model and Modules

The proposed model architecture and the modules are discussed below. Here the data owner is uploading the file to the cloud server. The client is ready to get file from the cloud so the data owner assigns a random audit check for the client. The cloud server in turn sends the files to the TPA and send request for integrity audit and getting back the response. The TPA sends the request and notification to the data owner for audit check. The owner verifies it to the TPA. After all the process has done the integrity is verified now the cloud data user is ready to use the data stored in cloud. The same process is repeated once if there is any modification done by the user and the integrity check is identified and the changes are updated to the server. Figure 4 represents the general architecture of the proposed system is given.

7.1   Data Owners

Data Owner Registration: For uploading the file in the cloud the owner have to start it with the registration. The owner will fill up the details and each of them is stored in the database and it is maintained by the cloud server.

   Owner Login: Whenever he wants to upload the data he needs to login in the cloud server for that the owner has to provide some user name and password for login.

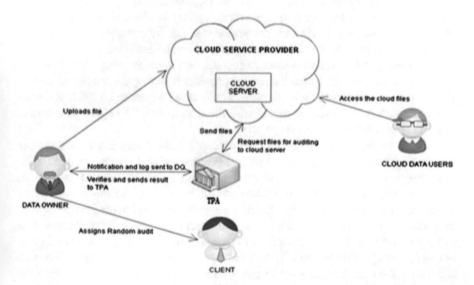

Fig. 4   Architecture of data integrity

User Registration: In the cloud storage if a user wants to access the data, the registration should be done first.

User Login: While accessing the data in the cloud storage the user have to login every time after the registration is done [21].

7.2   Third Party Auditor

Third Party Auditor Registration: Before wants to enter into the cloud the TPA needs to register first. Then he can able to do some integrity work in cloud on behalf of the owner. He also allots some cloud providers for providing some services to the users.

TPA Login: Once he logged into the cloud he can able to see the number of uploaded data and modified one which is done by the owner.

7.3   Data Sharing

The integrity of the shared data in the static groups available in the cloud is audited by us. The group is created previously before the data sharing and the members are not changed during the data sharing in the cloud. Only the original owner can decide who can able to view the shared data in the group while outsourcing. And the data integrity should be checked while sharing data in the dynamic group. Here some new user can be added to the existing group or the old one can also be revoked while sharing the data by preserving the identity privacy.

In the above module the basic registration and login details are explained. Who are all in the process and who have to register and login are discussed. Now the architecture is described. The cloud service provider is having the own cloud server. The data owner is ready to send the files in the cloud server and TPA is there for integrity check from the cloud server with the data owner. The cloud data user is ready to access the file in the cloud server.

Initially the data owner is sending the files to the cloud server. He only assigns priority of the data and files which is public, private, and both. In case of private the cloud data user able to access only if they are having the access permission. Then in case of public cloud the data is viewed by the entire user. So the possibility of modification of data is there. Here the data owner is responsible for the updating of data which is done by the cloud user. He only decides whether the user can access and the modifications are uploaded. For this every time the data owner need to be online for the updating process. Instead the data owner is now allows the TPA who is fixed by the CSP for doing the integrity check in the cloud server regarding the data updation. The data owner uploads the file to the cloud server. The CSP there send the files to the TPA for integrity check. The TPA sends the notification of the updation process done by some data user to the data owner. The data owner in return sends the result back to the TPA so that the server can able to upload the modified data. Now the TPA sends the integrity check result to the cloud server that the access permission is there for the updation from the cloud data user. The server in turn updates the file and sends another for the integrity check. The cloud data user is accessing the files in the cloud server.

The process is repeated again for the files. Here the entire data is stored in cloud is there for the process. The lots of data in the cloud are to be checked so the TPA is there for integrity check. It is not possible to check the integrity easily for that the entire data is divided into blocks using the fragmentation technique. Then the data is separated into each individual unit then the integrity check is done. For all this process the data owner needs much more time and the workload is heavy. The TPA is allotted by the CSP is many so the work is divided among them and easily done. This is the mainly for the public data access. Because for public data only the integrity check is main and for private data the data user cannot able to access the data it is well protected by some encryption algorithm. Only the authorized user can access the data and the integrity is maintained in the cloud server using the TPA.

In the previous section the general algorithm of TPA explains about the integrity check that is how the data in the cloud are considered. The data is taken and divided using the fragmentation technique and the data blocks are identified by the authorized user and integrity check is done. That concept is used in the architecture for the data integrity check.

7.4 Function of Multilevel Encryption and Decryption Technique

The proposed system explains about the combination of both the algorithms such as RSA and MD5 for encrypting and decrypting the data. Here the RSA is used for encrypting the data and MD5 is mainly for data integrity check and also for the securing the data. The algorithms use different key values for encrypting and decrypting. The multilevel encryption method is used for protecting the data integrity in cloud. The TPA checks the data integrity by using the cryptographic algorithms. The proposed method uses the RSA and MD5 for encrypting the data while the user is uploading the data into cloud database. The reverse process of MD5 and RSA is for decrypting the data and gives back to the user. Both the algorithms work great in securing the data in the cloud. Especially MD5 is used for data integrity purpose comparing with RSA. Figure 5 shows the Block diagram of encrypting the data. Figure 6 shows the Block diagram of decryption of data in cloud storage.

The cloud user uploads the file. The RSA algorithm is used for encrypting the file which is having the block size of 1024 bits. The user uses the public key for encrypting the data into cipher data as $c = m^\wedge e$ mod n, where m is the original message and c is the cipher text. It is having 1 round for encrypting the plain to

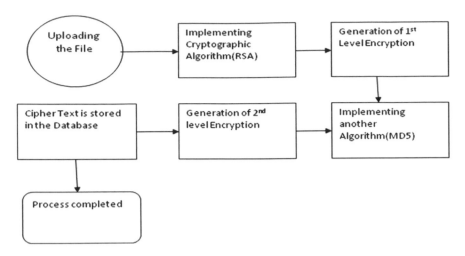

Fig. 5 Block diagram of encryption of data

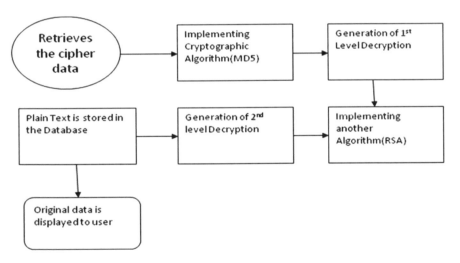

Fig. 6 Block diagram of decryption of data

cipher text using different key values. Then the 2nd level encryption takes place using the MD5 algorithm. Here D5 having the bit value of 128 bits and 4 rounds for converting the cipher text. Finally the process gets completed and the cipher data is stored in the cloud database.

Here the reverse process is used for decrypting the data. First the MD5 algorithm is used for decrypting the data and the first level of decryption is occurred. Then RSA algorithm is applied for the second level of decryption process. Here the cipher data c is converted to plain data using the secret key $\{d, n\}$, $m = c^{\wedge}d \bmod n$, where c is the cipher text and m is the plain text. Now the plain data is displayed to the user.

In the proposed system the RSA algorithm for encrypting the data as first level of encryption and MD5 is the second level of encryption for the output of RSA algorithm. Finally the encrypted data is stored in database. The reverse process is for decrypting the data in which first the MD5 is for decrypting the data and RSA is decrypting the output of the MD5 algorithm result. Finally we got the original result. Here the encryption and decryption are used in multilevel and the result data is stored in the database.

8 Conclusion

Recently Cloud computing is an unavoidable things of day today life. Because each and every moment we are using the cloud data in every thing so there is a possibility of security also an issue of cloud. Here the discussion is mainly about how the data are stored in the cloud and how the cloud provider and the owner of the data and the server are checking the integrity is a major concern. The existing system says about the drawback and proposed explains about the TPA integrity check and the data owner check using some cryptographic algorithms. In future the process is enhanced and will provide a highly secured integrity check with minimal cost and in an effective way.

References

1. Chang, V., & Ramachandran, M. (2016). Towards achieving data security with the cloud computing adoption framework (pp. 1–14).
2. Yu, Y., et al. (2017). Identity-based remote data integrity checking with perfect data privacy preserving for cloud storage. *IEEE Transactions on Information Forensics and Security, 12*(4), 767–778.
3. Ali, M., Khan, S. U., & Vasilakos, A. V. (2015). Security in cloud computing: Opportunities and challenges. *Information Sciences, 305,* 357–383.
4. Mahalakshmi, B., & Suseendran, G. (2016). Effectuation of secure authorized deduplication in hybrid cloud. *Indian Journal of Science and Technology, 9*(25).
5. Wang, B., Chow, S. S. M., Li, M., & Li, H. (2013). Storing shared data on the cloud via security-mediator. In *Proceedings of International Conference on Distributed Computing Systems* (pp. 124–133).
6. Chen, D., & Zhao, H. (2012). Data security and privacy protection issues in cloud computing. *2012 International Conference on Computer Science and Electronics Engineering, 1*(973), 647–651.
7. Aldossary, S., & Allen, W. (2016). Data security, privacy, availability and integrity in cloud computing: Issues and current solutions. *International Journal of Advanced Computer Science and Applications, 7*(4), 485–498.

8. Tate, S. R., & Mikler, A. R. (2003). *Performance evaluation of data integrity* Vandana Gunupudi thesis prepared for the Degree of Master of Science. December 2003 APPROVED.
9. Jayani, V. E. A., Ulasi, K. T., & Unitha, D. R. P. S. (2016). Public integrity auditing for shared dynamic cloud data with group user revocation. *IEEE Transactions on Computers, 8*(16), 3146–3152.
10. Subashini, S., & Kavitha, V. (2011). A survey on security issues in service delivery models of cloud computing. *Journal of Network and Computer Applications, 34*(1), 1–11.
11. Kumar, S., Ramalingam, S., & Buyya, R. (2016). An efficient and secure privacy-preserving approach for outsourced data of resource constrained mobile devices in cloud computing. *Journal of Network and Computer Applications, 64,* 12–22.
12. Botta, A., De Donato, W., Persico, V., & Pescap, A. (2014). On the integration of cloud computing and internet of things.
13. Raghavendra, S., et al. (2016). Survey on data storage and retrieval techniques over encrypted cloud data. *International Journal of Computer Science and Information Security, 14*(9).
14. Fabian, B., Ermakova, T., & Junghanns, P. (2015). Collaborative and secure sharing of healthcare data in multi-clouds. *Information Systems, 48*(May), 132–150.
15. Yang, K., Member, S., Jia, X., & Member, S. (2012). An efficient and secure dynamic auditing protocol for data storage in cloud computing (pp. 1–11).
16. Liu, C., Yang, C., Zhang, X., & Chen, J. (2015). External integrity verification for outsourced big data in cloud and IoT: A big picture. *Future Generation Computer Systems, 49*(August 2014), 58–67.
17. Ateniese, G., Di Pietro, R., Mancini, L. V., & Tsudik, G. (2008). Scalable and efficient provable data possession. In *Proceedings of the 4th International Conference on Security and Privacy in Communication Networks 2008* (p. 1).
18. Sanaei, Z., Abolfazli, S., Gani, A., & Buyya, R. (2014). Heterogeneity in mobile cloud computing: Taxonomy and open challenges. *IEEE Communications Surveys & Tutorials, 16*(1), 369–392.
19. Katkade, S., & Katti, J. V. (2017). Integrity check of shared data on cloud with various mechanisms. *International Journal of Computer Applications, 159*(6), 1–3.
20. Halevi, S., Harnik, D., Pinkas, B., & Haim, E. (2013). Proof of ownership in remote storage systems PoW solution : A general protocol security-efficiency tradeoff.
21. Wang, B., Li, B., & Li, H. (2015). Panda: Public auditing for shared data with efficient user revocation in the cloud. *IEEE Transactions on Services Computing, 8*(1), 92–106.

An Effective Hybrid Intrusion Detection System for Use in Security Monitoring in the Virtual Network Layer of Cloud Computing Technology

T. Nathiya and G. Suseendran

Abstract Security in the cloud computing environment is very important in the detection of intrusions into the virtual network layer. Denial of service (DoS) and distributed denial of service (DDoS) attacks are the main threats to cloud computing, and it is therefore crucial to protect against these types of intrusive attack. In this chapter, the effective monitoring of security by a hybrid intrusion detection system (H-IDS) in the virtual network layer of cloud computing technology is discussed and a detailed view of insider and outsider attackers in the virtual network layer is provided. This framework splits into four layers, namely virtual machine layer, node layer, cloud cluster layer, and cloud layer. Signature and anomaly techniques are used to detect known as well as unknown attacks and all virtual machine (VM) host systems which are available in the cloud computing environment are considered. The cloud cluster layer uses a correlation module (CM) to detect distributed attacks, and the Dempster-Shafer theory (DST) is employed in the final decision-making phase of the intrusion detection system (IDS) in order to improve its accuracy.

Keywords Hybrid cloud computing · Intrusion detection system
Network security · Dempster-Shafer theory

T. Nathiya (✉)
Department of Computer Science, School of Computing Science,
VELS University, Chennai 600117, Tamil Nadu, India
e-mail: tnathiya17@gmail.com

G. Suseendran
Department of Information Technology, School of Computing Sciences,
VELS Institute of Science, Technology & Advanced Studies (VISTAS),
Chennai 600117, Tamil Nadu, India
e-mail: suseendar_1234@yahoo.co.in

© Springer Nature Singapore Pte Ltd. 2019   483
V. E. Balas et al. (eds.), *Data Management, Analytics and Innovation*,
Advances in Intelligent Systems and Computing 839,
https://doi.org/10.1007/978-981-13-1274-8_36

1   Introduction

Modern cloud computing can be defined as the renovation of information technology infrastructure. Resources are deployed and hosted in the virtual environment via the internet for the benefit of end users. This type of development is termed an internet-based computing environment. A cloud data center has a virtually shared server in multiple location points worldwide, which provides software, hardware, infrastructure, and many resource tools [1, 2]. In Jan 2011, the National Institute of Standards and Technology (NIST) SP 800-145 proposed three services. First, infrastructures as a service (IaaS) allows the customer to create and access their own virtual machines. The platform as a service (PaaS) development deploys tools, frameworks, an application programming interface (API), and languages which are used to build and run applications, such as the Google App Engine and Microsoft Azure. The third type of service is software as a service (SaaS) and this employs a fully online application process that can be directly installed by the user, in a similar way to email. The most important aspect is the different levels of cloud computing involved, such as cloud service provider (CSP) data centers, the internet (network level) and the end user [3]. CSP describes shared pooled computing resources which provide on-demand self-services where the user only pays for the resources they actually use. Cloud computing saves on cost, saves on energy, and is rapidly developing and empowering customers. These are the main influential factors that are leading to customers being increasingly likely to adopt this technology.

Service providers access services in four ways. These are private, public, community, and by hybrid cloud. This chapter primarily focuses on the hybrid cloud as a combination of the private cloud and the public cloud. Most organizations are currently moving to the hybrid cloud [4]. Data and infrastructure management in the cloud is provided by a vendor [5]. There are serious risks involved in the handing over of sensitive data from providers and cloud computing is easily targeted by attackers. Malicious clouds can be classified into three types, namely network level, end-user level, and CSP level. The internet (network level) is a highly sensitive level and is subject to domain name system (DNS) attacks, distributed denial of service (DDoS) attacks, ransomware attacks, IP spoofing, port scanning, and routing information protocol (RIP) attacks [6].

There are many security issues affecting cloud computing services on a daily basis. Hackers employ illegal actions in order to disturb services. Two types of hacking attack occur and these are termed insider threats and outsider threats. In 2011, a total of 1041 instances of data loss occurred, and in 2012 a total of 1047 data breach incidents occurred, both in the first nine months of the year.

Epsilon leaked millions of names and email addresses from its customer database, for example, and from Stratfor in the United States, 75,000 credit card numbers and 860,000 usernames and passwords were stolen [7]. In 2017, ransomware attacks affected many banks, National Health Service hospitals in the United Kingdom, large telecom companies, and natural gas companies, while in 74 countries tens of thousands of systems were hacked [8]. Intrusions in the network layer of the cloud can be classified as either of two types (see Fig. 1). The first type is malicious insider attackers. These

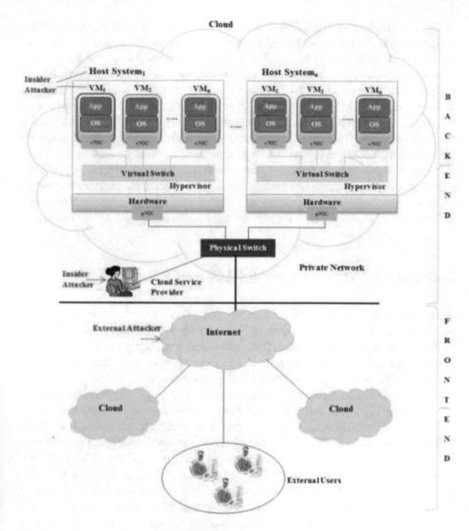

Fig. 1 Architecture of insider and outsider attackers

originate from the internal cloud and give unauthorized access to internal users or service providers. The second type, known as malicious outsiders originate from outside the cloud. Both attackers affected integrity, availability, and confidentiality. In 2014, the Amazon cloud server was affected by a DDoS attack and hosting services was suffered. On that occasion, the attacker deleted important data backups and affected machine configuration resulting in systems being completely down for approximately 12 h.

The malicious insider enters the cloud infrastructure using either admin access, the host system of the administrator or the administrator creates its own virtual machine (VM). An administrator gives some advantages to be entered the kind of attack to launched [9]. So, the provider almost protects inside and outside attacks. The traditional network security layer used to prevent outsider attacks employs a firewall, but the case of insider attacks (back end) is a difficult problem to address. To overcome this problem, an intrusion detection system (IDS) is often used in the network. An IDS is essential in the detection of attacks to the network security layer [10], as it monitors network traffic and allows intrusion data to be detected. In this chapter, we discuss the design of an effective hybrid intrusion detection system (H-IDS) for security monitoring in the network security layer of cloud computing technology. H-IDS makes the efficiency of signature and anomaly based technique. These H-IDS challenges are discussed later in this chapter. In addition, discussed on Dempster Shafer theory (DST) which is used in distributed intrusion. Using this new algorithm, it is possible to detect a network security layer attack and satisfy the cloud IDS requirements of a virtual network.

The rest of this chapter discusses related work and network attacks, and provides a brief description of network attacks. The following sections present clear details of the security framework, details on analyses of the proposed security framework, a discussion of future work, and references are given at the end of the chapter.

2 Intrusion Detection System (IDS) and Related Work

In the early days, security methods such as encryption, firewalls, virtual private networks, and gateways etc. were employed but didn't provide a high enough degree of security. Due to the nature of such static techniques it wasn't possible to depend on them completely. As a result there was a need to improve and increase the use of dynamic techniques to monitor systems and provide notification of illegal activities. A real-time deduction approach was introduced, known as an intrusion detection system (IDS) [11]. These systems are software applications that monitor networks and work to detect malicious activity and report it to either an administrator or SIEM (Security Information and Event Management). In the cloud computing environment, IDS are classified depending on the detection technique employed. There are three technique, namely signature-based IDS, anomaly-based IDS, and hybrid IDS. The signature-based IDS is a means of detecting all known attacks using a rule from a signature database. It produces a low incidence of false alarms and uses a fast multi-pattern matching (FMPM) algorithm to detect Denial of service (DoS) attacks. Signature-based IDS continuously update new attacks and these not detect the novel attack. Mainly problems of misuse IDS is every new type of attack it's can't be updated [12]. The second technique is anomaly-based IDS. This is employed to detect novel attacks and to identify unknown attacks. It has two main advantages over signature-based IDS. The first advantage is the ability to detect an unknown attack (or zero-day attack) and in order to detect new types of attack there is no need to update the database because the constantly update the normal profile database but implementation part is very difficult that is the main drawback of anomaly based IDS. The third technique involves the use of hybrid IDS which apply signature detection along with anomaly techniques to improve the performance and potential of current IDS.

IDS: Host-based IDS (HIDS) and network-based IDS (NIDS). HIDS are set at each host machine and monitor movement from both inside and outside the host machine. HIDS is dynamic sensor which trace the internal state of the system.

It investigates network packets of the specific host and is supposed to detect the program which accesses resources in the host. HIDS can't detect network attacks because they only operate within the host system. However, NIDS are placed at routers and switches and can observe network traffic and all traffic through the NIDS monitors. NIDS cannot encrypt data and can only inspect packets to see whether they contain malicious data. NIDS are better than HIDS because HIDS work to protect from within a system but NIDS work to protect all the hosts that are connected to the network [4, 13].

Le Dang et al. [14] proposed a multiple-pattern exact matching algorithm that reduced character comparisons and memory space based on a graph transition structure and using a dynamic linked list and searching techniques, such as the Wu-Manber algorithm (WM), Aho-Corasick algorithm (AC) and Comments Water Algorithm (CW) to compare various algorithm to be created multi-pattern exact matching algorithm. In this algorithm to produced high efficient space and time.

Puri et al. [15] proposed a novel security framework which employed statistical learning-based approaches to detect attacks in real time. Support vector machine (SVM) and regression tree (RT) algorithms were used to classify the attacks in HIDS. 90% available training set from KDDCUP 99 dataset is used and 10% is execution time and produced high accuracy of detection and lowest false rate. The proposed algorithm is used to obtain a realistic result by adding a network analytic tool to capture the packet to be transferred over the network, such as Snort etc.

Vieira et al. [16] proposed grid and cloud IDS (GCIDS). These systems combine both signature- and anomaly-based techniques and use artificial neural networks to detect unknown attacks. If any attack is intruding to detect and alert system informs to another node. This is an efficient means of finding unknown attacks. The outcome of this paper was a reduction in the duplication of data. However, the techniques need greater training time and increased sample detection accuracy.

Singh et al. [17] proposed a combined IDS-related architecture for the cloud. In this structure, a NIDS is inserted into each host system in order to observe network packets. The cloud cluster controls a correlation unit (CU) in light of the heap on the bunch. Grunt programming is used to identify the mark based known assaults. A combination of decision tree (DT) and support vector machine (SVM) statistical approaches is applied to an anomaly technique to detect unknown attacks. The use of this approach has improved detection accuracy. However, this concept is high network traffic, CU may leads maximize the success.

Kamatchi and Modi [18] proposed and implemented Bayesian theory in the virtual network layer to detect both known and unknown attacks, and a Snort tool was also used to detect known attacks. The proposed framework involved a signature-based technique with an anomaly detection technique, and resulted in an improvement in detection time but required the use of training samples. An overview of the above approaches to H-IDS and proposed frameworks is shown in Table 1.

Table 1 Analytical study of H-NIDS for cloud based

Reference	Feature				
	IDS type	Is the data in real time?	Positioning	Advantages	Challenges
Multi-pattern matching extra algorithm in IDS in the cloud, 2016 [14]	Network-based	Yes	On each host machine	Produced highly efficient memory space and time	Generates a high number of false positive alarms
Novel security in the cloud Hybrid IDS, 2017 [15]	Network-based system	Yes	On each node	Execution time and produced Accuracy detection rate and false rate	Requires more training samples
IDS for grid and cloud, 2010 [16]	Host system	Yes	On each node	Reduces false positive and false negative alarm rates	Needs more training data and samples Detection rate
Collaborative IDS framework for the cloud, 2016 [17]	Network-based	Yes	On each cluster node	Protects against DDoS attacks	Needs a number of training samples
NIDS in the cloud, 2016 [18]	Network-based	Yes	At the executing server	Detection percentage is very high	Needs a lot of training data samples
Fuzzy clustering IDS in the cloud, 2016 [19]	Host-based	Yes	On each node to be tested	Removes the problem of a sharp boundary	Low detection accuracy time

3   Proposed Security Framework

3.1   Design Goals

The proposed security architecture has been used to detect attacks on virtual networks in the cloud and it provides a minimum number of false alarms, a greater degree of accuracy, minimum communication, and lower cost which are all important requirements in the cloud environment. Under the proposed security framework design, all important factors, namely a large-scale dynamic system, ability to identify a variety of attacks, scalability and synchronization of IDS sensors are addressed.

3.2 Design of Proposed Security Framework

Distributed environments, such as that of cloud computing, are targeted by attackers for identified organization. As a result, IDS are employed to protect both the inside and the outside of the cloud as shown in Fig. 2. The diagram consists of four layers with the lower layer sending the alert signal to the upper level. These four layers are the cloud layer, cloud bunch layer, node layer, and VM (virtual machine) layer. In this level consisting of module via that are management module, the correlation module, NIDS. Cloud users communicate with the front end of the cloud and this in turn communicates with the external internet audience and the internal cloud network. The back end relates to the physical hardware and software used in cloud network services.

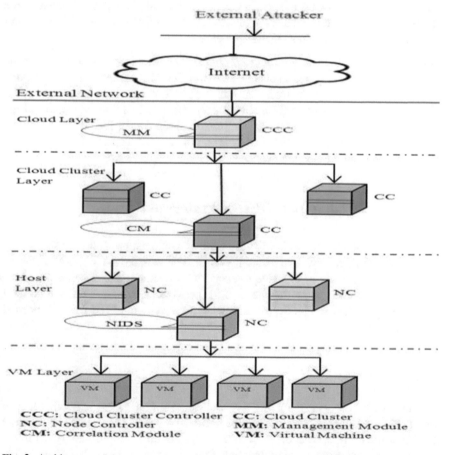

CCC: Cloud Cluster Controller   CC: Cloud Cluster
NC: Node Controller   MM: Management Module
CM: Correlation Module   VM: Virtual Machine

Fig. 2 Architecture of the proposed security framework

Virtual Machine (VM) Layer: This is the lowest layer of the framework and contains VMs allocated to the cloud user. To protect the VM from an unauthorized malicious attacker, the VM is controlled by the node layer.

Host Layer and NIDS: This layer is connected to several node controllers (NC), which are connected to a number of VMs. Each NC implements a NIDS which monitors the virtual network traffic generated in virtual switch on all host machine. Reports and alerts are sent to the correlation module (CM) of the cloud cluster layer. The NC has server processing of NIDS. Single instance of the Virtual network contains multiple VMs which could be observed NC deployment in NIDS. It safeguarded with insider and outsider network attack.

Cloud Cluster (CC) Layer and Correlation Module (CM): This acts as a back end for clustering in the cloud computing environment. The cloud cluster communicates with the NC and cloud controller (CC). The CM collects the evidence of the attack from the NIDS. Following this, a DST combination rule is applied to identify the evidence of distributed attacks at the cluster layer. The CM manages VM execution and service level agreements (SLAs) per cluster.

Cloud Layer and IDC Management Module (MM): This is the highest layer of the design and contains the edge to outside network. The MM receives intrusion reports from the cloud cluster network in this layer. The Cloud Cluster Controller (CCC) manages and controls the cloud clusters. Here, access control is allowed the privileged cluster to send alert to CCC. The Dempster-Shafer theory (DST) is then applied to the correlation module (CM) this combination of the rule to identifying the collusion attacks in the cloud network.

3.3 Functioning of the H-NIDS Framework

The major characteristics of H-NIDS are their dynamic nature, ability to self-adapt, scalability, and efficiency. As a result, it was decided to make the proposed system a real-time system that was adaptive to scope with dynamic attack series of developments. A detailed view of the hybrid network intrusion detection system framework is shown in Fig. 3. The following steps are employed for both normal and threat packets.

3.3.1 Firewall

The firewall is responsible for filtering traffic according to the network security policy. First, arriving packets having the correct source IP address and destination IP address are tested and are either allowed to pass or are blocked. The packet filters are very fast accessing to implementing the end routers. Since it decreases the number of unauthorized packets coming from an external network, NIDS is only used to detect insider attacks.

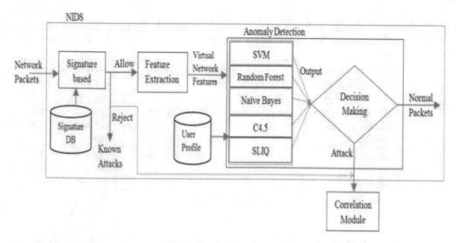

Fig. 3 Functioning of the H-NIDS framework

3.3.2 H-NIDS

H-IDS are mounted in the cloud virtual network layer (virtual switch) which is connected to the available hosts. It monitors network traffic to/from the VM. H-NIDS combine both signature- and anomaly-based intrusion detection methods.

(a) **Signature-Based Intrusion Detection**: This contains a set of rules used to identify the intrusion. This type of technique is used to detect known attacks and is termed signature-based intrusion detection. It passes attack packets through to the CM. Normal packets are sent to feature extraction. Use of this technique leads to a high degree of accuracy, rapid detection of known attacks, and a minimal false positive rate. The signature database simultaneously updates with information about the packets. Snort software is used in signature-based intrusion detection.

(b) *Feature Extraction*: Traditional features, such as protocol, port number, and extra virtual network features are combined with virtual private IP, VLAN identifier etc. and packets details are sent anonymously to the intrusion detection module. If reduces these features to improve the performance of the instruction detection technique [20].

(c) *Anomaly-Based Intrusion Detection*: This method can be classified into three types. First, the statistical-based approach acquires organize movement action and then creates a client profile relating to them. This client profile is continuously checked for deviation between their conduct and the ordinary one. Unilabiate, multi-variations, time arrangement, and self-comparative techniques are all used. Second, knowledge-based methodologies incorporate limited state machines, portrayal dialects, and master frameworks, all with predefined rules [20]. The third method is a machine learning technique involving connected hereditary calculations, neural systems, Bayesian systems, fluffy rationale, and

Table 2 Comparative study of anomaly-based intrusion detections

S. No	Method/Approach used	Rate of deduction	Disadvantages
1	Semi-supervised fuzzy clustering algorithms [19]	To prevent reaching of optimum clustering quality	Model does not apply in the real world
2	Decision tree and support vector machine [17]	Overall detection rate 99.40, 0.60% intrusion missing, 1.69% of alarms are false. Accuracy rate of 98.92%	Decision tree does not include the forest tree
3	Artificial neural network algorithm [22]	100% true positive rate with only 2 false alarms	Efficiency can be improved by using feature reduction
4	Neural network [23]	95% accuracy achieved	Model does not discuss with real time
5	Genetic algorithm [24]	57% of attacks detected from a random set with no false alarms	Method using feature selection techniques only

bunched locations. All these techniques are used in the detection of an attack [21]. Table 2 shows a comparative study of anomaly-based intrusion detection techniques, and refers to various papers using anomaly-based approaches to calculate training time and testing time in order to provide a better deduction rate, increased accuracy, and more frequent alarms.

(d) **Decision Module**: DST blend manage is makes the keep going basic leadership on obscure assaults. To constructing the DST combination rule is as follows:

Bayesian statistics assign probabilities to each new dataset, which can be derived from the elementary probability

$$P\left(\frac{A}{B}\right) = \frac{P\left(\frac{A}{B}\right)P(A)}{P(B)}$$

where, $p\left(\frac{A}{B}\right)$ represents the posterior probability of the model, $p\left(\frac{B}{A}\right)$ represents the probability function of the data, $P(A)$ denotes the prior possibility of the model, which is both a strength and weakness of Bayesian statistics and $P(B)$ represents an evidence of probability [25].

Dempster-Shafer Theory: This theory of evidence was first formulated by Shafer in 1976 and originates from Bayesian theory. Our examination the DST has one of a kind favorable circumstances in dealing with in interruption investigation. The technique was implemented with an IDS-ready correlation module (CM). For instance, DST supports the genuine and false case. Data moves to CM for false case. It is earlier likelihood of assault, by utilizing DST we can allocate a 0.1

certainty to "assault" or "genuine", 0 certainty to "no-assault" or "false", and 0.9 certainty to "{true, false}" [26].

Frame of Discernment and Belief Function: The DST approach permits three types of response: assault, no assault and don't know—the last alternative of permitting numbness in evidential thinking. In this way, that the DST just an arrangement of separate theories of intrigue, i.e., {attack, no-attack} is called edge of wisdom. The basic probability assignment (BPA) work is the power set of the casing of acumen. The BPA ranges from 0 to 1 with a goal defined as

$$m\theta = 2°$$
(1)

and

$$\sum_{A=0}^{1}(A) = 1$$
(2)

where BPA for null set Θ is m $(\Theta) = 0$, a function Bel: $2° \rightarrow [0, 1]$ is a belief function over θ if it is given by BPA.

$$Bel(A) = \sum_{B \subseteq A} m(B)$$
(3)

For all $A \in 2°$, $Bel(A)$ represents a measure of aggregate convictions to the confirmation of A. the few conviction works over a similar casing of acumen and in light of particular of Evidence, Dempster combination Rule, which is given by condition (4).

$$m(c) = \frac{\sum Ai \cap Bj = C^{m1}(Ai)^{m2}(bj)}{1 - \sum Ai \cap Bj = \theta^{m1}(Ai)^{m2}(Bj)}$$
(4)

For all non-empty $C \subseteq m(c)$ cases is a basic likelihood assignment of the combination rule. Anywhere $m1, m2$ is a basic probability. The belief function is over the same frame θ.

3.3.3 Correlation Module (CM)

The CM represents the cloud cluster layer and its role is to detect all types of attack and corresponding clusters for all node layers. It gathers the evidence of intrusion from all of the HNIDS. The authorization controls only allow alerts from HNIDS located in the same clustering of nodes. The CM deploys the DST combination rule in order to detect complicity of intrusion from various HNIDS. Whenever a distributed attack is detected by the CM, an intrusion report is sent to the management module (MM) for further action.

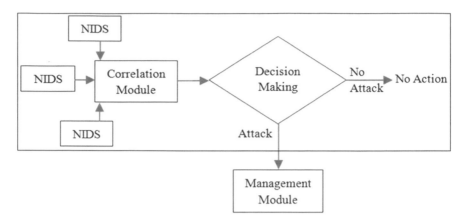

Fig. 4 Detailed view of the management module

3.3.4 Management Module (MM)

The MM is the main part of the proposed framework. A view of management module processing in the cloud is presented in Fig. 4.

Using the DST combination rule, the MM receives an alert from the CM in the cloud cluster layer. If an attack is identified by the MM, then the administrator is informed. The signature database is then updated with the identification of the new type of attack in order to improve its future efficiency.

4   Analysis of the Proposed Security Framework

Cloud computing is a powerful tool which uses the resources as a service through internet. The ability to provide security for cloud resources is therefore a major asset. The cloud suffers from traditional attacks such as address resolution protocol (ARP), DoS, and DDoS etc. These attacks affect both cloud resources and services offered. Many cloud providers use firewalls like Microsoft Azure and Eucalyptus etc. However, firewalls only prevent outsider attacks, and insider attacks and some particularly tricky outsider attacks are not detected by firewalls. For example, port 25 is common port used by mail servers. If there is an attack on any common port, it is not possible for the firewall system to separate normal traffic from attack traffic. To overcome this problem, we propose an effective HIDS for use in security monitoring in the virtual network layer of cloud computing technology. This framework is deployed at the front end as well as at the back end VMM in order to detect external and internal attacks.

A hybrid IDS is a combination of signature-based and anomaly-based techniques and is used to locate distributed attacks in the cloud infrastructure. The known attack detects attacks fast which only compare the signature matching using signature DB. For unknown attacks, the CM and MM use a DST combination rule, which provides fast intrusion detection and updates all changes in the signature database.

The CM functions to reduce the number of intrusion alarms which are forwarded to the MM. As a result, it decreases mathematical calculations and communication cost. This new projected security framework satisfies the maximum number of cloud IDS requirements under the virtual network security layer.

Intrusion is handled by NIDS in virtual machines which are connected in the virtual network. NIDS may exist either in each virtual machine or node controller. Both the MM and CM are capable of controlling NIDS, and NIDS themselves are scalable since they can adjust to either the NC or VM node.

4.1 Security Analysis of the Proposed Framework

In our proposed framework, security monitoring of IDS is deployed across the entire host machine in a virtual network that minimizes attacks in VM. NIDS is being provided for pass the sign of the detecting intrusion in Correlation Module. The CM provides authorization to find distributed and unknown attacks over virtual networks and runs on controlled access which permits authorized NIDS transaction alert alone.

The management module (MM) provides the authorization to find distributed intrusions in the cloud network layer. Node performance is affected by any attracter, and in order to avoid this problem the MM only provides access to authentic requests verified through the signature database at various levels. The authentication mechanism fetch rule is applied to each network transaction.

5 Conclusion and Future Work

The principle security concern in cloud computing is one of finding interruptions in the virtual system layer. In this chapter, an effective NIDS for use in security monitoring to identify virtual network layer related attacks and distributed attacks is discussed. The use of a HID technique improves accuracy and reduces false alarms. The alert mechanism at various layers of the cloud is to capture distributed intrusion. As per our analysis, the proposed framework is easily achievable in current virtual network security and this study offers a great deal of encouragement and inspiration to that end.

In future, security frameworks will be validated using a classification algorithm implanted via a Weka tool in order to demonstrate real-time analysis of location rate along with the low rate of false alarms in the private cloud. The analysis of our framework will improve training and testing times and help to provide 100% accuracy.

References

1. Turab, N. M., Abu, A., & Shadi, T. (2013). Cloud computing challenges and solutions. *International Journal of Computer Networks & Communications (IJCNC)*, *5*(5), 209–216.
2. Lock, H.-Y. (2012). InfoSec reading room. *Reading*.
3. Raghav, I. (2013). Intrusion detection and prevention in cloud environment: A systematic review. *International Journal of Computer Applications*, *68*(24), 7–11.
4. Mahalakshmi, B., & Suseendran, G. (2016, July). Effectuation of secure authorized deduplication in hybrid cloud. *Indian Journal of Science and Technology*, *9*(25).
5. Amudhavel, J., et al. (2016). A survey on Intrusion Detection System: State of the art review. *Indian Journal of Science and Technology*, *9*(11), 1–9.
6. Potteti, S., & Parati, N. (2015). Hybrid intrusion detection architecture for cloud environment. *International Journal of Engineering and Computer Science*, *4*(5), 12146–12151.
7. Chou, T.-S. (2013). Security threats on cloud computing vulnerabilities. *International Journal of Computer Science and Information Technology*, *5*(3), 79–88.
8. G. #132 Ismael Valenzuela—Global Director, and Foundstone Consulting Services. *Targeted ransomware attacks in the cloud*. [Online]. Available: https://files.sans.org/summit/healthcare2016/PDFs/Prediction-2017-I-Survived-a-Ransomware-Attack-in-my-Cloud-Ismael-Valenzuela.pdf. Accessed June 18, 2017.
9. Pitropakis, N., Anastasopoulou, D., Pikrakis, A., & Lambrinoudakis, C. (2014). If you want to know about a hunter, study his pray: Detection of network based attacks on KVM based cloud environments. *Journal of Cloud Computing: Advances, Systems and Applications*, *3*(1), 20.
10. Kene, S. G., & Theng, D. P. (2015). A review on intrusion detection techniques for cloud computing and security challenges. In *IEEE Sponsored 2nd International Conference on Electronics and Communication Systems (ICECS)* (pp. 227–232).
11. Ansari, G. (2016). Framework for hybrid network intrusion detection and prevention system. *Interational Journal of Computer Technology & Application*, *7*(August), 502–507.
12. Barabas, M., Homoliak, I., Drozd, M., & Hanacek, P. (2013). Automated malware detection based on novel network behavioral signatures. *IACSIT International Journal of Engineering and Technology*, *5*(2).
13. Kumar, U. (2015). A survey on intrusion detection systems for cloud computing environment. *International Journal of Computer Applications*, *109*(1), 6–15.
14. Le Dang, N., Le, D., & Le, V. T. (2016). A new multiple-pattern matching algorithm for the network intrusion detection system. *IACSIT International Journal of Engineering and Technology*, *8*(2).
15. Puri, A., & Sharma, N. (2017). A novel technique for intrusion detection system for network security using hybrid SVM-cart. *International Journal of Engineering Development and Research (IJEDR)*, *5*(2), 155–161.
16. Vieira, K., Schulter, A., Westphall, C., & Westphall, C. M. (2010). Intrusion detection for grid and cloud computing. *IT Professional Magazine*, *12*(4), 38–43.
17. Singh, D., Patel, D., Borisaniya, B., & Modi, C. (2016). Collaborative IDS framework for cloud. *International Journal of Network Security*, *18*(4), 699–709.
18. Kamatchi, A., & Modi, C. N. (2016). An efficient security framework to detect intrusions at virtual network layer of cloud computing. In *19th International ICIN Conference-Innovations in Clouds, Internet and Network* (pp. 133–140).
19. Thong, P. H., & Son, L. H. (2016). An overview of semi-supervised fuzzy clustering algorithms. *IACSIT International Journal of Engineering and Technology*, *8*(4), 301–306.
20. Sondhi, J. (2014). A review of intrusion detection technique using various technique of machine learning and feature optimization technique. *International Journal of Computer Applications*, *93*(14), 43–47.
21. Jeong, H. D. J., Hyun, W. S., Lim, J., & You, I. (2012). Anomaly teletraffic intrusion detection systems on Hadoop-based platforms: A survey of some problems and solutions.

2012 15th IEEE International Conference on Network-Based Information Systems (pp. 766–770), *NBIS*, September, 2012.

22. Jones, C. B., & Carter, C. (2017). Trusted interconnections between a centralized controller and commercial building HVAC systems for reliable demand response. *IEEE Access, 5,* 11063–11073.
23. Mann, A. S., & Kumar, V. (2016, November). An efficient method for estimation of cost in cloud computing using neural network. *Indian Journal of Science and Technology, 9*(44).
24. Singh, P., & Hazela, B. (2016). Design & Development of a new hybrid system to Prevent Intrusion at cloud using genetic algorithm. *International Journal of Advance Research in Computer Science and Management Studies, 4*(6).
25. De Vos, A. F. (2000). A primer in Bayesian Inference. Web ref: http://personal.vu.nl/a.f.de.vos/primer/primer.pdf
26. Phule, S. G., & Chavan, G. T. (2015). Intrusion response with Dempster Shafer theory of evidence to detect and overcome routing attack in Mobile Ad hoc Networks. *International Research Journal of Engineering and Technology (IRJET), 2*(2), 410–416.

IALM: Interference Aware Live Migration Strategy for Virtual Machines in Cloud Data Centres

V. R. Anu and Sherly Elizabeth

Abstract In IT industry most of the real-time online services are cloud based. Large data centres have been widely used to allocate and establish these services through virtual machines (VM) in physical servers. Live migration is a mandatory feature of all modern hypervisors in these cloud data centres for implementing key services like load balancing, server consolidation, high availability, etc. without much delay. Performance optimization of live migration is an active are of research now. But there has been little attention given by research community on the area named VM migration interference or resource contention among co-located VMs as a by-product of live migration, which cause performance degradation and may lead to SLA violation. Here we present an interference aware VM live migration strategy IALM which manages effectively the issue generated by VM interference while live migration. Extensive experiments and large scale simulation are done with CPU intensive and network intensive workloads on Xen platform to feature out performance gain in terms of network throughput and CPU consumption. VM migration interference like nature of workloads, system properties, characteristics including intensity and length of interference are also analysed in this work.

Keywords Cloud computing · Live migration · Performance interference
Virtualization

V. R. Anu (✉)
Mahatma Gandhi University, Kottayam, Kerala, India
e-mail: anuvraveendran@gmail.com

S. Elizabeth
IIITMK, Kazhakkoottam, Kerala, India
e-mail: sherly@iiitmk.ac.in

© Springer Nature Singapore Pte Ltd. 2019
V. E. Balas et al. (eds.), *Data Management, Analytics and Innovation*,
Advances in Intelligent Systems and Computing 839,
https://doi.org/10.1007/978-981-13-1274-8_37

1   Introduction

Live migration of virtual Machines (VM) is inevitable feature of cloud data centres. Virtualization is the revolutionary technology which supports live migration to implement key infrastructure features of cloud like effective utilization of resources, improved energy efficiency, consolidation of workload and load balancing. A considerable amount of energy is saved with virtual machine migration from under loaded servers by putting them in idle/sleep mode. Several mechanisms are introduced and adapted to improve the effectiveness of live migration and to reduce the power consumption of data centre effectively. The mapping between the physical machines (PMs) and VMs are done in such a way that total amount of energy consumed by PM while running maximum number of VMs, must be minimized. It is reported that, in worldwide the data centre consumes about 35 billion kilowatts of electricity which is equivalent to output of 30 nuclear plants and this number consistently growing around 12% every year. Minimizing energy consumption in datacenters is globally challenging issue [1, 2].

Virtualization allows diverse applications to run in isolated environments of VMs with in the Physical Machine. In recent studies and research it is clear that the isolation provided by virtualization through VMs in PMs are not effective all the time, in fact there is a performance degradation in cloud online services which are running in VMs. With the analysis of various performance matrices of virtual machines it is clear that this performance degradation is mainly due to interference among multiple VMs running on the same hardware platform or physical machine. This performance degradation sometimes leads to SLA violation also. Since isolation provided by virtualization causes disturbance in certain entities of source as well as target physical machines, which lead to the performance degradation of both VM and PM. This interference issue plays a vital role in the effectiveness of VM migration and it also increase migration cost in multitier application (Fig. 1).

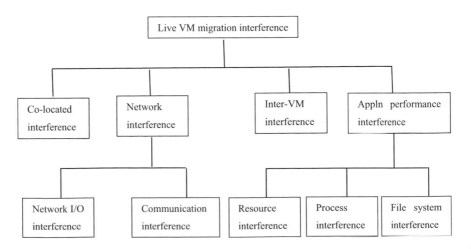

Fig. 1   Classification of interference due to VM migration

Live migration is a network centric and CPU moderate application [3, 4]. Since many categories interference are present in the above hierarchical structure most influenced divisions of interference on VM migration performance are inter VM interference and network interference. In this paper we study the effect of inter VM interference and network interference on performance degradation of VM before and after migration. Majority of cloud services running on these VMs demands little time delay. In this work we introduced an algorithm to perform interference aware live migration. We conduct extensive experiments and large scale simulations to evaluate the performance of our algorithm in terms of network throughput, CPU consumption and inter process communication. We also compare this algorithm with traditional interference unaware VM migration strategies.

2 Background Details and Related Work

Novakovie et al. [1] proposed a system called Deep-dive for transparently identifying and managing performance of IaaS providers. Deep-dive discussed co-located interference in a deep level by introducing modules for detecting interference. Also a warning system to reduce overhead of interference and a migration mechanism to transfer the VMs which cause interference in the infrastructure platform. Liu and He [5] introduced a coordination system named VM buddies to solve the issues created by co-related VMs in multitier applications. They identified a synchronizing protocol which will reduce the time delay and cost of transfer for a large set of co-related VMs. Also added a bandwidth allocation algorithm for effective usage of network in cloud data center. Xu et al. [6] presented a light weight interference aware live migration strategy called iAware. It analyses and mitigate co-located and VM migration interference. Yang et al. [7] analysed and identified the characteristics of disk I/O scheduler in a hypervisor using distributed I/O performance measured system called VExplorer. Pu et al. [8] done an extensive experimental study of interference caused by CPU bound and network bound virtual machines by the same physical infrastructure. They put forward eight metrics to analyses the interference on VMs. Zhang et al. [9] presented live migration of virtual machines with less cost and fewer application interference. They studied the cost evaluation model of performance interference. Liu [10] described about a robust virtual network which will not break during live migration process. Takouna et al. [11] propose PVA (Peer VM Aggregation) to introduce a communication pattern dynamically during VM migration. They identified that non breakage virtual network and dynamic communication pattern reduce the communication interference. They evaluated this method at the time of VM placement, network utilization of each link during performance degradation and results show that the total amount of network traffic significantly reduces network utilization by 25%. Shim [12] introduced a static VM consolidation algorithm which minimize the energy consumption in data center and reduce performance degradation of jobs due to inter VM interference on VMs on same physical machine. Wood et al. [13] argued the effectiveness of virtualization in cloud data centers by eliminating hotspots through VM migrations. Here they discussed black box strategy which is fully OS and application agnostic and grey-box approach that exploit OS and application level statistics.

2.1 An Overview of VM Interference

Live migration of VMs contributes advantages like power consumption management, high availability, fault tolerance, workload balancing and server consolidation. However, migrating VM may cause some disturbances in certain entities of source and destination PMs before and after migration. As a result performance degradation and thus SLA violation also happens in cloud environment and this is termed as VM migration interference [13–15].

To explain performance interference between jobs running on the same physical machine we first measures the execution time of a job1 without any other job in a PM. Now introduce same kind of job job2 into the PM and measure the execution time of job1. Now take the ratio of execution time of job1 in second scenario to first scenario and is greater than 1. Repeat the same for 'n' number of jobs. From the observation it is clear that the ratio getting bigger in each time and larger the performance degradation.

During live migration while transferring VM from one PM to another its network connections get interrupted and time required to resetting network connection. This degrades the performance of VM even though most of the running services on cloud demands little latency [16]. Running network intensive workloads in isolated environments on a shared physical machine lead to high overheads due to extensive context switches and accumulated events in driver domain and VMM. In most of systems for running applications, the same network path is used to transfer migration related information and other routine data transfer between the VMs. This cause network contention and affected the execution of both this type of processes and sometimes severely degrade the performance of the system.VM Networking interference will become worse, when migrating a VM with a high memory dirtying rate. Migration of small idle VM instances took up to 55 s in this scenario. To deal with network interference due to contention in network, several commercial virtualized platforms recommended a dedicated network for live migration. Additionally, to reduce the ill effects of interference during migration many vendors suggest a no-break virtual network during migration. Isolation of network bandwidth between VMs is another mechanism to deal with network interference; but in many situations it sacrifice the performance of some VMs. The victim VMs has to under gone a long period for migration process and its downtime also get prolonged and resulted a performance degradation [17, 18].

In inter VM interference it is observed that the interference issue highly co-related to number of VMs in each physical machine and type of application using CPU, memory-intensive or I/O-related application. In [10] authors tries to migrate VMs which are more aggressive in terms of resource utilization. The VM placement manager tries to find out a non-interference PM that will be best match for this culprit VM. In [19] they use a hotspot detector and migration manager to avoid VM contention issue which cause interference while utilizing the PM resources. They had done a black box screening by external observation into the VM management to identify hotspot. Also done a grey-box monitoring since it is feasible to gather OS level statistics and application logs which also helps to take quality decision to identify hotspots [16, 20]. Due to larger number of VMs in PM, heavy contention in cache space occurs which lead to interference. The number of cache miss ratio of PM while running VMs in that basically show their performance degradation due to interference issue. From the observation it is clear that utilizing round robin strategy for allocating resources is a fair strategy to deal with interference issue; which ensures availability of resources to all VMs equally in the common infrastructure [10].

2.2 System Model of VM Interference

Live migration is a network intensive and CPU moderate application; which logically implies that migrating VM to and from PMs with heavy I/O and CPU contention cause performance degradation due to interference. So VM migration interference can be represented as the function of network I/O contention and CPU resource contention on source and destination PM [19].

Our root cause analysis represents the overall performance degradation due to VM migration interference in the following way.

$$\text{Overall performance degradation } M_i = M_{cp} + M_{off\text{-}cp} + M_d + M_n, \qquad (1)$$

where M_{cp} represents CPU contention time, $M_{off\text{-}cp}$ represents memory access contention time, M_d disk access contention time, M_n network contention time.

The first three components can be considered as contention due to the presence of co-located VMs.

$$\text{Overall performance degradation } M_i = M_c + M_n \qquad (2)$$

The network interference can be defined as the ratio of total number of network interrupt in VM to the number of interrupts handle without lagging. Severity of interference increases with higher ratio. Moreover network throughput of VM fluctuated widely under interference. By considering both factors network interference can be represented as,

$$M_n = N_p + N_t/N_l \qquad (3)$$

N_p is fluctuated network throughput, N_t is total network interrupt raised and N_l is capacity to handle interrupts without lagging.

Next we estimate the CPU resource contention due to the presence of co-located VMs. If more number of VMs are coming into the PM, the contention issue increases because each VM requested for CPU time. Representing it in the ratio form,

$$M_c = C_d/C_l \qquad (4)$$

C_d is total number of VMs demands CPU, C_l is total number of VMs to CPU can be allotted without lagging.

Severity of interference increases as the ratio increases.

$$C_d = C_c + C_q \qquad (5)$$

C_c is VMs currently used CPU and C_q is VMs queued to get CPU.

In the diagram it is mentioned that other type of contention issues may happen. For example co-located interference can be explain as the function of network I/O contention, CPU contention along with cache and bandwidth contention on migration destination PMs.

Overall performance degradation can be represented as the sum of network contention and co-located interference [21]. It can be represented as,

$$M_i \approx f(M_n, M_c) = a.M_n + b.M_c \qquad (6)$$

where a, b are constants which regulate other factors.

3 Interference Aware Live Migration Strategy (IALM)

Input: n number of candidate VMs for live migration

Let available VMs are $V = \{V_1, V_2, V_3, \ldots, V_n\}$, Available PMs are $\{P_1, P_2, P_3, \ldots, P_m\}$

Output: Selected VMs to be migrated to destination PM.

Method:

Step 1: for selecting candidate VMs with least estimated migration interference on source PM.

    1. Initialize M_{min}= infinity, $S = \emptyset$

    2. for all VMs V_i in V do

    3. $M_s = M_n + M_c$; computing migration interference on source VM with network interference value M_n and co located interference value M_c

    4. if $M_s < M_{min}$ then

        a. $M_{min} = M_s$; S= $\{V_i\}$

    5. else if $M_{min} == M_s$ then

        a. S= S U $\{V_i\}$

    6. for all VMs V_i in S do

        a. if $M_n < M_{n\,min}$ then

            i. $M_{n\,min} = M_n$; $S_n = \{V_i\}$

        b. else if $M_{n\,min} == M_n$; then

            i. $S_n = S_n$ U $\{V_i\}$

        c. cnd if

        d. end for

    7. for all VMs V_i in S do

        a. if $M_c < M_{n\,min}$ then

            i. $M_{n\,min} = M_c$; $S_p = \{Vi\}$

        b. else if $M_{n\,min} == M_c$ then

            i. $S_p = S_p$ U $\{V_i\}$

        c. end if

        d. end for

    8. end if
    9. end for
Step 2: For finding out all PMs capable of hosting S_n and S_p
    1. for all V_i in S_p do
    2. for all PM P_i in P capable of hosting V_i do
            a. $M_{c\,min}$= infinity
            b. M_{cs}= M_{cn} +M_{cp}; computing migration interference on destination PM
               with network interference value M_{cn} and co located interference value
               M_{cp}
            c. if M_{cs} < $M_{c\,min}$ then
            d. $M_{c\,min}$= M_{cs}; R= {P_i}
            e. else if $M_{c\,min}$ == M_{cs} then
            f. R= RU {P_i}
            g. end if
            h. end for
            i. for all PMs P_i in R do,
                    1. To avoid migration interference Move as chronological order
                            a. VMs with smallest S_n to P_i with smallest M_{cn}
                            b. VMs with smallest S_n to P_i with largest M_{cp}
                    2. To avoid migration interference Move as chronological order
                            a. VMs with smallest S_p to P_i with smallest M_{cp}
                            b. VMs with smallest S_p to P_i with largest M_{cn}
            j. end for.

Here the algorithm for implementing interference aware VM live migration is executed in two steps. Step 1 executed at source PM to find out capable VMs for live migration without interference and Step 2 to find out list of PMs capable of including the VMs from source without interference. Here we do not mention specifically any migration strategy for live migration, but delta compressed live migration or pre-copy method will work more effectively. The algorithm works as per Eq. (2). Step 1 collects list of VMs which cause least interference for migration. This list of VMs again divides into two, one contains VMs with least network interference and another with least co-located interference. In Step 2, it find out favourable PMs collection which can hold the coming VMs without interference. To avoid interference at destination, the set of PMs also divided into two, set of PMs with least network interference and least co located interference. Then perform the live migration of VMs into matched PM using any of the live migration strategy.

4   Experimental Setup and Results

We designed our experiments to exercise both net I/O traffic and CPU utilized processes for evaluating performance interference. The following diagram shows the experimental set up in this paper. Many isolated domains V_1, V_2, V_3, ... , V_n

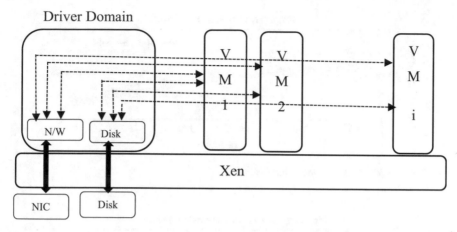

Fig. 2 Xen test bed

with resource allocated are hosted in the Physical platform. Here we are trying to live migrate network VMs with least interference to destination PM. Also ensures that the destination PM will not have interference issue after receiving this VMs. The same scenario follows in the case of CPU bound VMs also. Apache services provide HTTP communication between the VMs. Client uses httperf as load-generator are designed to access remote virtualized servers. PMs used in this scenario are connected by high speed Ethernet switch. The Linux kernel 2.6.32.10-xen is recommended for Xen 4.0.2 and Kernel 2.6.18.8-xen is recommended for Xen 4.0 (Fig. 2).

Interference is a big concern when we consider latency sensitive cloud applications. Interference occurred in cloud environment in different intensity and different duration as per workload, contention in physical resource, CPU allocation or network I/O [22]. As an example, if the resource contention for CPU exist in a cloud environment under low application load results a negligible impact whereas same resource content at heavy application load cause severe performance degradation. Another factor which determines the severity of interference is length of interference; how long interference exist in the given scenario. It is not at all a big deal if it last only for small period of time but the same will affect the performance of the system if it last for a long time [13].

In most of prior works on interference analysis, the study done on application performance under fixed workload and system conditions. But this sort of analysis doesn't give a real-time solution since applications often experience different workloads and service requirement of an application can change over time to time. Hence result infer from interference analysis under specific workloads might not hold true for all conditions.

From Fig. 3 it is clear that due to interference resource contention occurs in high request. A backlog of unserved requests are generated during severe interference

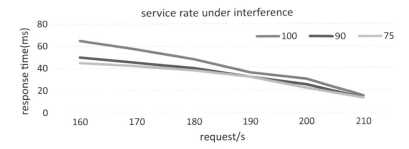

Fig. 3 Effect of service rate under interference

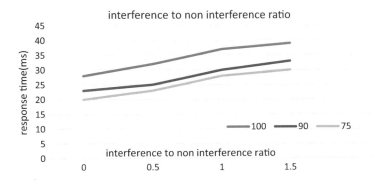

Fig. 4 Ratio of time spent in interference to non-interference

which degrade the performance of the system. Due to high interference resource contention increases which cause reduced service rate (Fig. 4).

If interference is periodic we can compare the time spent by VM under interference to that under non-interference [23, 24]. The graph shows that response time of cloud application as a function of ratio of interference to non-interference length. As the ratio increases response time increases sub linearly. When interference rate increases response time of a service goes up and it take a large amount of time to serve a request. When interference is low; response time is also negligible. Since service capacity is reduced due to interference then response time will be higher than non-interference condition. In other ways interference reduce the service rate.

To analyse interference in detail we collected the following system level characteristics of running VMM with VMs.

- VMM events per second: This metric gives the number of VMs are in the event channel. In other words it reflects the inter VM communication interference.
- VM switches per second: The number of VMM switches controls VMs per second. This metric put light on control interference among VMs.
- I/O count per second: During one second how many pages are transferred through I/O channel.

Table 1 System level characteristics of I/O workloads with running in VM

Work load (KB)	CPU %	Event/s	I/O execution (pages/ execution)	Waiting stage % Dom 0	Blocking stage % Dom 0	CPU % domain 0
1	97.50	224,100	4.97	4.85	10.55	46.83
5	97.1	242,210	6.14	4.51	10.48	45.71
15	68.5	294,754	8.45	3.18	13.74	37.75
30	55.45	357,842	3.75	1.87	17.71	35.87
50	48.74	354,587	3.21	1.25	19.38	33.54
100	42.54	331,254	3.01	0.75	20.95	31.04

- Execution per second: The number of execution periods per second for each VM.
- CPU utilization: Measured CPU utilization for each VM including Com 0. Total utilization is the summation of all.
- VM state in (waiting & Blocked): In both this stages VM is not utilizing CPU.

From Table 1 it is clear that 1 and 5 KB are CPU intensive workload and utilize CPU maximum. For these loads CPU utilization in domain 0 comparatively high as compare with other workloads. Their blocking state (waiting for I/O to complete) is relatively low and waiting stage is high as compared with other workloads. Event/s shows the number of VMs in the channel. Here network I/O processing workloads hold more number of VMs and this lead to interference. The workloads in 30, 50, and 100 KB utilize CPU low level shows that they are networking I/O processes and number of VMs involving in these workloads are more as compared with CPU bound processes. Their Blocking state value in Dom 0 is high since they demand more network I/O related processes to complete while completing their execution. The event and switch number are increasing gradually and event per second are also related to request rate of workload.

Analysing IALM algorithm, it performs double verification on source and destination PM to ensure there is no interference during live migration. As per Eq. (2) IALM algorithm divides the group of VMs in source PM into two categories, VMs with least network interference and VMs with least co-located interference. The same category VMs then transferred into the destination PM ensures that there no interference and thus performance degradation due to this migration. The algorithm select destination PM for set of VMs from source PM in reverse order also i.e., VMs with smallest network interference value to PM with largest co-location interference value.

5 Conclusions

In this paper we presented a VM interference aware live migration strategy IALM, which analyses and mitigates the issues caused by VM migration interference which ultimately leads to performance degradation and SLA violation. This method overcome ill effects of VM migration interference by analysing VM network interference and VM CPU interference. The extensive experiments and simulation done on Xen reveals intensity and performance degradation issues of interference on real-time cloud applications. The algorithm ensures VM grouping and VM communication in physical environment in such a way that it will nullify the effect of interference by balancing different categories of workloads and always ensures to run VMs below interference threshold. The algorithm performs well in simulated environment and we would like to implement in any of real-time cloud environment.

References

1. Novakovic, D., Vasic, N., Novakovic, S., Kostic, D., & Bianchini, R. (2013). Deepdive: Transparently identifying and managing performance interference in virtualized environments. In *Proceedings of the 2013 USENIX Annual Technical Conference*, no. EPFL-CONF-185984.
2. Votke, S., Javadi, S. A., & Gandhi, A. Modeling and analysis of performance under interference in the cloud.
3. Bloch, M. T., Sridaran, R., & Prashanth, C. (2014). Analysis and survey of issues in live virtual machine migration interferences. *International Journal of Advanced Networking & Applications.*
4. Amannejad, Y., Krishnamurthy, D., & Far, B. (2015). Detecting performance interference in cloud-based web services. In *2015 IFIP/IEEE International Symposium on Integrated Network Management (IM)* (pp. 423–431). New York: IEEE.
5. Liu, H., & He, B. (2015). Vmbuddies: Coordinating live migration of multi-tier applications in cloud environments. *IEEE Transactions on Parallel and Distributed Systems, 26*(4), 1192–1205.
6. Xu, F., Liu, F., Liu, L., Jin, H., Li, B., & Li, B. (2014). iaware: Making live migration of virtual machines interference-aware in the cloud. *IEEE Transactions on Computers, 63*(12), 3012–3025.
7. Yang, Z., Fang, H., Wu, Y., Li, C., Zhao, B., & Huang, H. H. (2012). Understanding the effects of hypervisor i/o scheduling for virtual machine performance interference. In *2012 IEEE 4th International Conference on Cloud Computing Technology and Science (CloudCom)* (pp. 34–41). New York: IEEE.
8. Pu, X., Liu, L., Mei, Y., Sivathanu, S., Koh, Y., Calton, P., et al. (2013). Who is your neighbor: Net i/o performance interference in virtualized clouds. *IEEE Transactions on Services Computing, 6*(3), 314–329.
9. Zhang, W., Zhu, M., Gong, T., Xiao, L., Ruan, L., Mei, Y., et al. (2012). Performance degradation-aware virtual machine live migration in virtualized servers. In *2012 13th International Conference on Parallel and Distributed Computing, Applications and Technologies (PDCAT)* (pp. 429–435). New York: IEEE.

10. Liu, C. (2016). A load balancing aware virtual machine live migration algorithm. In *Proceedings of the International Conference on Sensors, Measurement and Intelligent Materials* (pp. 370–373).
11. Takouna, I., Rojas-Cessa, R., Sachs, K., & Meinel, C. (2013). Communication-aware and energy-efficient scheduling for parallel applications in virtualized data centers. In *Proceedings of the 2013 IEEE/ACM 6th International Conference on Utility and Cloud Computing* (pp. 251–255). IEEE Computer Society.
12. Shim, Y.-C. (2015). Inter-VM performance interference aware static VM consolidation algorithms for cloud-based data centers. In *Recent advances in electrical engineering* (Vol. 18).
13. Wood, T., Shenoy, P. J., Venkataramani, A., & Yousif, M. S. (2007). Black-box and Gray-box strategies for virtual machine migration. In *NSDI* (Vol. 7, pp. 17).
14. Kesavan, M., Gavrilovska, A., & Schwan, K. (2010). On disk I/O scheduling in virtual machines. In *Proceedings of the 2nd Conference on I/O Virtualization* (pp. 6–6). USENIX Association.
15. Barker, Sean, Yun Chi, Hyun Jin Moon, Hakan Hacigümüş, and Prashant Shenoy. "Cut me some slack: Latency-aware live migration for databases." In Proceedings of the 15th international conference on extending database technology, pp. 432–443. ACM, 2012.
16. Lu, P., Barbalace, A., Palmieri, R., & Ravindran, B. (2013). Adaptive live migration to improve load balancing in virtual machine environment. In *European Conference on Parallel Processing* (pp. 116–125). Berlin, Heidelberg: Springer.
17. Zheng, J., Ng, T. S. E., Sripanidkulchai, K., & Liu, Z. (2014). Comma: Coordinating the migration of multi-tier applications. In *ACM SIGPLAN Notices* (Vol. 49, no. 7, pp. 153–164). New York: ACM.
18. Piraghaj, S. F., Calheiros, R. N., Chan, J., Dastjerdi, A. V., & Buyya, R. (2015). Virtual machine customization and task mapping architecture for efficient allocation of cloud data center resources. *The Computer Journal, 59*(2), 208–224.
19. Mishra, A., Jain, R., & Durresi, A. (2012). Cloud computing: Networking and communication challenges. *IEEE Communications Magazine, 50*(9).
20. Sabina, S. (2014, July). Multiple correlation coefficient approach for VM migration. *International Journal of Advanced Research in Computer and Communication Engineering, 3*(7).
21. Yang, Y. (2016). *On optimizations of Virtual Machine live storage migration for the Cloud.* Ph.D. diss., The University of Nebraska-Lincoln.
22. Caglar, F., Shekhar, S., & Gokhale, A. (2011). Towards a performance interference-aware virtual machine placement strategy for supporting soft real-time applications in the cloud.
23. Casale, G., Kraft, S., & Krishnamurthy, D. (2011). A model of storage I/O performance interference in virtualized systems. In *2011 31st International Conference on Distributed Computing Systems Workshops (ICDCSW)* (pp. 34–39). New York: IEEE.
24. Zhang, Q., Liu, L., Ren, Y., Lee, K., Tang, Y., Zhao, X., et al. (2013). Residency aware inter-VM communication in virtualized cloud: Performance measurement and analysis. In *2013 IEEE Sixth International Conference on Cloud Computing (CLOUD)* (pp. 204–211). New York: IEEE.

Artificial Bee Colony Optimization—Population-Based Meta-Heuristic Swarm Intelligence Technique

Anand Nayyar, Vikram Puri and G. Suseendran

Abstract Swarm Agents are known for their cooperative and collective behavior and operate in decentralized manner which is regarded as Swarm Intelligence. Various techniques like Ant Optimization, Wasp, Bacterial Foraging, PSO, etc., are proposed and implemented in various real-time applications to provide solutions to various real-time problems especially in optimization. The aim of this paper to present ABC algorithm in a comprehensive manner. The ABC-based SI technique proposed has demonstrated that it has superior edge in solving all types of unconstrained optimization problems. Many researchers have fine-tuned the basic algorithm and proposed different ABC based algorithms. The result show that still lots of work is required mathematically and live implementation in order to enable ABC algorithm to be applied to constrained problems for effective solutions.

Keywords Swarm agents · Artificial bee colony optimization · Honey bees
Waggle dance · Optimization · Artificial bee · Swarm · Swarm intelligence

1 Introduction

The word "Swarm" [1] denotes animal aggregation like Birds, Fishes, Elephants, Bats as well as the colonies of insects like Ants, Bees, termites and Wasps performing collective behavior. Every agent in the swarm works without any sort of

A. Nayyar
Graduate School, Duy Tan University, Da Nang, Vietnam
e-mail: anandnayyar@duytan.edu.vn

V. Puri (✉)
eLearning Center and R&D, Center for Visualization & Simulation,
Duy Tan University, Da Nang, Vietnam
e-mail: vikrampuri03@gmail.com

G. Suseendran
Department of Information Technology, School of Computing Sciences,
VELS Institute of Science, Technology & Advanced Studies (VISTAS),
Chennai 600117, Tamil Nadu, India
e-mail: suseendar_1234@yahoo.co.in

© Springer Nature Singapore Pte Ltd. 2019
V. E. Balas et al. (eds.), *Data Management, Analytics and Innovation*,
Advances in Intelligent Systems and Computing 839,
https://doi.org/10.1007/978-981-13-1274-8_38

513

supervision and each of the agent has stochastic behavior because of its perception in the neighborhood. Swarm Intelligence (SI) is primarily concerned with collective behaviors that results from local interactions of swarm agents among each other along with their environment [2, 3]. Swarm Intelligence is basically regarded as discipline concerned with the study and research of collective behavior of swarm performed in nature like Building of nest, Foraging, sorting of items in colonies of insects, flocking, herding, schooling behaviors, etc. Considering the engineering view, it is regarded as bottom up design of distributed systems that demonstrates novel behavior at upper level due to results of varied actions of number of units interacting among each other along with their environment.

Swarm Intelligence [4] is regarded as one of the most important areas of research which is applied by different researchers for problem solving, computations and optimizing solutions. Swarm Intelligence as a discipline can be applied to a wide range of disciplines like Mathematics, Robotics, Telecommunications, Computer Science, Mechanical Engineering for performing combinatorial and continuous optimizations. Results show that Swarm Intelligence based techniques have given excellent results as compared to other techniques like Fuzzy Logic, Genetic Algorithm, Discrete Mathematics, etc.

The Algorithms in Swarm Intelligence are utilized to solve various real-time problems. The most classical algorithm in Swarm Intelligence is Genetic Algorithm (GA) [5, 6]. With the passage of time, various other swarm intelligence algorithms are proposed by various researchers like: Ant Colony Optimization (ACO) [7–10], PSO [11–13], Cat Swarm (CSO), Artificial Immune System, Elephant Swarm, Bat Swarm Optimization, Differential Evolution (DE) [14, 15] Wasp Optimization and many more.

Bees swarm around their hive and the behavior can be extended to other systems. Some other approaches based on Bee Colony Optimization are proposed to depict the same behavior of honey bees to solve various combinatorial problems [16].

In general, bee colony based algorithms [17] don't have any sort of unified foundation, and are based on different behavioral concepts. The algorithms can be distinguished on the basis of: Mating behavior of Honey bees; Foraging behavior of honey bees. In addition to this, recent research being conducted [18] and proposed algorithm suggested that selecting nest-site behavior of honey bees can be used in the real world for proposing optimization solutions.

Mating-Inspired Algorithms are inspired from bee colonies. Genetic Diversity is regarded as foundation stone which drives the mechanism of bee towards ecological success [19]. It is due to Young Queen polyandrous behavior during her flight.

Foraging behavior inspired algorithms of honey bees are regarded as decentralized process which works on decision making of individual bees. These algorithms inspire the bee colonies to maintain appropriate ratio of exploration and exploitation of food sources. This behavior is highly adaptive; means can be changed depending on the resources if required [20–22].

In Honey bee colonies, Scout bees efficiently search for food and back to hive and tell the source of food via waggle-dance. Waggle-dance also specifies the distance of food from source, direction as well as quality. On the basis of waggle-dance, the bees distribute themselves in terms of profitability.

In ABC Algorithm, bees are classified into two groups—Employed Bees (EB) and Onlooker Bees (OB). Employed Bees perform the task of finding and maintaining promising solutions; Onlooker bees perform the task of local search. The task of exploration and exploitation is done by Employed bees. If the solution of EB's does not improvise over a series of steps, it will be abandoned and then a new random solution will be used. EBs' are known as 'Scouts' when choosing a new random solution.

In this research paper, the comprehensive review of Artificial Bee Colony Optimization (ABC) is presented which was proposed by Karaboga [22].

2 Artificial Bee Colony

2.1 Features of Intelligent Swarms

The world consists of many different kinds of swarms, but considering every swarm as intelligent is not right, as every swarm has its own intelligence. The important characteristic which results in the collective behavior by means of local interactions among simple agents is "Self-Organization". The following are the four types of characteristics given by Bonabeau et al. [1]:

1. Positive Feedback: It is basically defined as the development of best structures. Trail laying, recruitment and reinforcement in ants forming Ant Colony Optimization are examples of positive feedback.
2. Negative Feedback: Counterbalancing positive feedback and stabilizing collective pattern. Negative feedback is utmost necessary in order to avoid saturation.
3. Fluctuations: Random walk, task switching, errors among individual swarm agents are termed as fluctuations which are highly important for creativity.
4. Multiple Interactions: One individual agent in complete swarm makes use of information coming from other agents and spread throughout the network for work completion.

Apart from the above four characteristics, tasks are performed simultaneously by specialized agents called "Division of Labor" which is equally important as Self-Organization to complete intelligence in swarm agents [23].

2.2 Real Honey Bees—Foraging Behavior

A model depicting the foraging behavior of honey bee colony based on reaction-diffusion equations [24–26] was proposed by Tereshko. The emergence of collective intelligence of honey bees comprises of three main components:

a. Sources of Food
b. Foragers—Employed
c. Foragers—Unemployed

Food Sources: In order to determine the effective food source, a forager bee has to determine various characteristics with regard to food sources like, distance between food source and hive, taste of nectar, energy richness, easiness in extracting the energy from nectar. Considering simplicity, the "Profitability" of food source can be represented with single quantity.

Employed Foragers: It is primarily designated at specific food source which is currently being exploited. The bee carries the information about the food source back to hive and share with other bee agents in the hive. The information being shared is all about: distance, direction as well as profitability of food source.

Unemployed Foragers: They perform the task of only searching the food source to exploit. Under bee colonies, there are two types of Unemployed foragers: Scouts and Onlookers. Scouts—search the food in the nearby environment surroundings; Onlookers: wait in hive for information by scouts to exploit the food source. The average number of scouts is near to 5–10% as compared to other bees in bee colonies or in hive.

Information exchange among honey bees is regarded as the most important event which occurs. While examining the entire hive, it is possible to distinguish between some parts that commonly exist in all hives. The most important aspect of hive is to exchange the information regarding food source via Waggle-Dance. As the information about different food sources is available to onlooker bees on the dance floor, onlooker bees watches various dances and then decides which is the most profitable food source to exploit. The most profitable source has the highest probability to be chosen as compared to other food sources.

Considering honey bees, the following characteristics can be defined on which the principle of Self-Organization is based:

- Positive Feedback: With the increase in the amount of food sources, more onlooker bees visits the food sources.
- Negative Feedback: Food source exploitation is stopped by bees.
- Fluctuations: The scouts perform search in environment to locate food source in random fashion.
- Multiple Interactions: The information of food sources is shared by employed bees with onlooker bees on the dance floor to exploit and extract food (Fig. 1).

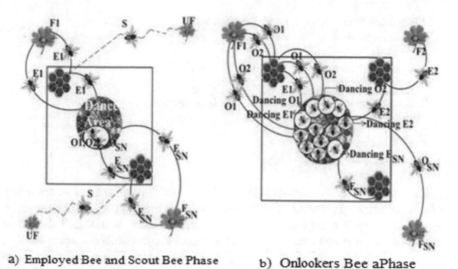

a) Employed Bee and Scout Bee Phase   b) Onlookers Bee aPhase

Fig. 1 **a** Phases of working of employed bee and scout bee **b** onlooker bee phase [37]

2.3 Artificial Bee Colony Optimization (ABC) Algorithm

2.3.1 Theoretical Foundation of ABC Algorithm

Artificial Bee Colonies can be categorized into three types of bee groups: Employed Bees; Onlooker Bees and Scout Bees.

Employed Bees are designated at specific food sources and comes back to hive and perform the waggle dance to aware the onlooker bees regarding food source. Onlooker bees watch the waggle-dance of employed bees to choose the most profitable food source among all waggle-dances. Scout bees search for food source randomly. Onlooker and Scout Bees are also termed as "Unemployed Bees".

All the food sources are discovered by scout bees in the initial phase. After that, the food sources nectar level is exploited by employed bees and onlooker bees, and this process remains continuous unless the bees become exhausted. Then, the employed bee which was exploiting the exhausted food source becomes a scout bee in search of further food source once again. In other words, the employed bees whose food sources are exhausted transforms itself to scout bee. In ABC Algorithm, the position of food source is regarded as a possible solution to the problem and the nectar amount in food determines the quality (fitness) of the problem. The number of employed bee is equal to a number of food sources (solutions) as every single employed bee is associated with only one food source.

Algorithm Structure—General Approach
Initialization Phase
REPEAT
        Employed Bees Phase
        Onlooker Bees Phase
        Scouts Bees Phase
        Memorize the best solution attained till now
UNTIL (Cycle = Maximum Cycle Number or a Maximum CPU Time)
ABC Algorithm- Pseudocode.

2.3.2 Phases of ABC Algorithm

Initialization Phase: In the first step of ABC Algorithm, generation of randomly distributed initial population of sn/2, where sn = size of population. Each solution x_i ($I = 1$, 2, sn/2) is regarded as D Dimensional vector. It is regarded as time consuming process and sometimes become impossible to estimate feasible solution in random fashion. In this phase, random values between the lower and upper limits of parameters are assigned to solution parameters. Failure i is regarded as non-improvement number of solution x_i.

The following Algorithm demonstrates Initialization phase of ABC Algorithm

> **for** $i = 1$ to $sn/2$**do**
>   **for** $j = 1$ to D**do**
>   Generate $\bar{x}_i$ solution
> $x_i^j = x_{min}^j + rand(0, 1)(x_{max}^j - x_{min}^j)$
>   where x_{min}^j and x_{max}^j are lower and upper bound of the
> parameter j, respectively.
>   **end for**
> $failure_i = 0$
> **end for**

After the phase of initialization, the evaluation of population is done with respect to the search iterations being performed by employed bees, onlooker bees and scout bees.

The following algorithm demonstrates the procedure deployed by Employed Bees to search the food in a random manner in an environment.

- Feasible Solution is: $violation_i \leq 0$ is best as compared to infeasible solution, i.e., $violation_j \geq 0$.
- The solution which is having best objective function value is selected between $violation_i \leq 0$ or $violation_j \leq 0$.
- The one having smaller constrained violation is preferred, i.e., $violation_i \geq 0$ and $violation_j \geq 0$.

When all employed bees complete the process of food search, they return to hive and share the information via Waggle-dance with onlooker bees on the basis of probability values.

The following Algorithm shows the Onlooker Bees—Probability calculation phase.

for $i = 1$ to $sn/2$ **do**
Calculate the probability values p_i for the solutions using fitness and/or violation of the solutions by

$$p_i = \begin{cases} \left(0.5 + \dfrac{fitness_i}{\sum\limits_{j=1}^{sn} fitness_j} \right) \times 0.5 & \text{if solution is feasible} \\[4ex] \left(1 - \dfrac{violation_i}{\sum\limits_{j=1}^{sn} violation_j} \right) \times 0.5 & \text{if solution is infeasible} \end{cases}$$

where $violation_i$ is the penalty value of the solution $\bar{x}_i$ and $fitness_i$ is the fitness value of the solution $\bar{x}_i$ which is proportional to the nectar amount of that food source. The fitness is determined by

$$fitness_i = \begin{cases} 1/(1+f_i) & \text{if } f_i \geq 0 \\ 1 + abs(f_i) & \text{if } f_i < 0 \end{cases}$$

where f_i is the cost value of the solution $\bar{x}_i$.
end for

ABC Algorithm—Onlooker Bees Phase-Probability Calculation

Onlooker bees compute the feasible solution on basis of employed bee's information and choose which food source has the highest amount of nectar.

for $i = 1$ to $sn/2$ **do**
  for $j = 1$ to D **do**
    Produce a new food source $\bar{v}_i$ for the employed bee of the food source $\bar{x}_i$ by using

$$v_{ij} = \begin{cases} x_{ij} + \phi_{ij}(x_{ij} - x_{kj}), & \text{if } R_j < MR \\ x_{ij}, & \text{otherwise} \end{cases}$$

where $k \in \{1, 2, \ldots, sn\}$ is randomly chosen index that has to be different from i and ϕ_{ij} is uniformly distributed random real number in the range of $[-1,1]$. R_j is uniformly distributed random real number in the range of $[0,1]$ and MR is a control parameter of ABC algorithm in the range of $[0,1]$ which controls the number of parameters to be modi?ed.
  end for
  If no parameter is changed, change one random parameter of the solution $\bar{x}_i$ by
$$v_{ij} = x_{ij} + \phi_{ij}(x_{ij} - x_{kj})$$
where j is uniformly distributed random integer number in the range $[1, D]$.
  Evaluate the quality of $\bar{v}_i$
  Apply the selection process between $\bar{v}_i$ and $\bar{x}_i$ based on Deb's method
  If solution $\bar{x}_i$ does not improve $failure_i = failure_i + 1$, otherwise $failure_i = 0$
end for

After giving a food source, ABC Algorithm [17] makes a selection among various food sources to select the best one. Another modification of ABC Algorithm is to solve constrained optimization problem using DEB's rule rather than using Greedy Algorithm [27].

By applying DEB's rule, the bee is able to memorize the new position by overwriting the old food position. Deb's method makes use of tournament selection operator, where two solutions are compared with the following conditions:

- Feasible Solution is: $violation_i \leq 0$ is best as compared to infeasible solution, i.e., $violation_j \geq 0$.

- The solution which is having best objective function value is selected between $\text{violation}_i \leq 0$ or $\text{violation}_j \leq 0$.
- The one having smaller constrained violation is preferred, i.e., $\text{violation}_i \geq 0$ and $\text{violation}_j \geq 0$.

When all employed bees complete the process of food search, they return to hive and share the information via Waggle-dance with onlooker bees on the basis of probability values.

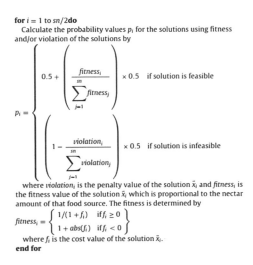

for $i = 1$ to $sn/2$**do**
    Calculate the probability values p_i for the solutions using fitness and/or violation of the solutions by

$$
p_i = \begin{cases} \left(0.5 + \left(\dfrac{fitness_i}{\sum\limits_{j=1}^{sn} fitness_j} \right) \times 0.5 \right) & \text{if solution is feasible} \\ \left(\left(1 - \dfrac{violation_i}{\sum\limits_{j=1}^{sn} violation_j} \right) \times 0.5 \right) & \text{if solution is infeasible} \end{cases}
$$

where $violation_i$ is the penalty value of the solution $\bar{x}_i$ and $fitness_i$ is the fitness value of the solution $\bar{x}_i$ which is proportional to the nectar amount of that food source. The fitness is determined by

$$
fitness_i = \begin{cases} 1/(1 + f_i) & \text{if } f_i \geq 0 \\ 1 + abs(f_i) & \text{if } f_i < 0 \end{cases}
$$

where f_i is the cost value of the solution $\bar{x}_i$.
end for

ABC Algorithm—Onlooker Bees Phase-Probability Calculation

Onlooker bees compute the feasible solution on the basis of employed bee's information and choose which food source has the highest amount of nectar.

$t = 0, i = 1$
repeat
    if $random < p_i$**then**
        $t = t + 1$
        for $j = 1$ to D**do**
            Produce a new food source $\bar{v}_i$ for the onlooker bee of the food source $\bar{x}_i$ by using
        end for
        Apply the selection process between $\bar{v}_i$ and $\bar{x}_i$ based on Deb's method
        If solution $\bar{x}_i$ does not improve $failure_i = failure_i + 1$, otherwise $failure_i = 0$
    end if
    $i = i + 1$
    $i = i \bmod ((sn/2) + 1)$
until $t = sn/2$

Onlooker Bee Phase—ABC Algorithm

After the distribution of all onlooker bees, food sources which are not rich in nectar are not exploited and left abandoned. The food sources which are left abandoned by bees is replaced by new food sources discovered by scouts. Different ABC Algorithms [28] proposed for constrained and unconstrained problems is the production of artificial scouts at a predetermined period of cycles for discovering food sources randomly. This time period id called Scout Production Period.

```
if cycle mod SPP = 0then
  if max(failure_i) > limitthen
    Replace x̄_i with a new randomly produced solution by
  end if
end if
```

SCOUT Bees Phase ABC Algorithm

```
place each employed bee on a random position in the search space
while stop criterion not met do
  for all employed bees do
    if # steps on same position = 1 then
      choose random position in search space
    else
      try to find better position
      if better position found then
        move from current position to found position
      end if
    end if
  end for
  for all onlooker bees do
    choose an employed bee and move to its position
    try improve position
  end for
end while
```

Complete ABC Algorithm

2.4 ABC Algorithm Under Constrained & Unconstrained Optimization Problems [29]

ABC Algorithm in Constrained Optimization Problems
Step 1: Initialize the Population of Solutions.
Step 2: Population Evaluation
Step 3: cycle =1
Step 4: Repeat
Step 5: Produce new solutions for the employed bees and evaluate the bees.
Step 6: Apply Greedy Selection process
Step 7: Calculate the Probability values for the solutions.
Step 8: Produce the new solutions for the onlookers from the solutions selected depending on conditions and evaluation of solutions.
Step 9: Apply Greedy Selection Process
Step 10: Determine the abandoned solution for the Scout. If exists, and replace with new randomly produced solution.
Step 11: Memorize the best solution attained so far
Step 12: Cycle= cycle +1
Step 13: until cycle = MCN

ABC Algorithm in Unconstrained Optimization Problems [29]
Step 1: Initialize the Population of Solutions.
Step 2: Population Evaluation
Step 3: cycle =1
Step 4: Repeat
Step 5: Produce new solutions for the employed bees and evaluate the bees.
Step 6: Apply selection process based on Deb's Method.
Step 7: Calculate the Probability values for the solutions.
Step 8: Produce the new solutions for the onlookers from the solutions selected depending on conditions and evaluation of solutions.
Step 9: Apply selection process based on DEB's method.
Step 10: Determine the abandoned solution for the Scout. If exists, and replace with new randomly produced solution.
Step 11: Memorize the best solution attained so far
Step 12: Cycle= cycle +1
Step 13: until cycle = MCN

3 Bee Colony Optimization

Bee Colony Optimization [30–36] was developed by Lucic and Teodorovic and is primarily based on the principle of collective bee intelligence. Artificial Bees are regarded as individual agents, and work collaboratively to solve complex combinatorial optimization problems. Every bee in the colony proposes a novel solution for each problem.

Bee Colony Optimization comprises of two phases: Forward Pass and Backward Pass. In forward pass, every artificial bee explores the area. It performs some predefined set of movements which either give a new solution or proposes improvements to existing solutions. After getting new results, the bees move back to the next and proceed to the next phase called backward phase. In backward phase, the bees share the information regarding their solution.

Applying Bee Colony Optimization to Travelling Salesman problem, the problem of TSP is decomposed into stages. In every stage, a bee selects a new node to be included in the partial Travelling salesman tour visited so far.

Considering mother nature, bees perform a waggle-dance which acts as an information base for other bees regarding quantity, quality and distance of the food. In BCO search algorithm, the artificial bees propose solution quantity, i.e., Objective function value. During the stage of backward pass, every bee decides which food source to consider whether to search again or go to some other food source or abandon the existing food source. Every bee, choose a new solution from recruiters by roulette wheel.

In second forward pass, bees expand the partial solutions created, with a pre-defined number of nodes, and after that perform backward pass and return to hive. On returning to the hive, bees again decide whether to choose, make a decision or perform third pass.

The two phases in BCO, i.e., Forward and Backward and performed until the best solution is determined.

Consider stages as st = {st1, st2, st$n$} which are regarded as a finite number of pre-selected stages, m is the number of stages. B stands for number of bees participating in the search process and I will be regarded as number of iterations to be performed.

Partial solutions at stage st$_j$ can be denoted as S_j (j = 1, 2, ...m).

(1) *Initialization.* Determine the number of bees B, and the number of iterations I. Select the set of stages ST = {st_1, st_2 ,..., st_m}. Find any feasible solution x of the problem. This solution is the *initial best solution*.
(2) Set i: = 1. Until i = I, repeat the following steps:
(3) Set j = 1. Until j = m, repeat the following steps:
Forward pass: Allow bees to fly from the hive and to choose B partial solutions from the set of partial solutions S_j at stage st_j.
Backward pass: Send all bees back to the hive. Allow bees to exchange information about quality of the partial solutions created and to decide whether to abandon the created partial solution and become again uncommitted follower, continue to expand the same partial solution without recruiting the nestmates, or dance and thus recruit the nestmates before returning to the created partial solution. Set, j: = j + 1.
(4) If the best solution x_i obtained during the i-th iteration is better than the best-known solution, update the best known solution (x: = x_i).
(5) Set, i: = i + 1.

Bee Colony Optimization Algorithm

4 Conclusion and Future Scope

Artificial Bee Colony Optimization is regarded as one of the most modern techniques discovered in the area of Swarm Intelligence. It is derived from honeybees foraging behavior. The paper presents the comprehensive concept of Artificial Bee Colony Optimization. Considering all the works performed and algorithms been proposed by several researchers, most of the work is performed theoretically and lots of work need to be updated and modified to fine tune the performance for better implementation in real-world applications. It is used for solving combinatorial optimization problems. In order to improvise the overall performance of Bee Colony Based Algorithms, new production mechanisms need to be worked out and proposed.

Future Scope

Considering future work, the prime focus would be on improvising the methodology and work structure of ABC Algorithm and implementation of ABC-based Routing Protocol for MANETS, WSN and other wireless communication networks.

References

1. Bonabeau, E., Dorigo, M., & Theraulaz, G. (1999). Swarm intelligence: From natural to artificial systems (No. 1). Oxford: Oxford University Press.
2. Blum, C., & Li, X. (2008). Swarm intelligence in optimization. In *Swarm Intelligence* (pp. 43–85). Berlin, Heidelberg: Springer.
3. Kennedy, J. (2006). Swarm intelligence. In *Handbook of nature-inspired and innovative computing* (pp. 187–219). US: Springer.
4. Garnier, S., Gautrais, J., & Theraulaz, G. (2007). The biological principles of swarm intelligence. *Swarm Intelligence, 1*(1), 3–31.
5. Goldberg, D. (1989). *Genetic algorithms in optimization, search and machine learning.* Reading. Boston: Addison-Wesley.
6. Guo, Y., Cao, X., Yin, H., & Tang, Z. (2007). Coevolutionary optimization algorithm with dynamic sub-population size. *International Journal of Innovative Computing, Information and Control, 3*(2), 435–448.
7. Dorigo, M., Birattari, M., & Stutzle, T. (2006). Ant colony optimization. *IEEE Computational Intelligence Magazine, 1*(4), 28–39.
8. Maniezzo, V., & Carbonaro, A. (2002). Ant colony optimization: An overview. In *Essays and surveys in metaheuristics* (pp. 469–492). US: Springer.
9. Stützle, T. (2009, April). Ant colony optimization. In *International Conference on Evolutionary Multi-Criterion Optimization* (pp. 2–2). Berlin, Heidelberg: Springer.
10. Nayyar, A., & Singh, R. (2016, March). Ant Colony Optimization—Computational swarm intelligence technique. In *2016 3rd International Conference on Computing for Sustainable Global Development (INDIACom)* (pp. 1493–1499). IEEE.
11. De Castro, L. N., & Von Zuben, F. J. (1999). Artificial immune systems: Part I–basic theory and applications. Universidade Estadual de Campinas, Dezembro de, Tech. Rep, 210(1).
12. Kennedy, J. (2011). Particle swarm optimization. In *Encyclopedia of machine learning* (pp. 760–766). US: Springer.
13. Xie, X., Zhang, W., & Yang, L. (2003). Particle swarm optimization. *Control and Decision, 18,* 129–134.
14. Karaboga, D., & Akay, B. (2009). A comparative study of artificial bee colony algorithm. *Applied Mathematics and Computation, 214*(1), 108–132.
15. Goldberg, D. E., & Deb, K. (1991). A comparative analysis of selection schemes used in genetic algorithms. *Foundations of genetic algorithms, 1,* 69–93.
16. Yang, X. S. (2005). Engineering optimizations via nature-inspired virtual bee algorithms. In *Artificial intelligence and knowledge engineering applications: A bioinspired approach* (pp. 317–323).
17. Akay, B., & Karaboga, D. (2012). A modified artificial bee colony algorithm for real-parameter optimization. *Information Sciences, 192,* 120–142.

18. Diwold, K., Beekman, M., & Middendorf, M. (2010). Bee nest site selection as an optimization process. In *ALIFE* (pp. 626–633).
19. Mattila, H. R., & Seeley, T. D. (2007). Genetic diversity in honey bee colonies enhances productivity and fitness. *Science, 317*(5836), 362–364.
20. Biesmeijer, J. C., & de Vries, H. (2001). Exploration and exploitation of food sources by social insect colonies: A revision of the scout-recruit concept. *Behavioral Ecology and Sociobiology, 49*(2), 89–99.
21. Teodorovic, D., Lucic, P., Markovic, G., & Dell'Orco, M. (2006, September). Bee colony optimization: Principles and applications. In *NEUREL 2006. 8th Seminar on Neural Network Applications in Electrical Engineering, 2006* (pp. 151–156). IEEE.
22. Karaboga, D. (2005). *An idea based on honey bee swarm for numerical optimization (Vol. 200)*. Technical report-tr06, Erciyes University, Engineering Faculty, Computer Engineering Department.
23. Millonas, M. M. (1994). Swarms, phase transitions, and collective intelligence. In *Santa Fe Institute Studies in the Sciences of Complexity-Proceedings Volume*—(Vol. 17, pp. 417–417). Massachusetts: Addison-Wesley Publishing Co.
24. Tereshko, V., & Loengarov, A. (2005). Collective decision making in honey-bee foraging dynamics. *Computing and Information Systems, 9*(3), 1.
25. Tereshko, V. (2000, September). Reaction-diffusion model of a honeybee colony's foraging behaviour. In *International Conference on Parallel Problem Solving from Nature* (pp. 807–816). Berlin, Heidelberg: Springer.
26. Tereshko, V., & Lee, T. (2002). How information-mapping patterns determine foraging behaviour of a honey bee colony. *Open Systems and Information Dynamics, 9*(02), 181–193.
27. Karaboga, D., & Akay, B. (2011). A modified artificial bee colony (ABC) algorithm for constrained optimization problems. *Applied Soft Computing, 11*(3), 3021–3031.
28. Karaboga, D., & Basturk, B. (2008). On the performance of artificial bee colony (ABC) algorithm. *Applied Soft Computing, 8*(1), 687–697.
29. Karaboga, D., Akay, B., & Ozturk, C. (2007). Artificial bee colony (ABC) optimization algorithm for training feed-forward neural networks. *MDAI, 7,* 318–319.
30. Lucic, P., & Teodorovic, D. (2001, June). Bee system: Modeling combinatorial optimization transportation engineering problems by swarm intelligence. In *Preprints of the TRISTAN IV triennial symposium on transportation analysis* (pp. 441–445).
31. Lucic, P., & Teodorovic, D. (2002). Transportation modeling: An artificial life approach. In *Proceedings. 14th IEEE International Conference on Tools with Artificial Intelligence, 2002. (ICTAI 2002)* (pp. 216–223). IEEE.
32. Lučić, P., & Teodorović, D. (2003). Computing with bees: Attacking complex transportation engineering problems. *International Journal on Artificial Intelligence Tools, 12*(03), 375–394.
33. Lučić, P., & Teodorović, D. (2003). Vehicle routing problem with uncertain demand at nodes: The bee system and fuzzy logic approach. In *Fuzzy sets based heuristics for optimization* (pp. 67–82).
34. Teodorovic, D. (2003). Transport modeling by multi-agent systems: A swarm intelligence approach. *Transportation Planning and Technology, 26*(4), 289–312.
35. Teodorovic, D., & Dell'Orco, M. (2005). Bee colony optimization—A cooperative learning approach to complex transportation problems. In *Advanced OR and AI methods in transportation* (pp. 51–60).
36. Teodorović, D. (2009). Bee colony optimization (BCO). In *Innovations in swarm intelligence* (pp. 39–60).
37. Shah, H., Ghazali, R., & Hassim, Y. M. M. (2014). Honey bees inspired learning algorithm: Nature intelligence can predict natural disaster. In *Recent Advances on Soft Computing and Data Mining* (pp. 215–225). Springer, Cham.

Enhanced VSDL Hash Algorithm for Data Integrity and Protection

G. Karthi and M. Ezhilarasan

Abstract In the modern world, securing information is a challenging task. Cryptography is the field of building information to an unintelligible form in order to secure it. Cryptography provides various services such as data integrity, confidentiality, and access control. Data integrity is the assurance of reliability and completeness of data throughout its lifetime. In this paper, modified VSDL hash algorithm is proposed which provides data integrity. Various cryptographic hash algorithms are available to provide data integrity. Among the various cryptographic hash algorithms existing, an analysis based on different metrics such as algorithm strengths and weaknesses is performed. The experimental results of those algorithms show the overall comparative performance with the existing systems. The proposed algorithm provides 2% better results compared with the existing algorithms.

Keywords Cryptography · SHA · Data integrity · VSDL · Discrete logarithms

1 Introduction

Cryptography is an art of rewriting the information on the applications into disorganized or in unintelligible format to protect the data from unauthorized access. It relates to the study of mathematical techniques associated with the aspects of knowledge security like the confidentiality, integrity, and authentication of data. Data integrity is the mechanism of identifying the completeness of the received data, whether it is modified or not. The integrity of a message could be provided by

G. Karthi (✉) · M. Ezhilarasan
Department of Computer Science and Engineering, Pondicherry Engineering College,
Puducherry, India
e-mail: karthi.govindharaju@gmail.com

M. Ezhilarasan
e-mail: mrezhil@pec.edu

© Springer Nature Singapore Pte Ltd. 2019
V. E. Balas et al. (eds.), *Data Management, Analytics and Innovation*,
Advances in Intelligent Systems and Computing 839,
https://doi.org/10.1007/978-981-13-1274-8_39

cryptographic hash functions. The given message m is applied to the hash function f which produces a fixed length hash code such that $f = h(m)$.

Then, the message is concatenated with the hash code generated and sent it across to the receiver. The receiver separates the message and the hash code and produces the new hash code for the message received. Finally, the newly generated hash code h' is compared with the hash code h; if equal then the message is complete. Otherwise, the message is modified.

Different hashing algorithms used for data integrity are MD2, MD4, MD5, MD6, SHA, ECOH, VSH, and FSB.

A hash function f can take any arbitrary length of string and computes fixed length of message called hash code as output. The hash function f can be defined in Eq. (1) as follows:

$$f : \{0, 1\}^* \rightarrow \{0, 1\}^d. \tag{1}$$

Hash functions [1] have the following properties:

The properties of one-wayness: Given the hash value $h(s)$, it is computationally hard to convert the hash value to the original message s. This is known as one-way property. An arbitrary message $\{0,1\}^* \rightarrow \{0,1\}^d$ is one-wayness, if from the given hash value $\{0,1\}^d$, it is highly impractical to produce the message $m\{0,1\}^*$ such that $h(m) = d$.

Second preimage resistant: There should not exist two different messages for a single hash value. For a given message s, it is highly impractical to uncover another message s' such that $h(s) = h(s')$ and $s = s'$.

Collision resistant: For any two messages s_1 and s_2, it is impossible to find the same hash values, i.e., $h(s_1) = h(s_2)$.

The organization of the paper is as follows: the first section gives the introduction of the work, the section two covers the related works done, and the section three discusses the algorithms needed for integrity. The evaluation parameters for the algorithms are defined in the section five. The section six gives the proposed system, and sections seven and eight give the results and conclusion of the work.

2 Related Works

In [2], the hash algorithms MD4 and RIPEMD were analyzed and discussed. Their strength against collision attacks was found. In [3], various attacks like inversion attacks and differential attacks against the hash functions are discussed. Sasaki and Kazumaro [4] propose a hybrid system which uses MD5, AES, and RSA for securing digital data.

Geethavani et al. [5] give an outline of the performance of hash algorithms based on the preimage attack. In [6, 7], MD4 and MD5 hash algorithms are developed in the year 1990 and 1992, respectively, by Rivest. SHA-1 algorithm [8] was developed in the year 1995 by NIST. The input is the arbitrary length and produces an *s* fixed size 160-bit hash code. An SHA-1 version is an improvised algorithm than the message digest. In the year 2002, the extended version of SHA-1 is developed and named as SHA-2 [9] which has three variants called SHA-256, SHA-384, and SHA-512. This SHA version is popular until the next version of SHA series is developed. SHA-3 [10] the modified version of SHA series was introduced in the year 2005, and later various modified versions of SHA-3 were released. In [11], SHA-512 algorithm was designed by Kahri et al. which provides fault coverage of 99.9%. The combined SHA algorithm with chaotic-based neural networks [12] gives good resistant against differential attacks and successfully defends against collision attack. A symmetric algorithm is combined with SHA [13] to produce a variable length hash value, and it is secure against various linear and differential attacks.

The FSB [14] hash algorithm was invented by Augot et al. in the year 2003. The complexity of these hash functions depends on the syndrome decoding problem. This work is later modified by Finiasz et al. [15] by adding sponge construction function. A more efficient version of FSB was proposed by Meziani et al. [16] which work 30% more faster and its security depends on the regular syndrome decoding (RSD) problem, which is an NP-complete problem. A new hash algorithm, very smooth hash (VSH), developed by Contini et al. [17] is a provably secure algorithm. VSH produces 1024- and 2048-bit hash value and is effective against collision attacks. The new variant of VSH is very smooth number discrete logarithm (VSDL), a modified hash algorithm [18], and the security of the hash algorithm depends on the discrete logarithm problem.

3   Integrity Algorithms

Integrity, in general, is the property of safeguarding the accuracy and completeness of data. The following are the algorithms used for the integrity of the data.

3.1   MD4

The MD4 message digest algorithm is a cryptographic hash work created by Ronald Rivest in the year 1990. The process length is 128 bits. The work has impacted later plans, for example, the MD5, SHA-1, and RIPEMD algorithms. The message is cushioned to guarantee that its length is distinct by 512. A 64-bit paired padding of the first length of the message is then linked to the message. The message is handled in 512 blocks in the Damgard/Merkle iterative structure, and each square is prepared in three particular rounds.

3.2 MD5

The message digest 5 (MD5) is a cryptographic hash algorithm invented by Ron Rivest in the year 1992. The MD5 works with an arbitrary input, and it produces a fixed length of 128 bits. The MD5 algorithms are efficient on the properties of the hash functions.

3.3 SHA-1

SHA-1 160-bit hash function is an extension of MD5. SHA-1 was developed by the National Security Agency and is widely thought about the successor to MD.

3.4 SHA-2

SHA-2 could work with different variants of outputs. The input could be a varying length of the input string. Secure hash algorithm 2 comes with additional security parameters.

3.5 SHA-3

SHA-3 has many variants of 224, 256, 384, and 512 bits fixed hash code output. The complexity of the algorithm lies with the compression function. The complex structure of the computing function gives SHA-3 more reliability and security.

3.6 FSB

The fast syndrome-based hash functions (FSB) were developed in the year 2003. FSB algorithms are proven to be secured compared with the other hash algorithms. The strengths of the algorithms depend on the regular syndrome decoding (RSD). The algorithms start with the compression task F, with the parameters n, r, and w such that $n > w$ and $w \log(n/w) > r$. The parameters n, r, w, and s should be natural numbers, where n is the length of the input word size, r is the output hash size, w is the weight of the word, and s is the input message size. The FSB algorithm is defined as follows:

　　Step 1: An arbitrary message of size s is taken as input.

　　Step 2: Convert the message s into a regular sequence of word of length n and weight w.

Step 3: Add the corresponding columns of the random matrix H according to the value obtained from the n and w from the formula $w \log(n/w)$ from previous step.

Step 4: Output the hash r.

3.7 VSH

The very smooth hash (VSH) algorithm is a secure and efficient hash algorithm. The strength of VSH algorithm depends on the nontrivial modular square root (NMSR) problem. The NMSR problem is defined by the following: let "q" be the product of two prime numbers and given "q".

Find $b \in Z_n^*$, such that

$$b^2 = \prod_{i=1}^{k} p_i^{e^i}, \tag{2}$$

where p is the prime number and e is the integer.

By solving Eq. (1), we solve the NMSR problem which is the complexity of the hash algorithm. For computing the hash, the following steps are defined as follows:

Step 1: The given input message m is split into l number of blocks of length k, such that $\prod_{i=1}^{k} p_i < n$, where p is the prime number and n is a very large integer number.

Step 2: Let L be the smallest integer such $L = l/k$ that $l = \sum_{i=1}^{k} l_i 2^{i-1}$.

Step 3: Next, the compression function F is applied as given in Eq. (2)

$$F_{j+1} = F_j^2 \prod p_j^{e_j} \bmod n, \tag{3}$$

for $j = 0, 1, \ldots, L$.

Step 4: The output hash r has been obtained as F_{L+1}.

3.8 VSDL

Very smooth discrete logarithm based hash (VSDL) is a variant of VSH algorithms. In this algorithm, the strength depends on solving the discrete log problem. The discrete logarithm is defined in Eq. (3)

$$y = b^{x_i} \bmod e, \tag{4}$$

where b is the base integer and x is any integer between $i = 1, 2, \ldots e\text{-}1$, and e is the prime number. The values of b, x, and e are given, and it is easy to find y but finding the value of x is difficult when the values of y, b, and e are known. This modification is done in the compression function of the VSH. The modified compression is as follows: compression function F is applied as given in Eq. (4)

$$F_{j+1} = F_j^2 \prod b_j^{m_j} \bmod e, \tag{5}$$

for $j = 0, 1, \ldots, L$.

3.9 ECOH

Elliptic curve only hash (ECOH) is an elliptic curve-based algorithm developed in the year 2008. The ECOH algorithm has many variants based on the size of the message digest. Each variant will work on different parameters. The ECOH algorithm is defined as follows:

Step 1: The given input message M is separated into equal block size of n.

Step 2: Last block message is padded with 0 and 1, if it is not of equal length.

Step 3: Then, each block of message is applied with transformation function F, as given in Eq. (6)

$$F = F(M_i, i). \tag{6}$$

The transformation function F transforms the message block to elliptic curve points.

Step 4: In addition to transformation function, two more elliptic curve points $T1$ and $T2$ are added to the transformation function as given in Eqs. (7) and (8). First point transformed from the length of the message and the second point is taken based on the message length and XORing all the messages.

$$T_1 = P(n), \tag{7}$$

$$T_2 = P(M_i, n). \tag{8}$$

Step 5: Summation of all elliptic curve points is done and it is passed to an output function R, as given in Eqs. (9) and (10).

$$E = \sum_{i=0}^{n-1} P_i T_1 + T_2 \tag{9}$$

$$R = f(E), \tag{10}$$

where R is the output hashed value.

4 Analysis of Hash Algorithms

The security of hash algorithms depends on the ability of the algorithms to defend against cryptanalytic attacks. The most popular attacks against hash algorithms are collision attack and inversion attack. In Table 1, the elaborate results of the attacks were given and with the help of the analysis, it is observed that VSDL algorithms provide good security against the attacks.

5 Evaluation Parameters for Hash Algorithms

Each of the hash algorithms has its own strength and weakness. In this section, various hash algorithms are analyzed based on the following evaluation parameters.

Table 1 Security strength of various integrity algorithms

Algorithm	Output size (bits)	Security strength in bits	
		Collision	Inversion
SHA-1	160	Collision found	Inversion possible
SHA-2-224	224	112	32
SHA-2-256	256	128	Inversion possible
SHA-2-384	384	192	128
SHA-2-512	512	256	Inversion possible
SHA-3-224	224	112	224
SHA-3-256	256	128	256
SHA-3-384	384	192	384
SHA-3-512	512	256	512
FSB-160	160	100	82
FSB-224	224	135	115
FSB-256	256	190	131
FSB-384	384	215	196
FSB-512	512	285	264
VSH-1024	1024	587	641
VSH-2048	2048	987	1154
VSDL-1024	1024	641	1024
VSDL-2048	2048	1047	2048
ECOH-224	224	112	143
ECOH-384	384	192	206
ECOH-512	512	256	287

5.1 Processing Time

The hash processing time is the processing time for converting the input message into hash message and vice versa. The processing time depends upon plaintext block size. In our experiment, the processing time is measured in milliseconds per bit. The performance of the systems depends on the processing time.

From Figs. 1 and 2, as the size of the output hash code increases, the processing time also increases. The processing time of VSDL and SHA-3 algorithms is very high when compared to other algorithms.

5.2 Entropy

Entropy is a measure of randomness or uncertainty of the information. The hash algorithms should yield high randomness in hashed message, so that there is less or

Fig. 1 Average processing time of other hash algorithms

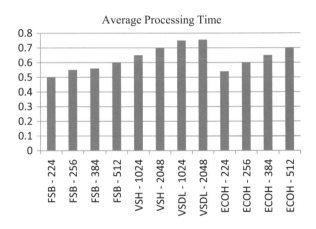

Fig. 2 Average processing time of SHA algorithms

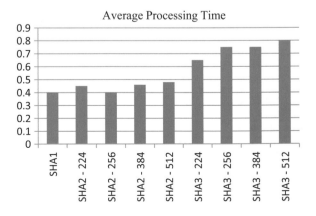

Fig. 3 Entropy of SHA algorithms

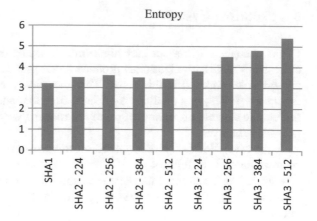

Fig. 4 Entropy of other hash algorithms

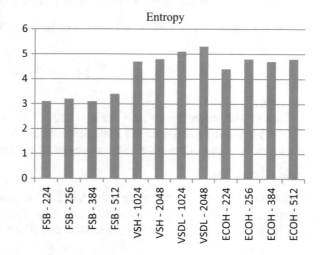

no dependency between input message and hashed message. With high randomness, the relationship between input message and hashed message becomes complex. This property is called confusion. A high degree of confusion is desired to make it difficult to guess to an attacker.

Figures 3 and 4 show the average entropy per byte of information for hash algorithms. From Fig. 4, it is clear that the SHA-3-512 VSDL hash algorithms produce better results than the rest of hash algorithms.

6 Proposed System

From the analysis of various hash algorithms based on the security parameters, it is clear that VSDL hash algorithms produce better security. A modified VSDL hash algorithm is proposed in this paper. This modified *very smooth number discrete logarithm (VSDL)* is a modified *S*-bit hash function, and the hardness depends on finding the discrete logarithms of primitive roots of very smooth numbers modulo, an *S*-bit composite.

A modified VSDL algorithm based on the primitive roots of the discrete logarithm problem is as follows:

Step 1: The given input message *m* is split into *l* number of blocks of length *k*, such that $\prod_{i=1}^{k} p_i < n$, where *p* is the prime number and *n* is a very large integer number.

Step 2: Let *L* be the smallest integer such $L = l/k$ that $l = \sum_{i=1}^{k} l_i 2^{i-1}$.

Step 3: Next, the modified compression function *F* is applied as given in Eq. (11)

$$F_{j+1} = F_j^2 \prod p_j^{e_j} \bmod n, \tag{11}$$

for *j* = 0, 1, ..., *L*.

Step 4: The output hash *r* has obtained as F_L.

The proposed system used the modified version of VSDL hash algorithm, and the main functions of the proposed system are as follows:

Padding function:

The padding function makes sure that all the blocks of the message are of equal length and multiples of 512, 1024 bits, etc. The input message could be of any length, and the padding function makes sure that all the blocks of message will be in multiples of equal length.

Initialization function:

There are seed values which should be initiated at the start of the compression function. The base value and the power value of the discrete logarithm problem should be initialized so that it will provide the randomness.

Compression function:

This is the most important function in the hash function, as in the proposed the compression function uses primitive roots of the discrete logarithm problem. Here, the primitive root gives more randomness to the compression function.

7 Results and Discussion

In this section, the experimental results of the proposed work on various parameters are discussed. The evaluation parameters used are randomness, efficiency, security against collision attacks, and avalanche effect.

7.1 Efficiency

In this experiment, the measured time consumption of hash algorithms in milliseconds was given. Figure 5 shows the timing chart for various hash functions in milliseconds for an input message size of 64 bytes. The proposed algorithm is more efficient than the other VSH variant hash algorithms. The processing time shows that the proposed system is efficient than the other versions of VSDL.

The processing time shows that the proposed system is efficient than the other versions of VSDL.

7.2 Analysis of Inversion and Collision Attacks

The security of hash algorithms depends on the ability of the algorithms to defend against cryptanalytic attacks. The most popular attack against hash algorithms is collision attack and inversion attack. In Table 2, the hash algorithms are compared against collision attack and inversion attack. From the values obtained, it is clear that the proposed system and VSDL-2048 provide better security against attacks. In general, for the hash algorithms, the collision resistance should be $2^{n/2}$. For example, for a 512-bit hash value, the collision result should be 2^{256}, i.e., it requires 2^{256} operations to find the collision for that hash value. In our experiment, the hash value of a message is generated and it is converted to decimal value. Then, a bit in

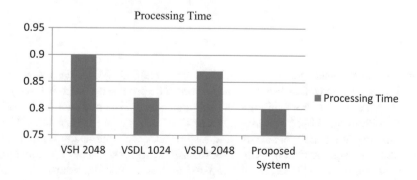

Fig. 5 Computational time of VSDL algorithms

Table 2 Security against collision and inversion attacks

Algorithm	Output size (bits)	Security strength in bits	
		Collision	Inversion
VSH-1024	1024	587	641
VSH-2048	2048	987	1154
VSDL-1024	1024	641	1024
VSDL-2048	2048	1047	2048
Proposed system	2048	1255	2048

Table 3 Absolute difference mean value

Algorithm	Output size (bits)	Mean/Byte
VSH-1024	1024	79.25
VSH-2048	2048	87.24
VSDL-1024	1024	84.62
VSDL-2048	2048	90.52
Proposed system	2048	94.76

the message is randomly chosen and it is changed, and then a new hash value is generated and it is converted to decimal value.

Now the two hashed decimal values are compared and the absolute difference (AD) is calculated using Eq. (12), which is the resistance against the collision attack:

$$AD = \sum_{i=1}^{n} |m_i - m_i'|, \qquad (12)$$

where m_i is the ith decimal value of the message and m_i' is the ith decimal value of the modified message. This experiment is done for many times and the mean value per byte of the message is calculated and tabulated in Table 3.

8 Conclusions

From above analysis, we have found that the modified VSDL hash algorithm provides security as well as faster hashing speed. The proposed system is having all the requirements. So, from this review and analysis, VSDL and modified VSDL hash algorithms have been performing better than the remaining algorithms. These two hash algorithms are more secure and fast to work with and in future, there is a wide scope of improvement in both of these hash algorithms. These two algorithms will provide integrity effectively as compared with the other algorithms.

References

1. Damgard, I. (1989). A design principle for hash functions. In *The Proceddings of CRYPTO'89, LNCS* (Vol. 435, pp. 416–427).
2. Wang, X., et al. (2005). Cryptanalysis of the Hash Functions MD4 and RIPEMD. In *Annual International Conference on the Theory and Applications of Cryptographic Techniques.* Berlin, Heidelberg: Springer, 2005.
3. Stevens, M. M. J. (2012). Attacks on hash functions and applications. Mathematical Institute, Faculty of Science, Leiden University, 2012.
4. Sasaki, Y., & Kazumaro, A. (2009). Finding preimages in full MD5 faster than exhaustive search. In *Annual International Conference on the Theory and Applications of Cryptographic Techniques.* Berlin, Heidelberg: Springer, 2009.
5. Geethavani, B., Prasad, E. V., & Roopa, R. (2013). A new approach for secure data transfer in audio signals using DWT. In *2013 15th IEEE International Conference on Advanced Computing Technologies (ICACT)* (pp. 1–6), September 2013.
6. Rivest, R. (1990). The MD4 message-digest algorithm, 1990.
7. Rivest, R. (1992). The MD5 message-digest algorithm, 1992.
8. NIST. (1995). Secure Hash Standard (SHS), federal information processing standards 180–1, 1995.
9. NIST. (2002). Secure Hash Standard (SHS), federal information processing standards 180–2, 2002.
10. NIST. (2005). Secure Hash Standard (SHS), federal information processing standards 180–3, 2005.
11. Kahri, F., Mestiri, H., Bouallegue, B. & Machhout, M. (2017). An efficient fault detection scheme for the secure hash algorithm SHA-512. In *2017 International Conference on Green Energy Conversion Systems (GECS)*, Hammamet (pp. 1–5) 2017.
12. Abdoun, N., El Assad, S., Taha, M. A., Assaf, R., Deforges, O., & Khalil, M. (2016). Secure Hash Algorithm based on Efficient Chaotic Neural Network. In *2016 International Conference on Communications (COMM)*, Bucharest (pp. 405–410) 2016.
13. Aggarwal, K., & Verma, H. K. (2015). Hash_RC6—Variable length Hash algorithm using RC6. In *2015 International Conference on Advances in Computer Engineering and Applications*, Ghaziabad (pp. 450–456) 2015.
14. Augot, D., Finiasz, M., & Sendrier, N. (2005). A family of fast syndrome based cryptographic hash functions. In E. Dawson & S. Vaudenay (Eds.), *Mycrypt 2005* (Vol. 3715, pp. 64–83)., LNCS Heidelberg: Springer.
15. Finiasz, M., Gaborit, P., Sendrier, N. (2007). Improved fast syndrome based cryptographic hash functions. In *ECRYPT Hash Workshop 2007*.
16. Meziani, M., Dagdelen, Ö., Cayrel, P. L., & El Yousfi Alaoui, S. M. (2011). S-FSB: An improved variant of the FSB hash family. In *Communications in Computer and Information Science* (Vol. 200). Berlin, Heidelberg: Springer 2011.
17. Contini, S., Lenstra, A. K., & Steinfeld, R. (2006). VSH, an efficient and provably collision resistant hash function. In *Lecture Notes in Computer Science* (Vol. 4004, pp. 165–182) 2006.
18. Halunen, K., Rikula, P., & Roning, J. (2008). On the security of VSH in password schemes. In *Third International Conference on Availability, Reliability and Security (ARES 2008)* (pp. 828–833) 2008.

Author Index

© Springer Nature Singapore Pte Ltd. 2019
V. E. Balas et al. (eds.), *Data Management, Analytics and Innovation*,
Advances in Intelligent Systems and Computing 839,
https://doi.org/10.1007/978-981-13-1274-8

Printed in the United States
By Bookmasters